Nature Remade

CONVENING SCIENCE: DISCOVERY AT THE MARINE BIOLOGICAL LABORATORY

Nature Remade

Engineering Life, Envisioning Worlds

EDITED BY
LUIS A. CAMPOS,
MICHAEL R. DIETRICH,
TIAGO SARAIVA, AND
CHRISTIAN C. YOUNG

The University of Chicago Press Chicago and London

The University of Chicago Press, Chicago 60637
The University of Chicago Press, Ltd., London

Published 2021
Printed in the United States of America

30 29 28 27 26 25 24 23 22 21   1 2 3 4 5

ISBN-13: 978-0-226-78326-0 (cloth)
ISBN-13: 978-0-226-78343-7 (paper)
ISBN-13: 978-0-226-78357-4 (e-book)
DOI: https://doi.org/10.7208/chicago/9780226783574.001.0001

Library of Congress Cataloging-in-Publication Data

Names: Campos, Luis A., editor. | Dietrich, Michael R., editor. | Saraiva, Tiago, editor. | Young, Christian C., editor. | Marine Biological Laboratory (Woods Hole, Mass.)
Title: Nature remade : engineering life, envisioning worlds / edited by Luis A. Campos, Michael R. Dietrich, Tiago Saraiva, and Christian C. Young.
Other titles: Convening science.
Description: Chicago ; London : The University of Chicago Press, 2021. | Series: Convening science: discovery at the Marine Biological Laboratory | Includes bibliographical references and index.
Identifiers: LCCN 2020058521 | ISBN 9780226783260 (cloth) | ISBN 9780226783437 (paperback) | ISBN 9780226783574 (ebook)
Subjects: LCSH: Bioengineering.
Classification: LCC TA164 .N39 2021 | DDC 660.6—dc23
LC record available at https://lccn.loc.gov/2020058521

♾ This paper meets the requirements of ANSI/NISO Z39.48-1992 (Permanence of Paper).

And last—and this is the main thing—when I hear people speak of reshaping life it makes me lose my self-control and I fall into despair.

Reshaping life! People who can say that have never understood a thing about life—they have never felt its breath, its heartbeat—however much they have seen or done. They look on it as a lump of raw materials that needs to be processed by them, to be ennobled by their touch. But life is never a material, a substance to be molded. If you want to know, life is the principle of self-renewal, it is constantly renewing and remaking and changing and transfiguring itself, it is infinitely beyond your or my obtuse theories about it.

Boris Pasternak, *Doctor Zhivago*

CONTENTS

Introduction: Engineering Life, Envisioning Worlds

LUIS A. CAMPOS, MICHAEL R. DIETRICH,
TIAGO SARAIVA, AND CHRISTIAN C. YOUNG

"Engineering" has firmly taken root in the entangled bank of biology even as proposals to remake the living world have sent tendrils in every direction, and at every scale. *Nature Remade* explores these complex prospects from a resolutely historical approach—tracing cases across the decades of the long twentieth century that span the many levels at which life has been engineered—molecule, cell, organism, population, ecosystem, and planet.

When biologists have (re)made nature, either they have gained knowledge of life that has entailed its engineering, or their attempts to engineer life have posited new ideas of what can be known. Both efforts have been intertwined with visions of a better world and efforts to remake living systems. But, like all reconstructions of the world, the engineering of life was never simply a technical question. Every effort at remaking nature around us—even those most seemingly future-oriented efforts extending to the Earth itself and beyond—inescapably occurs in a particular social and political milieu. Tracing material practices of the engineering of biology through concrete historical cases ranging from the development of field sites for experimentation of new test organisms to the hybridization of game species and wildlife conservation, or from the development of genetic modification to the new frontiers of synthetic biology, highlights how tinkering with life entails the (re)making of both biological and social order.

Divided into three parts, the cases considered in *Nature Remade* interrogate the control technologies channeling life's expression, suppression, and interaction; offer examples of the material practices and infrastructures involved in (re)making living forms in the laboratory and in the field; and envision the possible futures produced as a result. The three parts of *Nature Remade* focus on themes of *control and reproduction, knowing as making,* and *envisioning.*

While there are many scholarly works on the technological dimensions of biological research, none deliberately contrast different historical approaches to engineering life at different scales, in different biological systems, and in varied social contexts.[1] More than a mere metaphor for a kind of biological practice, engineering ideals and practices have had multiple realizations across generations of biologists in different subdisciplines.[2] Paying attention to these many dimensions of the engineering of life makes it possible to write the history of biology in ambitious ways, which are also, importantly, more accessible to general historians.

As engineers, biologists have manipulated model organisms in elaborate laboratory experiments to unveil hereditary mechanisms; they have produced viruses at industrial scale to discover vaccines; they study and conserve wildlife using surveillance technologies. Discovery and understanding in the history of the life sciences are inseparable from manipulation, modification, industrial production, control, and maintenance—all activities within the realm of engineering. Hans-Jörg Rheinberger's "experimental systems" (1997) or Hannah Landecker's "living matter as technological matter" (2010) are exemplary cases of scholarship exploring the material dimensions of knowledge production in the life sciences.[3] Paul Rabinow (1992), when discussing the Human Genome Project, had already called attention to the importance of considering its technological features beyond the simple recognition that it made intensive use of machinery. He preferred to emphasize how the project exemplified a form of knowledge that produced its own object of research, a seminal approach for many science studies scholars.[4] In this volume, we seek to understand not just the importance of making use of technologies in biological research, but also the wider historical significance of considering the engineering framework in modern attempts at (re)making life. Taking the engineering of life seriously—as motivator, ideal, and lived reality—highlights how engineering approaches enact new forms of knowing life as well as new sociotechnical imaginaries.[5]

Conceptualizing organisms as technical systems—reworking their physiology, their mode or manner of reproduction, their genetic legacies—has been at the forefront both of new biological discoveries and of new means for entraining biological systems for human-desired purposes for over a century. As the Nobel Prize–winning geneticist H. J. Muller once noted, "the duty of biology" was not only to "make us all healthy, vigorous, and happy" but also "to study, to understand, and to reach into the heart of the organic world and refashion this radically to man's own advantage."[6] Philip Pauly (1987) famously referred to the "biological modernism" of the early twentieth century, in which life scientists' work was determined by "the framework of engineering."[7] Building on Muller's claims, Pauly's insight, and adding cues from engineering studies and history of technology, we attempt an updated and fuller exploration of how attention to engineering can open new and ambitious ways for writing histories of biology. In addition to the more obvious dialogue with the vast body of literature dedicated to the material dimensions of knowledge production in biology referred above, this engineering focus allows us to engage more unexpected literatures such as those crossing environmental history and science and technology studies (STS). As suggested by Sara B. Pritchard (2013), the "broad temporal and spatial scales" typical of environmental historians' narratives have challenged STS scholars who tend to be more site-bounded to engage with topics like imperialism, slavery, or industrialization.[8] Our full embracing of engineering in this volume aims at producing analogous effects for history of biology.

Fourteen contributions from the history of science, STS, environmental history, and art and design cohere in this volume, engaging the centrality of engineering for understanding and imagining modern life. This collection grew from two workshops on "Engineering Life" held at the Marine Biological Laboratory at Woods Hole in 2017 and 2018, organized by seminar leaders Jane Maienschein, James Collins, John Beatty, and Karl Matlin. These week-long workshops provided the contributors an opportunity to discuss many of their ideas with each other before undertaking the work of writing their chapters. Of course, the selection offered here is not exhaustive, but rather represents a sample of perspectives on this topic.

Three main themes of engineering life structure our endeavor: control and reproduction, knowing as making, and envisioning.

The emphasis on *control* as part of the engineering ideal pursued by Jacques Loeb (1859–1924) was one of Pauly's central productive insights. In fact, anxieties over the putative hubris of scientists promising to control every feature of life has led to a prolific critical literature following in the footsteps of Michel Foucault. The used and abused notion of "biopolitics" that Foucault first put forward denoted the modern transition from state power as control over territories into more insidious interventions targeting whole populations and the bodies of citizens.[9] If the emergence of engineering as a profession in seventeenth-century France is directly linked to the defense and administration of the monarch's territories, the later emphasis on governing life meant redirecting engineers' attention to control of urban spaces, the workplace, or the household. The historical moment in which Loeb placed the study of life under the "framework of engineering" coincided with engineers' fixation on control, which they translated into a fascination with scientific management in the early twentieth century.[10] In the realm of the life sciences, as all the chapters of this section make clear, control of life translated overwhelmingly into control of reproduction.

Detailing varied processes—controlling reproduction of rats on New Zealand islands or in major urban centers elsewhere, pigs in the southeastern United States, or oranges in Mandatory Palestine—highlights the productive gesture of writing histories of biology informed by engineering. At the most generic level, engineering involves transforming nature to serve human needs. More interestingly, to render such service "engineering" redefines both what nature is and which human needs deserve consideration.[11]

Making sense of nature through mechanistic explanations was for Loeb a necessary first step for controlling life. In Christian Ross's contribution to this volume we see the present implications of Loeb's mechanistic nature in attempts at controlling life through gene drives. MIT scientists occupied with gene-editing techniques to control the population of rats in Aotearoa New Zealand willfully attempted to involve Māori communities in their undertakings. However, the mechanistic view of nature accounting for the success of gene editing did not have space for Māori notions of care and stewardship of humans and nonhumans developed through the weaving of kinship relations. As Ross notes, aspirations to scientifically know and technologically control life are bound up with a vision of the world and how it should be.

Rats are also the protagonists of Anita Guerrini's chapter. She convincingly makes the case that no other animal in human history was as subject to different regimes of control as rats. From poisoning rats to the breeding of standardized laboratory animals, they embody a long history of control. Importantly, their enduring presence close to us in our cities in spite of many eradication campaigns, reminds us of the obvious limits of control dreams. Such limits tend to get lost in too many narratives of engineering life, taking at face value the rosy futures and doomsday scenarios put into circulation by the promoters and detractors of disruptive innovation in the life sciences. Guerrini's narrative, although focused on the twentieth century, hints also at historical chronologies less captured by the glare of novelty. As is also clear with Tiago Saraiva's Jaffa oranges or with Abraham Gibson's feral pigs, to fully appreciate the historical significance of the promises and limits of control of life the historian of biology needs many times to engage with periodization that takes into account time scales of the actual living thing being engineered.

In Tiago Saraiva's text, engineering nature in Palestine to serve Jewish settlement in the interwar period involved both the obsessive control of citrus reproduction through cloning and the orientalist identification of Arab horticultural practices with an ahistorical premodern past. Arabs were relegated to a state of nature waiting to be transformed by Jewish settlers who asserted their own Jewish identity through the way they reproduced oranges. Reproduction control was not just a means to an end, a more efficient way of producing a commodity, but a practice sustaining a specific understanding of nature and humans. As eloquently demonstrated by Abraham Gibson's chapter, the history of interactions between pigs and humans goes well beyond artificial selection of the former to better serve the needs of the latter. Instead of understanding the proliferation of feral animals in the US South as a history of failure of taming the wild, of failure to control, one can see it as intended production of nature to sustain a certain type of contested human: the American hunter.

Considering engineering as a practice that blurs divisions between *knowing and making* was at the core of Pauly's studies of Jacques Loeb's infatuation with the technological possibilities of biology as living substrate. Loeb's biology, suggested by his experiments on sea urchin embryos of the late nineteenth century, viewed cells and tissues as manipulable technologies: "The idea is now hovering before me that man himself can act as a creator, even in living nature, forming it eventually according to his will. . . . Biologists label that the production of monstrosities; railroads, telegraphs, and the rest of the achievements of technology of inanimate nature are accordingly monstrosities."[12] Conceptualizing organisms as technical systems—reworking their physiology or metabolism, their mode or manner of reproduction, their genetic heredity—has been at the forefront both of new biological discoveries and of new means for entraining biological systems for human-desired purposes, from better breeding and genetic engineering to synthetic biology. The contributors to the second part of the book consider the detail of material practices in the life sciences aimed at (re)making life by reducing it to physical and chemical components and reassembling its building blocks, while studying the production of life-supporting environments as infrastructural projects.

The chapter by Dominic Berry highlights how productive it can be to go beyond the mere signaling of engineering life as metaphor in biological research or generic discussions of information technologies as a central theme in molecular biology. By exploring the detail of technological practices involved in synthesizing and analyzing DNA, he unveils two different roles of technology as providing access to nature and, alternatively, as enabling tinkering with nature. Such attention allows the historian to follow underappreciated genealogies of biological research at the intersection of "engineering, technology, chemistry, and beyond." Such attention to the various historical roles of technology in biological research is also at play in Edmund Ramsden's chapter, where he considers behavioral engineering alongside problems of misbehavior in animals. Again, the interest is in acknowledging not so much a mechanistic understanding of life but how different engagements with technologies lead to contrasting approaches in behavioral research. While B. F. Skinner's approach of considering organisms as machines responding to external stimuli pointed at malleability of animal behavior and total control of life, Marian and Keller Brand's involvement with the challenges of mass production of trained animals led them to delve into the biological significance of technological failure. Importantly for the purposes of this volume, Ramsden suggests that behavioral research became more fundamental the more engaged researchers became with the material conditions of the experiment. Counterintuitively, as knowledge of life becomes more fundamental, the closer it gets to engineering practices of explaining through making.[13] Or, in other words, explorations in the history of basic research in the life sciences deserve consideration as explorations in the history of engineering.

Picking up this theme at another scale, Joshua McGuffie examines the engineering of an entire landscape in the design and construction of the Pacific Proving Grounds in the Marshall Islands. His contribution aligns the priority of biological studies of radiation with the demand of creating infrastructure that would make those studies possible. While biologists prepared to focus on the effects of atomic tests on organisms and tissues, the larger human and natural systems needed to be remade. Knowing what changes occurred in tissues throughout the food web required labscapes engineered from the remains of military bases. Freezers and lab equipment powered by diesel generators took their place amid airstrips, piers, mess halls, dormitories, and communication towers. McGuffie opens a window into the interplay of making the world knowable and knowing the world that was made in these places.

Environmental reconstruction is also a significant theme of Lisa Onaga's study of the research and manufacture of artificial food for silkworm culture in Japan. Although at a different scale than that examined by McGuffie, Onaga's analysis demonstrates how the implementation of artificial food for silkworms allowed this form of sericulture to develop independently of the culturing of mulberry plants as a food source and allowed silkworm sericulture to be revived as an economic enterprise. This "modernization" of silkworm sericulture pulled it further across the already blurry boundary between nature and artifice, between industrialized organism engineered to serve economic purposes and model organism engineered to produce new biological knowledge.

Richard Fadok's essay on biomimesis and biomimicry also explores the trade across the boundary between nature and artifice as scientists use elements from living systems to guide or inspire the design of technological systems. Biologist-engineers are explicitly contrasted with those who consider natural selection as an engineering process in the production of biomimicry. The effect of Fadok's analysis is an inversion: instead of considering biologists as engineers, life itself is already engineered, and biologists are acting in response to the systems they find in nature.

At every level, the political scale of interventions always parallels or supervenes on the technical. Genes may be driven through populations, organisms may escape laboratories, populations may exceed their ability to be tracked. Biological engineering always requires social engineering—an entangled syncopation. In the state of nature, as Rousseau tell us, everything is political. The political projects associated with (re)making nature tend to be messier than those described in earlier paragraphs, transforming engineering of life into a practice of imagining new futures, of *envisioning.*[14] Rheinberger described "experimental systems" as "machines for making the future"—producers of knowledge in biology as well as markers of directions of future research. The issue is that the future is never easily contained inside laboratory walls, as all the chapters in the third section of the book highlight, and the distinction between scientific, literary, and political practices have a clear tendency to become blurred, leading to constant disputes on who is entitled to envision biological futures.

Luis Campos demonstrates the importance of following the unexpected transits between the futures engendered in science fiction and those produced in biology laboratories. The historical debates that emerged around the now mundane technique of recombinant DNA make sense only when detailing simultaneously the technicalities of "experimental systems" and the cultural significance of sci-fi narratives. Going beyond traditional descriptions of literature reacting to scientific innovations, Campos highlights the early relevance of Michael Crichton's novel *The Andromeda Strain* in framing the risks of recombinant DNA in terms of alien invasion of different forms of life. From Campos's chapter we learn that while many scientists resisted such framing and struggled to be in control of the future opened by their particular biological practices, they were never able to contain the proliferation of alternative futures envisioned not only by novelists but also by scientists turned into activists and even Congressional leaders under the influence of sci-fi scenarios.

While the catastrophic futures built on recombinant DNA invoked extinctions, infections, and invasions, those of twenty-first-century genomics promised lighter visions of resurrection, personalized healthcare, and endless entertainment. Nathaniel Comfort's chapter invites readers to consider neoliberalism as constitutive of current day genomics, unveiling the capitalist nature of futures put forward by techniques of gene editing (CRISPR) and genome-wide association studies (GWAS). Tracing the genealogy of these techniques to a *longue durée* history of separating nature from nurture, heredity from environment, it becomes apparent that current sophisticated understandings of multiple genes involved in complex traits did not eliminate what the author describes as a structural genetic determinism. As pervasive and insidious as structural racism, structural genetic determinism implicitly takes DNA as the core defining element of human nature, ready to be engineered to produce an allegedly brighter future. CRISPR and GWAS should certainly be understood as "machines for making the future" à la Rheinberger, but it should be acknowledged, Comfort reminds us, that the epistemic future they open is inseparable from and limited by its neoliberal imaginary.

The historical relevance of the distinction between structural racism and structural genetic determinism is particularly salient in dealing with what Ayah Nuriddin labels "black eugenics." The author describes in her chapter the surprising emancipatory agendas associated with the eugenic practices promoted by distinguished American black leaders such as W. E. B. Du Bois. Futures of racial equality were predicated on the possibility of engineering the black race by eliminating undesired traits that white eugenics took as fixed elements of black populations. In contrast with obsessions over degeneration produced by racial mixing characteristic of white eugenics, black eugenics emphasized instead the plastic nature of a population to be bettered through interventions targeting "chastity, temperance, education, and hygiene." Making use of Comfort's terminology, this could be described as an attempt in using structural genetic determinism to combat structural racism.

From all the contributions of this section of the book it is clear that dreamers were never modest. Independently of the scale of intervention—molecule, cell, organ, or human body—the scale of the future runs from the population, as in Comfort and Nuriddin's texts, up to nothing less than the entire planet. Tracing the genealogy of "planetary manipulation fantasies," James Fleming urges readers of this volume to embrace perspectives from the human and social sciences in the making of futures, thus enlarging the community of experts involved in such scaling-up operations. Such discussion is especially significant when writing the history of biology in (of?) the Anthropocene, our current era, in which humans have putatively become a geological force.[15] Adding to the gallery of biologists, sci-fi novelists, capitalists, activists, and politicians that populated previous chapters, and casting a critical eye on Anthropocene dreams of control, Fleming wants to incorporate scholars dealing with the social and cultural dimensions of biology in the common task of designing and imagining a common future.

Artists and designers did not wait for a formal invitation to join the conversation. Alexandra Daisy Ginsberg describes in this volume her artwork "Resurrecting the Sublime," in which she wanted "museum visitors to enjoy the total artifice of a resurrected smell" of extinguished plants produced through genetic engineering in a "simulated landscape." More than indulging in the technological hubris too common in Anthropocene literature, with geoengineering solving the harms caused by human activities, Ginsberg recovers the notion of the sublime as "expression of the unknowable." The engineering of life does not invoke in Ginsberg's work futuristic visions of total control, but instead feelings of loss in which life in the Anthropocene is better understood through explorations amidst the ruins of capitalism.[16]

Some readers might take such artistic intervention as a critique of the engineering motive in the life sciences, following the lines of traditional oppositions between what engineers do and artistic reactions to it. But if one embraces the methodological proposal of this volume of engaging seriously the history of engineering in order to rethink the history of biology one should acknowledge that romantic notions of the sublime that historically challenged simplistic versions of human progress were put forward not only by poets and musicians. As a prolific historiography dedicated to romantic science and technology of the first half of the nineteenth century has suggested, engineering practices should also be understood as romantic cultural interventions aimed at exploring new forms of feeling: steam engines were also romantic machines through which one could experience the underlying unity of movement and heat, the underlying unity of nature sung by romantic poets of the sublime.[17] More prosaically, engineers designed elaborate urban parks staging ruins of bygone civilizations in the middle of carefully cultivated nature. More than artists reacting to the futures put forward by biologists as engineers, we are faced with a continuum of practices in which art and engineering are braided in experimenting with our common future.

PART ONE

Control

1 * *Knowing and Controlling: Engineering Ideals and Gene Drive for Invasive Species Control in Aotearoa New Zealand*

CHRISTIAN H. ROSS

On the islands of Aotearoa, also called New Zealand, invasive species have been a prominent and persistent concern for local ecosystems. Traditional methods of biological control, though, can be difficult to implement and often have harmful side-effects for the environment and human health. Recent developments in genetic engineering have led to the creation of a new technology called gene drive, which some have suggested may provide a safer, easier alternative way to "restore damaged ecosystems and save endangered wildlife by genetically removing invasive species."[1] While the promises of gene drive for invasive species control have attracted the attention of many in Aotearoa New Zealand interested in preserving or restoring the islands' native environment, at the same time it has prompted calls for caution regarding their controllability and possible unintended consequences of their use.[2] However, the consideration of gene drive for the control of invasive species in Aotearoa New Zealand is more than just an issue of a controversial use of emerging biotechnology. At stake also are critical questions about what it means to know and control life. What are the kinds of knowledge that enable and underwrite the notions of controlling of life? If gene drive confers the power to control invasive species, who decides whether and how that control is exercised and with what responsibilities? And, crucially, what visions of the world are embedded in the aspirations of scientifically knowing and technologically controlling life?

Controlling life has long been a central aspiration of the biological sciences. In the early twentieth century, aspirations to greater control over life manifested in rigorous laboratory experimentation, attempts to engineer organisms to be more amenable to human purposes, and explanatory commitments to a mechanistic conception of life.[3] Mechanistic approaches have been ubiquitous in biological practice aimed at bringing living things and their functions into the purview of human intention and volition by isolating, manipulating, and better understanding the function of more fundamental parts.[4] The widespread mechanistic approaches in biology also helped create the conditions for the rise of genetics as the preeminent field in the biological sciences.[5]

In the decades following the close of World War II, experimental and molecular biology grew dramatically. Molecular biology increasingly displaced other fields of biology as a more physics-like science, a resemblance that lent a sense of hardness to biological theories and explanation.[6] The dominance of molecular biology laid the foundation for the epistemic status of genetics by providing a molecular articulation of fundamental processes of transcription and translation. Later, the development of recombinant DNA and similar biotechnologies further cemented genetics as the foremost scientific field for manipulating, changing, and controlling life for human uses.[7] The experimental and commercial successes of genetic engineering and its products entrenched a narrative of biological control as one of having mechanistic understanding of the underlying genetic components, demonstrated through genetic isolation, replication, and rearrangement in versatile, predictable, and modular ways.[8] By the close of the twentieth century, the genetic and mechanistic explanations of biology crescendoed in the promises of the Human Genome Project to provide greater control of human health through a putatively comprehensive understanding of human genetics that would provide means of intervening in fundamental biological processes.[9] More than just an ideological commitment to particular scientific methodologies, life as mechanism was an animating metaphor for what it meant to both know and control living things.[10] To control life was to mechanistically understand its underlying phenomena, and genetics provided the kind of causal, molecular-level explanations that mechanistic biology prized.

In the twenty-first century, aspirations to control life through genetics find new resonance in the technology of gene drive. Gene drive is a genetic engineering technology that embodies a mechanistic logic in which aspirations to control life are unified across biological scales, linking molecular genetic interventions with macroscale ecological outcomes. At its most basic level, gene drive enables the rapid spread of a desired genetic element through a species population by significantly increasing its inheritance in subsequent generations. An engineered gene drive employs genome-editing techniques and exploits innate cellular processes of genome repair to produce a self-propagating system of genetic modification such that, once introduced, the gene drive continues to spread across generations of a targeted population. The result is that the desired genetic element coupled to the gene drive is engineered into the genome of nearly every individual in a population over relatively few generations. The power and scalability of a gene drive to alter entire species populations has lent itself to many speculative ecological engineering projects ranging from the prevention of vector-borne disease to conservation efforts and invasive species control by inserting genetic elements that skew population sex ratios, cause sterility, introduce lethal mutations during development, or otherwise reduce the numbers of targeted species.[11]

Control over life extends beyond harvesting or harnessing of nature. It suggests a broadening of the horizon of humanity's power to make and remake nature ever more in our own image. Accordingly, what it means to control life is necessarily intertwined also with the control of those mediating technologies and the particular visions of the material and social world that they authorize.[12] Scholars studying the biosciences in society have long argued that knowledge making of biological sciences is not independent of the social and cultural worlds in which it takes place.[13] Accounts of biological objects like genes, DNA, and cells are inseparable from and co-produced with normative visions of social order and structures of power.[14] The entanglements of the biological and distinctive sociopolitical visions of the world become particularly evident in the ways that humans invest of themselves, their aspirations, and their ideologies in the nonhuman animals that they make and are made by.[15] Divergent biological understandings of the world, then, are not merely the result of a singular nature refracted through the lens of many cultures,[16] but are disparate social, political, and cultural constructions of life itself.[17] Put another way, the social precedes and is embedded within every part of biological projects to engineer life.[18] Therefore, at stake in questions of what it means to know and control life vis-à-vis gene drive are not only matters of technological precision, capacity, or predictability of technological outcomes but also of social order for the development and governance of science in society.

Scientific aspirations to engineer life also evoke questions about responsible science regarding a technology like gene drive. While the idea of scientific responsibility toward broader society is neither new nor unique to gene drive technology,[19] what it means to do science responsibly does find distinctive resonance with notions of restraint, precaution, and deliberate action associated with metaphors of engineering life. Scientific responsibility has been understood in terms ranging from adherence to particular norms of scientific practice,[20] individual "role responsibilities" of scientists so as to avoid research misconduct,[21] a "collective responsibility" of science as a knowledge-making profession,[22] or a "co-responsibility" between science and society.[23] However, others have problematized such accounts of responsibility in science as limited to the practices of scientists qua scientists.[24] Because science and society are not so neatly separable from one another, the responsibilities of science and scientists are implicated at every turn with epistemic, social, political, and moral stakes.[25] Those stakes are heightened all the more in projects of engineering and controlling life on potentially global scales. Thus, notions of responsible science are inherently wrapped up with the mechanistic rationales that have dominated the life sciences for a century and underwrite scientific claims to controlling life. Questions of responsible science regarding gene drive, then, are also questions about how one knows and attempts to control life.

Gene drive promises greater control by contrasting the mechanistic precision of genetic engineering to impress human designs upon nature with an implicitly more disordered and out-of-control state of nature. In doing so, gene drive highlights the ways in which technologies of control are integral to the aims of biological engineering and how such technologies render biological processes amenable to human intention and inescapably mediate human efforts to control life. As aspirations to bring nature under human control increasingly implicate a diversity of cultures and communities, mechanistic approaches common to Western biological engineering practice encounter alternative ways of understanding the natural world and what it means to control it.

This chapter examines discourse around the use of gene drive for invasive species control in Aotearoa New Zealand as an exploratory case of what it means to control life in the twenty-first century. It follows the research and community engagement activities of one of the technology's inventors and prominent advocates, MIT Media Lab scientist Kevin Esvelt, to trace his articulations of the function, capabilities, and stakes of gene drive technology. Analysis of the notions of control embedded in descriptions of gene drive technology itself shows how mechanistic biological control becomes intertwined with ideas of responsible scientific practice and reveals limitations of mechanistic biology in navigating social worlds and distinct practices of meaning-making. It also prompts reflective examination of the visions of remaking nature that are animated by particular approaches to knowing and controlling life.

Aotearoa New Zealand and Invasive Species Control

Introduced predatory species have been a longstanding feature of the island environments of Aotearoa New Zealand. The first such species introduced to the islands was the kiore, also known as the Polynesian rat (*Rattus exulans*), that accompanied the arrival of Māori peoples in the late thirteenth century. Other predatory rodent species came later to the islands as stowaways on the ships of European colonizers or as intentional additions to the environment to attempt to control the population of other introduced species brought by European colonists during the nineteenth century.[26] Though the designation of introduced species as "native" or "invasive" is itself a thorny issue,[27] by the mid-twentieth century, nearly all introduced predator species on the islands came to be widely considered a major ecological concern.[28]

During the second half of the twentieth century, sustained action toward invasive species control grew, especially targeting rodents. Early efforts were relatively modest removals. In 1959, a handful of volunteers began one of the first rodent removal operations on Ruapuke Island, a two-hectare islet off the eastern coast of Auckland in the Hauraki Gulf. The introduction of invasive rats in 1963 by commercial ships to Big South Cape Island, the largest island off the southern coast of New Zealand, galvanized national attention and resolve to eliminate the invasive species that had become increasingly out-of-control. Over the following decades, invasive species control in New Zealand grew as a national priority. Advances in toxin development in the late 1970s and 1980s led to increases in chemical control efforts that relied heavily on poison baiting of affected regions with poisons like Compound 1080 and brodifacoum. By the turn of the twenty-first century large-scale conservation efforts had become commonplace, with major projects declaring Campbell Island and Big South Cape Island cleared of invasive rodents in 2001 and 2006.[29]

Though the control of invasive species had enjoyed decades of broad public buy-in, by the early 2000s, that support began to wane. The chemicals used to eradicate invasive species were chosen because of their high toxicity. However, those toxins were also indiscriminate in what organisms they affected, which resulted in many cases of secondary poisoning of native and valued domesticated species. As a result, the chemical methods of invasive species control came under harsh criticism from environmental groups and communities that lived in areas where toxins and poison baiting were widely used. The dissonance of the broad use of toxic, nonspecific chemical controls for invasive species as part of environmental conservation became increasingly politically untenable.

One such group was the original Predator Free New Zealand (PFNZ), a grass roots movement founded by environmental activist Les Kelly in 2008. PFNZ had the ambitious aim of eliminating all invasive predatory species—including possums, stoats, feral cats, and rats—from all of New Zealand's more than 600 islands over the course of a couple decades. Importantly, the organization emphatically rejected the use of toxins and advocated for the consideration of alternative technologies to control invasive species.[30] The goals of groups like PFNZ gained support among the general public as well as the New Zealand government. In July 2016, New Zealand Prime Minister John Key announced the launch of Predator Free 2050, a government-supported initiative that put national resources and legitimacy behind the mission of eliminating invasive predator species from the island nation by 2050.[31] The vision for invasive species control presented was one of a New Zealand entirely devoid of their existence. Control of invasive species entailed not just a modulation of their effects on native ecosystems, but their complete removal from them.

Predator Free 2050 retained many of the priorities and commitments of PFNZ, in particular an emphasis on pursuing a wide range of technologies for biological control, including renewed consideration of genetic engineering. Genetic engineering technology had been controversial in past decades, but more recently, public attitudes had become more accepting of genetic engineering technologies as a means of controlling living things, particularly in the service of environmental protection.[32]

Gene Drive: A Technology of Control

In 2014, two years before the announcement of Predator Free 2050, MIT scientist Kevin Esvelt and collaborators from Harvard Medical School and the Harvard School of Public Health published a pre-print manuscript detailing the design, function, and speculative applications of gene drive.[33] The landmark paper presented gene drive as a possible tool for large-scale environmental engineering. The paper included a Venn diagram that summarized many possible applications of gene drive as ecological tools in human health, and conservation, specifically noting invasive species control. By driving a lethal or sterilizing genetic trait through invasive populations, like the rats in Aotearoa New Zealand, a gene drive could be used to effectively suppress and control their numbers and negative ecological impacts.

Though the paper did not specifically identify Aotearoa New Zealand as a good candidate region, it did explicitly state that gene drive might be used to "promote biodiversity by controlling or even eradicating invasive species from islands or possibly entire continents."[34] It did not take long for groups like Predator Free 2050 to connect the dots. Over the following years, gene drive grew to be one of the most high-profile technologies considered for invasive species control in Aotearoa New Zealand.[35]

Esvelt and his collaborators described gene drive as a technology for large-scale ecological engineering by causing a particular genetic element to be inherited in a sexually reproducing population at a much higher rate than that of typical probabilities of inheritance. Increasing the rate of inheritance results in the rapid propagation of that genetic element in a targeted population such that "over many generations, this self-sustaining process can theoretically allow a gene drive to spread from a small number of individuals until it is present in all members of a population."[36] Though there are naturally occurring instances of gene drive,[37] the 2014 proposal by Esvelt and colleagues relied on CRISPR-Cas9 genome-editing techniques to construct an engineered gene drive.

With CRISPR genome editing, new genetic elements are introduced by cutting the genome at a specific locus and providing an engineered DNA template that contains the desired genetic edits. First, an endonuclease protein is guided by custom-designed RNA molecules to specific, target locations in the genome. Once the endonuclease is guided to the target site, it cuts the DNA, introducing a double-stranded break into the genome. Upon detecting the double-stranded break, the cell's natural DNA repair processes attempt to repair the cut site, using the engineered DNA template as a guide. As a result, whatever desired genetic edits were included in the DNA template are copied and inserted into the genome.

To construct a gene drive, Esvelt and colleagues proposed inserting the genes for the CRISPR-Cas9 endonuclease, the guiding RNA molecules, as well as the genetic element to be driven into the genome. That way, any edited cells would also include the genes to produce more genome-editing CRISPR systems. Instead of a single genome-editing event, a CRISPR-based gene drive would continue to target, cut, and edit at the specified site in the genome indefinitely. For diploid species reproduction, that meant that if any offspring inherited an edited chromosome from one parent containing a gene drive, the analogous site in the chromosome inherited from the other parent would also be edited to contain the gene drive. Thus, if an organism edited with gene drive were to reproduce, even with a wildtype organism, the genomes of all resulting offspring would contain the gene drive and the desired genetic edits, ensuring its continued self-propagation.[38]

The paper's description of gene drive function and invasive species control was heavily mechanistic. Esvelt and colleagues appealed to the molecular biology of CRISPR genome editing as part of the "mechanistic reasons" why gene drive was an effective means of controlling invasive species.[39] Specifically, they argued that "if population-level engineering is to become a reality, all molecular factors relevant to [CRISPR genome editing] must be considered."[40] In doing so they employed a mechanistic notion of control in which biological engineering depended on sufficient knowledge of the underlying biochemical processes.

The paper also identified possible risks of gene drive use, namely the risk of a gene drive getting out-of-control. Esvelt and colleagues identified rare mating events and geographical leakiness of gene drive as some of main risks in which gene drive might spread to nontargeted species or regions. The strategies for mitigating those risks were to render gene drive more controllable through further acts of genetic engineering. Proposed solutions included introducing additional gene drive systems to counteract or outcompete the effects of a gene drive gone rampant, genetically engineering invasive species to be have unique genomic sequences for more precise targeting by a future gene drive, and the introducing of genetic sensitivities to less toxic molecular compounds for more traditional chemical control.[41] In the risk management strategies posited, keeping gene drive under control was possible through the mechanistic control of genetic engineering, closely linking the manipulation of molecular-level processes to the control of populations and ecosystems.

Gene Drive and Responsive Science

Part of how gene drive became preeminent among biotechnologies for invasive species control was its use of CRISPR genome editing. In the years before the announcement of Predator Free 2050, the science and popular media was inundated with news of the "CRISPR revolution" in biology with rippling implications across global societies.[42] The synchronous development of New Zealand's Predator Free 2050 initiative with the scientific fervor around CRISPR genome editing brought significant attention to gene drive and its applications—including invasive species control, particularly of invasive rodents. The hype and molecular precision afforded by CRISPR genome editing promised that gene drive would enable greater control of invasive populations through greater precision of biological engineering, sidestepping the negative environmental impacts of traditional methods of invasive species control.

Gene drive also benefited from the high-profile actions of one of its inventors, Kevin Esvelt himself. Esvelt assumed a prominent and presumptive leading role in the unfolding discussions around the applications of gene drive. His influence on perceptions and consideration of gene drive and genetic engineering more broadly was significant. Not only was Esvelt a scientist at one of the world's most academically and socially elite research institutions, he was also a charismatic media figure, sitting for numerous interviews and hosting public meetings to explain the science behind gene drive and to discuss the possible risks and benefits of its applications with a wide audience.[43]

Esvelt's emphasis on broader discussions about gene drive applications was not without principled cause, however. His participation in gene drive discourse, assuming a position as leading voice for the future of gene drive, followed directly out of his commitments to a particular vision of responsible scientific research, which he termed "responsive science."[44] For Esvelt, responsive science was "a way of conducting research that invites openness and community involvement" from those who stood to be affected by the applications of technologies like gene drive and entailed that scientists are "morally responsible for all consequences" of their research.[45] In doing so, Esvelt argued that scientists like himself were obliged to participate in and help inform discussions about the applications of his research on gene drive. Esvelt considered himself to have a personal, moral responsibility to ensure that if gene drive was to be developed at all, that it would be done so in a way that was sensitive to the interests and participation of the local communities where the technology would be used. In linking his sense of personal responsibility to the future uses and ultimate control of gene drive, Esvelt blended notions of individual responsibility and scientific responsibility. As a result, Esvelt began to frame responsible development of gene drive in terms of a mechanistic conception of life and notion of control.

Problems with Controlling Gene Drive

Esvelt came to regret suggesting that controlling invasive species was a possible application of gene drive. Despite initial enthusiasm behind gene drive for controlling invasive species, as his research group continued to study gene drive, they became increasingly aware of what they identified to be a significant problem.

The problem with the standard gene drive system that Esvelt's research group had described in 2014 as a tool for invasive species control was precisely that—control.

The standard version of gene drive that had been proposed was a self-propagating, globally-reaching system. If a standard gene drive were released into the environment, Esvelt and others argued that it would "likely to spread to every population of the target species throughout the world" as well as possibly to closely related species.[46] Given that one of the primary targets for gene drive control, invasive rodents, are also among some of the most globally distributed species, the possibility of unintentional gene drive spread was particularly alarming. Ironically, gene drive, the very technology designed and proposed to stem the spread of invasive entities and restore native, local environments, was itself an invasive threat to global ecosystems.

Even if the risk mitigation strategies they suggested were employed, it was increasingly clear to them that they could not ensure that a gene drive would remain localized geographically or to targeted invasive populations. Containment of gene drive to targeted invasive predator populations, even on the relatively isolated islands of Aotearoa New Zealand, was an unrealistic expectation given the realities of a highly interconnected global travel and trade systems. Unintended transmission of a standard gene drive could be facilitated by the same means by which rodents came to Aotearoa New

Zealand in the first place—as stowaways on commercial shipping vessels. Standard gene drive spread to some noninvasive populations globally was likely, if not unavoidable.

Standard gene drive simply was too unwieldly for the localized control needed for addressing invasive species in a particular region. Esvelt later said of his initial suggestion that it was "profoundly wrong of me to even suggest it, because I badly misled many conservationists who are desperately in need of hope. It was an embarrassing mistake."[47] In Esvelt's view, standard gene drive was untenable for conservation because it was not sufficiently controllable. Despite the emphasis in the initial paper on the molecular controllability of gene drive, Esvelt later wrote of standard gene drive systems that they "lack[ed] control mechanisms and are consequently highly invasive."[48] It did not allow for modulation of its effects once released, nor did it lend itself to biological containment measures. By early 2017, Esvelt had expressed significant skepticism as to how well a standard gene drive could be reliably contained. Ultimately, it was its own efficiency that made standard gene drive difficult to control, given the self-perpetuating way in which it propagated.[49]

Notably, their conclusion that standard gene drive proved to be less than controllable relied on the mechanistic logic and notion of control that had become integral to modern biological practice. If life was brought under control by the understanding and manipulation of more fundamental components, then the lack of control was made evident by virtue of knowledge about the genetic and biochemical functionality of gene drive as insufficiently amenable to producing the kinds of phenomena desired. Even argued in the negative, it was the knowledge about the self-propagating features of standard gene drive on a molecular level that was determinant in whether or not one had control over the entire function of the gene drive. Likewise, not only were invasive species populations to be controlled, but the technological means of that control also was a locus that required its own modes of control. The precise molecular manipulation that underwrote the claims to control of invasive species was also translatable to the control of gene drive itself. The various mitigating strategies presented indicated latent concerns about controlling the controls of life and further underscored the degree to which mechanistic approaches made controllability paramount. Standard gene drive was treated as a biological genie that, once released, could not be put back in the bottle.

Daisy Drive and Mechanistic Control

Beginning in 2016, Esvelt's research group published research on an alternative to standard gene drive, called daisy drive. Unlike standard gene drive, daisy drive promised greater innate controllability of invasive populations. Daisy drive was a variant of gene drive technology that Esvelt's group likened to a "daisy-chain" of serially linked gene drive systems. In a daisy drive, the molecular components of the gene drive were linked in "a linear series of genetic elements arranged such that each element drives the next in the chain," ultimately culminating in the driving of the desired genetic trait in the target population.[50] No genetic element drove itself, so the spread of the desired genetic trait was dependent on the successful driving of each of the prior, requisite parts. The daisy drive would function only if each of the necessary drive components were present.

Crucially, components earlier in the chain were designed to dissipate over a predictable number of generations. The result was that a daisy drive would have rapid initial spread in a targeted population but would eventually degrade and dampen in its effect and prevalence at a stable rate. The rate at which a drive might degrade could also be modified to be faster or slower, according to the amount of generational penetration and geographical spread desired for a particular application.[51]

The metaphor of a daisy-chain itself invoked a mechanistic accounting of complex biological function as the result of stepwise, causal relations between biological parts. Esvelt's research explicitly identified the manipulation of the molecular parts of daisy drive as the basis for the greater control, saying:

> By using **molecular constraints** to limit generational and geographic spread in a **tunable manner**, daisy drive approaches could expand the scope of ecological engineering by enabling local communities to make decisions concerning their own local environment.[52]

Esvelt's articulation of daisy drive invoked heavily mechanistic understandings of control. He analogized the function of molecular parts to control knobs altering population and ecosystem outcomes vis-á-vis invasive species. The mechanistic conceptions of control embedded in standard gene drive are only further reified in the notion of controlling life articulated in daisy drive. Daisy drive added more molecular and genetic parts, more mechanical relations between those parts, promising a greater degree and a finer resolution of control for both gene drive and invasive species. The metaphor of tunable control over daisy drive also suggested a wider range of possible intensities or frequencies of particular applications. Linking greater technical control to local decision making about gene drive implies there are more options on the table to choose from—not just whether to use gene drive, but how, to what extent, when, and by whom. It is worth noting, then, that in Esvelt's group's articulation of daisy drive, it was the increased controllability of daisy drive that was credited with the placing decision-making power in the hands of local communities. Esvelt's group explicitly linked the mechanistic account of control afforded by daisy drive to the empowerment of local communities to make decisions about the use of gene drive. Local co-governance and decision making about gene drive was enabled precisely because of the predictable control afforded by control of the molecular biology and genetics of gene drive, embedding a logic of mechanism in the articulation of the function of gene drive and the social ordering that they authorize.

In doing so, what it meant to do responsible science was linked with the mechanistic notions of control of daisy drive. Mechanistic control of gene drive was made out to be not only necessary for the technical challenges of conservation, but also a prerequisite for any practice of responsive science. The alignment of greater mechanistic control with values of Esvelt's paradigm of responsive science suggests that responsible scientific research also necessarily requires mechanistically controllable technologies. Mechanistic notions of controlling life, thus, become essentialized to responsible scientific practice. Such articulations of responsibility subtly reinforce mechanistic approaches to control as not only technologically practical, but also normatively necessary for biological engineering technologies like gene drive. More importantly, greater control of gene drive enabled the offloading of decision making to local communities while still maintaining a position of moral responsibility over gene drive by virtue of the fact that the mechanistic tunability of gene drive control remained within the purview of scientists like Esvelt.

Community Engagement and the Limits of Mechanism

Convinced of the inherent uncontrollability of standard gene drive, Esvelt began a campaign to push for the adoption of the daisy drive variant in considerations for the use of gene drive for invasive species control in Aotearoa New Zealand. Esvelt's advocacy included personally traveling to the island nation, interviews with local news media,[53] and a public meeting at the University of Otago in Dunedin in September 2017, inviting interested community members, policy makers, and conservation organizations to participate in an open dialogue about gene drive and invasive species control.[54] The meeting provided Esvelt an opportunity to articulate his case for the control advantages offered by daisy drive, if it were decided to move forward with the technology.

At the meeting, Esvelt met with individuals involved with the Predator Free 2050 initiative, the New Zealand government, and members of Māori-led organizations like Te Tira Whakamātaki (the Māori Biosecurity Network) with a record of representing Māori interests. Esvelt placed particular emphasis on including indigenous Māori groups in discussions about gene drive. In part, Esvelt was attempting to practice responsive science to include affected communities in discussions about the possible use of gene drive. Esvelt had stated, "Ultimately, the decision of whether to guide the development of daisy drive is up to the people and government of New Zealand."[55] Not only was doing so consonant with Esvelt's commitments to his "responsive science," but it was also a politically important maneuver for the consideration of gene drive in Aotearoa New Zealand given the complex relationship between the Māori and the New Zealand government as a result of the colonial history that continues to be felt.[56]

The attempts by Esvelt to be attuned to the cultural and political realities for Māori communities was evidenced in the content and tone of what he presented at the Otago meeting. He identified Māori co-governance of gene drive as a fundamental and necessary prerequisite for himself and his involvement with any future use of gene drive on the island nation, saying that "if Māori are not broadly supportive of a proposal [to use gene drive], our group will have no part of it."[57] Esvelt committed to partner with Māori communities in conjunction with New Zealand government and other organizations as part of broader discussion about the use of gene drive for invasive species control in attempts to ensure co-governance of the technology. Esvelt went as far as to incorporate Māori terms, language, and concepts into his descriptions of gene drive and the environment and acknowledged that Māori communities should have input about gene drive governance even if not put into scientific vocabulary.[58]

Yet, the account of gene drive for invasive species control he presented at the meeting still mirrored the mechanistic notions of control that Esvelt and his colleagues had published previously. The function of gene drive and human control of it retained the heavy emphasis on mechanistic descriptions and explanations, especially regarding daisy drive. Risks were framed as pertaining to the geographic spread and biological effectiveness of gene drive to control invasive populations, and they were accordingly addressed by the control advantages of daisy drive.[59] The focus remained on the molecular tools implemented, the genetic changes they caused, and the resultant impact on invasive populations.

However, despite his attempts to be culturally sensitive and inclusive of Māori communities in discussions about the use of gene drive, Esvelt nevertheless did not recognize and attend to an entangled history of science, indigenous peoples, and meaning.[60] The reductive, mechanistic approaches to knowing and controlling life that he emphasized themselves were not neutral but were wrapped up in the alignment of science with projects of colonization, systemic displacement, and epistemic violence toward indigenous people and ways of knowing. Māori communities in particular had a long history with extractive, colonial science regarding biotechnologies like genetically modified organisms and approaches to environmental conservation.[61] The longstanding and continued actions of scientific and colonizer actors had led to skepticism and suspicion of reductive, mechanistic bioscience and the ways that such technoscientific practices impact local environments and cultural values.[62]

Esvelt's lack of engagement with the complex, colonial relationships between science and Māori communities alongside his advocacy for the potential use of gene drive was yet another instance in a history of epistemic dispossession of indigenous people of lands, rights, and identities enabled and underwritten by scientific practice. By not attending to that history as part of his engagement with Māori communities, Esvelt demonstrated that he was unaware not only of the local politics and ecologies of the Aotearoa New Zealand but also of the historical context and stakes for engaging responsibly in the shadow of scientific colonization. Esvelt's inattentiveness to histories of science and empire was not just a case of an individual scientist's missteps or irresponsibility, but also symptomatic of the broader limitations of mechanistic conceptions of controlling life to consider the historical, social, political, and cultural aspects implicated in aspirations to engineer life and practice responsible science.

Furthermore, by describing control for gene drive primarily as a matter of the interactions of mechanistic biological parts, Esvelt also reinscribed a particular vision of nature as being shared between himself and Māori communities. Esvelt's mechanistic approach brought with it commitments to reductive, materialistic explanations as sufficient for knowing organisms, that organisms themselves were reducible to more basic biochemical and genetic parts. Moreover, Esvelt's engagement with Māori communities about gene drive presumed that the issue was primarily one of culture and cultural difference, not epistemic or ontological difference. Taken for granted in that translation was an understanding that the features of gene drive and invasive species control in Aotearoa New Zealand were correspondent to the same entities, that there was a singular, shared nature that was perceived through different cultures.[63]

However, the constructions of what constitutes nature and what constitutes life were precisely what was at stake in those disjunctures in ways of knowing the world. Distinctions such as what constitutes "living" and "nonliving" entities and how one has particular kinds of knowledge or beliefs about the features and significance of organisms, their environments, and broader contexts of place relating to biodiversity and geography are not always shared between Western and indigenous approaches to knowledge.[64] Likewise, significant scholarship has demonstrated that sharp differentiations between epistemology and ontology in Western knowledge systems are also not necessarily shared by indigenous worldviews.[65]

While Māori perspectives toward genetic technologies are not monolithic regarding gene drive or biotechnologies more broadly, Māori scholarship has demonstrated that Māori understandings of nature are not the same as, reducible to, or necessarily incoherent with understandings common in mechanistic biology.[66] Māori visions of the world situate humans within an ancestral lineage that draws and binds together human and nonhuman, animate and inanimate in a genealogical heritage of historically, culturally, and spiritually significant relations rather than as causally linked, mechanical parts.[67] Scientific conceptions of mechanistic biology and indigenous Māori knowledges were not merely culturally distinct ways of understanding a shared natural world, but fundamentally different ways of constructing nature itself.

By not attending to the multiplicity of constructions of nature and how one knows it, Esvelt naturalized his own mechanistic approaches and presumptions about gene drive and control. Esvelt's commitments to mechanistic explanations of the controllability of gene drive implicitly asserted Western, scientific constructions of nature as the normatively right way of understanding the world. As the epistemic basis of making gene drive predictable and controllable, mechanistic understandings of life were also how, for Esvelt, one would "responsibly" ensure Māori co-governance of the technology. In doing so, however, Esvelt pre-framed mechanistic approaches as the animating vision of engineering life in ways that would reflect the limits of biological mechanism and his inattention to the historical and political stakes for Māori communities.

Relational Missteps and Gene Drive Co-governance

The deficiencies in Esvelt's mechanistic approaches to controlling gene drive were soon made manifest in the fallout surrounding his controversial publication of new gene drive research. In November 2017, Esvelt and Neil Gemmell, a scientist colleague from New Zealand, published a scientific manuscript that described the deployment of daisy drive as a potentially preferable, more controllable solution for invasive species control.[68] The paper built on the previous months of community engagement discussion about the use of gene drive in Aotearoa New Zealand for invasive species control, and though the paper did not overtly prescribe gene drive for addressing invasive species concerns, it did specifically identify New Zealand as a possible candidate location for it. Rather than gene drive being one of many possible pathways forward, some perceived the message of the paper to be that the daisy drive variant was a tailored fit for New Zealand's invasive species problems.[69]

The day after the publication of the manuscript, some of Esvelt's Māori collaborators responded, wanting to understand why they had not been made aware of the forthcoming publication mentioning their home as a possible gene drive test site and why they had not been included in the process of reviewing the manuscript given the emphasis on community involvement that had been so prevalent during Esvelt's push for daisy drive over standard gene drive. One Māori scholar and the then chief executive of Te Tira Whakamātaki (the Māori Biosecurity Network), Melanie Mark Shadbolt, criticized the paper, saying it "unashamedly pushes for the use and consideration of their research" and failed to "build respectful relationships with indigenous peoples/communities."[70] Shadbolt further critiqued the paper, saying that because of the exclusion of Māori partners from every step of the gene drive development process "it is now possible that Māori may never get co-governance in the discussion and/or development of gene-drive technologies."[71] Such critiques from Māori scholars like Shadbolt characterized the stakes of gene drive for Māori communities as involving not only their biological function but also how they would be developed and governed. By undermining Māori participation in the process of gene drive development, Esvelt and his research group—despite intention to the contrary—were seen to have potentially irreparably jeopardized Māori access to meaningful co-governance of gene drive for invasive species control.

Shortly after the critical reception of the paper, Esvelt published a public apology on his research group's website. He openly acknowledged his failure to include his Māori collaborators in the publication of the manuscript as a mistake, saying, "I singularly failed to uphold the ideals of Responsive Science."[72] Despite his overt attention and emphasis to do otherwise, Esvelt's failure to fully include his Māori partners further added to a history of harms and injustices against indigenous people under the banner of science. In addition to bringing into sharp relief the longstanding failures of science to shed its own institutionalized, colonial habits, it highlighted the limitations of biological mechanism. Mechanistic approaches circumscribed the kinds of futures imagined in engineering aspirations to those that similarly characterize control of life in terms of stepwise, causal relations among biological components. Alternative conceptions of life, as well as how one knows and controls it, were displaced by Esvelt's presumption of the epistemic supremacy of mechanistic science.

Esvelt accepted responsibility for his actions and their possible consequences for the Māori and the governance of gene drive, and he committed to a redoubling of efforts to restore trust and make amends. Amends-making would include meeting again with his Māori partners in Te Tira Whakamātaki (the Māori Biosecurity Network) and the Te Herenga Māori (the Māori National Network) early in the following year to continue discussing their interests, concerns, and ideas for the future of gene drive in Aotearoa New Zealand.[73] In those meetings, Esvelt stated that he hoped to improve his understanding of both whether and how gene drive might be a help to local communities and ecosystems, as well as how he might "best fulfill [his] broader responsibilities to the world beyond."[74] However, how that process would unfold and with what impact for the future of gene drive and the role of the Māori in its governance remained unclear.

Conclusions

The controversy around Esvelt's manuscript publication without Māori input was not only a case of the continued relational missteps of scientific research amplified by Esvelt's own vigorous assertions to do otherwise. It also was suggestive of the disjunctures in what is salient to controlling life. The disconnect between in-depth discussion of the differences of standard gene drive versus daisy drive technologies and considerations of the relational realities for potentially affected communities reflected the emphasis of mechanistic approaches on reductive, technical explanations of controlling life. The oversight of not including his Māori partners was indicative not only of Esvelt's naiveté of the contested political situation of Aotearoa New Zealand but also of the limitations of mechanistic notions of control to consider alternatives.

The discussions and controversy around the use of gene drive for invasive species control in Aotearoa New Zealand show how a heterogeneity of notions of control, both of biological entities and of the technologies employed to control them, simultaneously exist with aspirations to remake the natural world. However, such aspirations are not merely a technological expression of environmental engineering or design. Ascertaining what forms of life are sufficiently native or invasive, and therefore ought to be protected or eradicated, is inherently value-laden and normative, as are decisions about if and how gene drive might be used to do so. Questions of who is empowered to make such determinations are fundamentally social and political matters. Likewise, control of invasive species is also simultaneously an exertion of control over the kinds of cultural memory and narrative about the history of native and invasive species in that environment and humans' relationships with it.[75] What it means to control life has implications not only for particular instances of biological engineering, but also for what, or whose, visions of nature remade are instantiated. In the making of particular epistemic visions of an engineered nature, so, too, are particular social formations of the world constructed.[76] To repurpose the words of Philip Pauly, the concepts of control employed are consequential for the kinds of "engineering ideals" they realize.[77]

Implicitly, mechanistic approaches to understanding and controlling life embody an engineering ideal akin to that of Jacques Loeb in the early twentieth century. His engineering ideal entailed a privileging of biochemical explanations for more complex biological phenomena and embodied a notion of control as the repurposing of biological processes according to human volition. Control over the natural world was the chief purpose of scientific practice.[78] Contemporary biological approaches are more than reminiscent of Loeb's mechanistic conception of life. They are a direct continuation of a scientific tradition that locates control of life in the ability to understand and manipulate biological parts to produce intended phenomena. In so doing, they authorize a vision of the world in which non-biochemical dimensions of controlling life are systematically occluded. Such elisions of history and context are a serious lapse, especially so because the motivations for engineering nature are inextricably intertwined with human values and desires for a reimagined world. It is critical, then, in the undertaking of engineering life not only to reflect on what it means to control life, but in so answering, to also recognize whose ultimate visions of nature and society remade are thereby privileged and made real.

2 * *A Tale of Two Rats*

ANITA GUERRINI

Humans have been trying to "remake" rats for at least two centuries, attempting to transform nature to fit our needs. These rodents have been our intimate companions for millennia. Most attempts at manipulating rats to enable their peaceful coexistence with humans entailed finding ways to kill them or at least exclude them from human presence. Remaking the rats themselves, whether by selective breeding, by training, or most recently by genetic modification, began only in the mid-nineteenth century. While killing rats remains a goal of urban dwellers, public health officials, and ecologists, at the same time rats have become highly valued tools of life science as indispensable laboratory animals. These incommensurable and indeed incompatible goals highlight the unique and peculiar relationship that has developed between rats and humans. They also highlight limits—social, biological, and political—to engineering nature.

Although there are many species of rats in the world, this essay focuses on two: the brown or Norway rat (*Rattus norvegicus*) and the black or ship rat (*Rattus rattus*). Among the five species of rats considered to be commensals, these two species are the most common, with global distribution.[1] Commensalism is usually defined as a relationship between two organisms in which one organism benefits, and one is unaffected. Although rats certainly benefit from their association with humans, many humans do not believe themselves to be "unaffected" by rats. Rats are urban pests, carriers of disease, and invasive species. As poet Robert Browning declaimed in 1843,

> Rats!
> They fought the dogs and killed the cats
>   And bit the babies in the cradles,
> And ate the cheeses out of the vats,
>   And licked the soup from the cooks' own ladles,
> Split open the kegs of salted sprats,
> Made nests inside men's Sunday hats

And even spoiled the women's chats
By drowning their speaking
With shrieking and squeaking
In fifty different sharps and flats.[2]

In addition, rats prey on vulnerable species. Nevertheless, they are second only to mice as laboratory animals, and have been kept as pets since the eighteenth century. At least half a million American households own a pet rat or mouse, and breeding "fancy rats" is a recognized hobby with its own membership society and annual shows.[3] These multiple functions of rats seem separate, but they are as closely intertwined as the tails of the semi-mythical "rat king."[4] Nonetheless, the engineering of rats has proceeded in very different directions, depending on where rats are: in the lab, in human living spaces, or in nature.

Brown and black rats are among the most manipulated creatures on the planet. Their usefulness as experimental animals was recognized early but not fully exploited until the nineteenth century; in 1621, physicians Theophilus Müller and Johann Faber, members, with Galileo, of the Accademia dei Lincei, dissected a female rat, which they described as a hermaphrodite. A century later, English surgeon William Cheselden mentioned investigating the testicles of a rat, but the first documented use of Norway rats as experimental animals occurred in the 1850s. They were the first mammalian species bred specifically for scientific use.[5] Yet because of their roles as urban and suburban vermin and as invasive species, draconian extermination programs aim at engineering out of existence rats that are not enclosed in a lab. While the bioengineering of lab rats is an obvious intervention into their biomechanics, extermination is the most complete intervention of all. Nevertheless, the knowledge of rat biology that scientists have gained in the lab has been applied only intermittently to rats in other settings, although as we learn from Christian Ross's chapter in this volume, gene drive has recently been proposed as a method of population control. Black and brown rats originated in Asia; the black rat most likely in India, and the brown farther north in central Asia.[6] The black rat had already reached Europe in antiquity and played a role in the spread of plague in the Middle Ages. It traveled, as its alternative name indicates, on ships.[7] It is the "big domestic mouse vulgarly called rat" in the 1551 *Historiae animalium* of Conrad Gessner.[8]

However, Gessner added at the end of this volume an entry for a "big country mouse" described as "somewhat red" in color that may have been the brown Norway rat. The brevity of his description (although he did include a picture) indicates that he did not know much about this animal.[9]

The name "Norway rat" originated in eighteenth-century England, when naturalist John Berkenhout surmised that this apparently new species of rat arrived on British shores in the 1720s from Norway.[10] That was not true, but the name stuck. Because brown rats are bigger and smarter than black rats, as well as more tolerant of cold weather, they soon took over Europe and moved on to the Americas.[11] However, black rats retained their place on ships and as Western trade circled the globe, so did they.

Gessner expressed no fondness for rats, describing black rats as "aflame with lust" and claiming that their urine could decay human flesh.[12] The arrival of brown rats in Europe led to increased invective. Fewer than two decades after Browning's account of the unfortunate town of Hamelin, "Uncle James" Rodwell's popular *The Rat: its History and Destructive Character* (1858) related the viciousness and cannibalism of rats, including tales of them feeding on infants and even on adults. City rats or country rats, wrote Rodwell, it made no difference. They will eat anything: "nothing dead or alive, on water or on land, is safe from their rapacity."[13] By the late eighteenth century, brown rats had become commensal species with humans, sharing their living spaces and benefiting from the association. They are ubiquitous in most major cities; the *New York Times* recently declared, "Rats are taking over New York City."[14] The urge to modify rats, whether to a constrained life or to an early death, has never been stronger.

From Rat Baiting to the Lab

When Unity Mitford, English aristocrat and friend of Hitler, made her society debut in 1932, she carried on her shoulder her pet white rat Ratular. She took Ratular "to dances and even to a Palace garden party."[15] In the nineteenth century, rat catchers in big cities in North America and Western Europe captured rats by the thousands to be thrown into pits for the popular sport of rat baiting, in which a dog would kill as many rats as he could within a given time. Rat catchers such as Jack Black, self-proclaimed rat catcher to Queen Victoria, set aside any unusually colored rats he caught—particularly the albino ones—and bred them for sale as pets. By handling rats in the course of breeding, rat fanciers also tamed them.[16]

In Asia, Japanese fanciers had kept rats as pets since the eighteenth century. A guidebook from 1775 referred to breeding them selectively for particular colors.[17] Even Rodwell in 1858 admitted that rats were smart and easily tamed. Alongside his stories of rats nibbling the toes off human babies were stories of rat courage and motherly love. An 1880 article in the British *Boy's Own Paper* described rats as "the most amusing of pets" as well as the most "cleanly animal in the world." A later article added,

"They are very funny."[18] Unity Mitford's Ratular was the consequence of a century or more of selective breeding, a form of engineering that remade Norway rats into pets—and into laboratory animals.

Scientists used the same Norway rats that Jack Black bred to sell as pets: mainly, albino rats. Nutrition and metabolism were the topic of the first published scientific work employing albino rats, published in 1856.[19] By the 1890s, scientists recognized rat intelligence and added neuroanatomy and, later, behavioral studies. Their small size made them easy to transport and to house, and by 1900, albino rats regularly crossed the Atlantic from European laboratories to American ones. Meanwhile, selective breeding continued, first in Europe and then in the US. Brown rats are impressively fecund: they can produce up to five litters a year, each litter consisting of four to eight or more rats. Those rats in turn reach sexual maturity in about three months.[20] With several generations over the course of a year, selective breeding—for color or behavior, for example—can proceed very quickly. Darwin's evolutionary theory added a theoretical basis for such breeding. In 1909, Henry H. Donaldson at the Wistar Institute in Philadelphia, who had initiated a plan to standardize the albino rat three years before, commented, "It would be hard to find another animal that combined so many virtues in so compact and pleasing a form."[21] The Wistar's colony of albino rats became the model for scientists around the world.

Scientists at the Wistar Institute engineered "Wistar rats," a uniform and consistent animal model that would yield reliable experimental results, and the Institute successfully marketed its rats globally. Intensive breeding practices, which included inbreeding—that is, mating siblings—served both experimental goals and this creation of a model animal. The institute also developed protocols for housing and diet that formed the basis for all subsequent animal experimentation on rats.[22] Donaldson, the institute's director of research from 1906 until 1938, assembled a mass of data on the Norway rat and its albino variant, including anatomical and metabolic information. This data proved to be critical to establishing the biological baseline for the "normal" rat and for setting breeding parameters. The animals' long use as experimental organisms has validated their scientific utility and provided, as historian Bonnie Clause put it, "a reference library of results upon which researchers can continue to build."[23]

The distance between urban rats nibbling on babies' toes and sleek uniform laboratory rats seems very long, but it took place over the span of less than one human lifetime. Helen Dean King at the Wistar Institute, who bred several of its best-known strains, began with Donaldson's albino rats. She carried one line of these rats to the 135th generation, and it became known as the "King Albino" or PA strain. She also captured wild Norway rats in Philadelphia, the basis of the brown Norway (BR) strain.

The Long-Evans strain began as a cross between a Wistar albino rat and a wild Norwegian rat caught in Berkeley. At least twenty-five modern strains of experimental rats descend from Wistar rats, and another fourteen, including the much-used Sprague-Dawley rat, arose from crosses of Wistar rats with other strains.[24]

Specific strains of rats, named for their breeders, including Osborne-Mendel and Sprague-Dawley, dominated rat research at midcentury, as rats became model organisms for biological research, defined in terms of their amenability to laboratory life and the body of data that allowed them to be proven experimental models for humans. They became "technological matter" as well as living matter. Rats were bred to be prone to cancer, or hypertension, or obesity. William E. Castle at the Bussey Institution at Harvard, established in 1908, began to study the genetics of rats, particularly the inheritance of coat colors, and eventually, knowledge of the rat genome revolutionized the breeding of laboratory rats.[25] Meanwhile, other sorts of engineering, such as the development of the germ-free rat in the 1950s, led to major advances in research. "Gnotobiology," the study of animals in a germ-free environment, depends on technology rather than breeding to keep newborn rats completely free of germs.[26] According to publication data, rats became increasingly important in the 1950s and 1960s. By 1966, when President Johnson signed the Animal Welfare Act into law, rats and mice were by far the most common experimental animals. However, even though subsequent revisions extended the AWA to cover all warm-blooded animals, rats and mice have continued to be explicitly excluded, along with birds. It seems that the identity of these rodents as "vermin" excludes them from regulatory consideration, even as new technologies have made them ever more valuable; another example of the redefining of nature to fit human needs.[27]

For the past century, scientists have continued to use descendants of the original Wistar rats and other twentieth-century strains; in the early 2000s, *Nature* estimated that more half of all experimental rats in use at that time descended from Wistar rats.[28] Sprague-Dawley and other popular strains continue to be employed in a variety of experiments. Since the sequencing of the brown Norway rat genome in 2004, however, new technologies have radically changed how scientists deploy experimental rats, and many of the animals themselves are now custom-made via the use of genetic modification, as new knowledge of life leads inevitably to its engineering. The use of assisted reproductive technology, including the recovery of rat oocytes, in vitro fertilization, embryo culture and transplantation, and microinsemination, have allowed unprecedented manipulation of rat reproduction. Researchers can manipulate oocytes (immature egg cells) and embryos in several ways, including somatic cell nuclear transfer in the oocyte (the first step in cloning), and establishing embryonic stem cells. These techniques in turn open the door to creating genetically modified rats.[29]

The first transgenic rat dates to 1990 and was created by injecting a single gene into a newly fertilized egg. The 2004 sequencing of the BN genome led to an explosion of new strains of rats and new uses for rats in the laboratory. The rat genome showed just how far apart rats and mice are, but also showed why the rat has been such a good model for the human body.[30] Scientists created the first "knockout" rat in 2008, in which a single gene sequence is "knocked out" or turned off. Genes can also be "knocked in" or added. Such mutations are heritable, creating new strains of transgenic rats. The technique known as CRISPR-Cas9 allows for finer tuning of DNA sequences, creating ever more specialized rats.[31]

By means of these techniques, the rat has become an even more valuable research animal. While genetic manipulation of mice preceded that of rats, so that earlier studies of genetically modified organisms focused on mice, the development of gene-editing technologies for rats has led to their comeback in the lab. Because they are bigger, more easily tamed, and more intelligent than mice, as well as genetically and physiologically more similar to humans, rats have a big advantage over mice as model organisms in many disciplines.[32] A recent report, however, noted that rats are still "playing catch-up" with mice, in part because of the expense of transgenic rats.[33] The monetary and scientific value of genetically engineered rats is enormous: a recent catalog from Cyagen, one of many providers of animal models for science, lists a group of three CRISPR-Cas9 knockout rats for prices starting at $17,000. More complex knockout models list for as high as $36,000 for three; these would then become the basis for a breeding colony.[34]

While other organisms, such as zebrafish, are now also coming into use, mice and rats still constitute at least 90 percent of experimental animals. The exclusion of rats from the US Animal Welfare Act means that scientists are not required to report the numbers of these animals that they use. While policies of the US Public Health Service cover experimental uses of these animals, these policies do not include reporting requirements. Estimates of the numbers of rats and mice used each year vary widely, from 12 million–27 million (according to a US government estimate in 2016) to 100 million (according to the animal rights organization PETA).[35]

Although the laboratory rat has long been a tool of science, transgenic technologies have carried lab rats farther and farther away from their origins, beyond the selective breeding of the past century. In 1999, Oxford zoologist Manuel Berdoy released his documentary film "The Laboratory

Rat: A Natural History," which followed a group of lab rats released into an outdoor enclosed environment. They quickly reverted to "wild" behavior, including nesting, hunting, and the creation of social hierarchies typical of rats in the wild.[36] It is not clear whether the present generations of genetically modified rats would be able to shed their laboratory life as easily.

Rats as Pests

In 1914, the "Extermino" Chemical Company of Dundee, Scotland, announced

> EXO for RATS. Tons have been supplied Government use against Plague, a recommendation beyond dispute. "The World's greatest." Every Rat destroyed. . . . No smell or mess.[37]

Rats outside of the laboratory continued to be pests. In the twenty-first century, country rats eat rice and other stored grains; recent losses in Asia are estimated at 5 to 10 percent of total output. In cities, rats chew on electrical cables and pipes. They carry noxious bacteria, viruses, and fungi, as well as serving as intermediate vectors for a variety of insect- and arthropod-borne diseases.[38] They sometimes bite babies. They ruin food supplies with their feces and urine. Because over half of the world's population now lives in urban areas, urban rats are even more of a problem than they were in the nineteenth century. Urban rats include both *Rattus rattus* and *Rattus norvegicus*, and Norway rats almost always displace black rats. The Norway rat is dominant in US cities. Urban rats tend to be bigger and mature earlier than rural rats because of the abundant availability of human-provided food. Owing to environmental hazards and fierce intraspecies competition, they have short life spans, and usually survive for only a year, but their fecundity means that their populations generally remain stable. In cities, streets form significant barriers, so the home ranges of urban rats tend to be confined.[39]

We know surprisingly little about the ecology of urban rats.[40] A 2013 review essay by Alice Feng and Chelsea Himsworth revealed some interesting facts, but also revealed large areas of ignorance. One salient fact, which also explained the difficulty of trapping urban rats for study, is that they are "neophobic"—they avoid new things in their environments, such as traps or unfamiliar food. Ecologist Charles Elton, who studied rats in the 1940s as part of the war effort in Britain, had earlier described this behavior as "new object reaction."[41] Journalist Robert Sullivan, who observed a colony of rats in New York City in 2000–2001, described the difficulty public health officials had in trapping and examining urban rats. Rats, moreover, have a finely tuned sense of taste that allows them to avoid tainted food or other harmful substances.[42]

Humans have tried to control urban rats—to engineer them out of existence—for hundreds of years. The migration of the bigger and smarter Norway rat to the west has only intensified these efforts, and with the discovery in the 1890s of the role of rats in spreading diseases such as plague, their eradication became an urgent issue of public health.[43] Early methods of control included pouring boiling water down rat holes, beating them, shooting them, and employing predators such as dogs and cats (although modern adult urban rats are so big, most cats avoid them). But the rise of the chemical industry in the mid-nineteenth century led to a new solution: poisoning them.[44]

Early rat poisons in the late nineteenth and early twentieth centuries included arsenic, strychnine, and the powdered bulbs of a Mediterranean plant known as red squill.[45] Enemy occupation of the South Pacific and southern Europe during World War II cut off Allied supplies of strychnine (derived from the seed of a tropical plant) and red squill. The eradication of rats was a wartime priority because they destroyed food supplies and carried diseases. A new rodenticide was urgently needed. Although DDT, the miracle pesticide of the 1940s, was fatal to rats in high doses, they soon learned to avoid it.[46] The US government therefore funded research into new rodenticides, and two came into use: sodium fluoroacetate, also known as Compound 1080 (about which we will hear more in the next section), and alpha napthyl thiourea, marketed as ANTU.[47] The city of Baltimore tested ANTU in an experimental field trial between 1942 and 1946, led by biologist Curt P. Richter. Richter's behavioral research at Johns Hopkins had used many Norway rats, but he had no experience with rats outside the laboratory. Through work with a combination of lab and wild rats, he developed ANTU.[48] Although the field trial in Baltimore was deemed successful, ANTU proved to be less of a panacea than Richter originally thought. Rats who ingested a sub-lethal dose were unlikely to try it again, and such doses led to resistance. Richter's goal in Baltimore was 100 percent eradication, but this could never be confirmed, since no one knew how many rats lived at a given site.[49] Richter described a death rate of 60–80 percent as "next to useless," but even at an estimated 95 percent success, rats returned to pre-ANTU levels within a few years. Public health officials in Baltimore pointed out that environmental controls that deprived rats of living spaces and access to food was equally, if not more, essential in keeping apart rats and people.[50]

Meanwhile, in 1948, another rat poison hit the market. Unlike ANTU, which affected the capillaries of the lungs, leading to respiratory failure, warfarin inhibited the clotting function of the blood, causing rats to die of internal bleeding. Discovered by the University of Wisconsin biochemist Karl Link, it became the most widely used rodenticide for the next half-century, even though rats began to exhibit resistance to it by 1960.[51] Scientists have since developed many other anticoagulants and other rodenticides, in a never-ending arms race against resistance. Also of concern is the collateral damage that ever more powerful poisons cause in other animals and in children who happen to ingest rat bait. Urban rats signify both poverty and affluence; the poor suffer most from their effects, but affluent societies provide the ample amounts of garbage that enable urban rats to flourish. Urban rat control programs urge both environmental controls that keep rats out of dwellings and keep food away from them, and the use of rodenticides. However, pest control is hardly ever the highest priority for urban governments, so no matter what lethal tools scientists may develop, their use is always constrained by local politics. Currently the rats seem to be winning this battle.

Rats as Invasive Species

In his 2011 novel, *When the Killing's Done*, T. C. Boyle explored the eradication of invasive rats on an island in the Santa Barbara Channel.[52] In the winter of 2001–2002, helicopter-borne ecologists dropped bait laced with the potent anticoagulant brodifacoum onto Anacapa Island, actually three linked islands. Their purpose was to eliminate black rats, and another application followed the next winter. According to the ecologists, this was "the first-ever invasive rodent eradication from an entire island where an endemic rodent was present and the first aerial application of a rodenticide in North America."[53] As Boyle explained, the rats most likely arrived with an 1853 shipwreck. Finding an island with few other mammals but abundant seabirds, they flourished. This combination of species turns out to be a common denominator among island invasions. Other nonnative species on Anacapa Island, introduced by ranching in the twentieth century, included cats, sheep, and rabbits, but by 2000 all of these species were gone. Only the rats and the small native Anacapa deer mouse—the endemic rodent—remained among mammals.[54]

The technique of dropping poison-laced bait had been perfected in New Zealand. The Anacapa Island ecologists took elaborate precautions to prevent poisoning other species, including staggering the applications among the east, middle, and west islands with accompanying translocation of affected species, and capturing some deer mice and birds to assure that breeding populations would survive. Nonetheless, many birds died from the poison, ranging from raptors to sparrows. From an ecological standpoint, the rat eradication was an undoubted success. Populations of endemic species, including the deer mouse, salamanders, and several species of birds, rebounded.[55]

Although urban rats, at least in North America, are nearly all *Rattus norvegicus*, "invasive rats" usually refer to *Rattus rattus*; the official list of the 100 worst invasive species does not include the Norway rat.[56] Rats have been particularly attracted to islands, and three species of invasive rats—black, brown, and Pacific (*Rattus exulans*) occupy 80 percent of the world's island groups. Their voracious appetites continue to prey upon numerous species and disturb global ecosystems. Rats, together with cats, dogs, and pigs, are responsible for the large majority of predator-caused extinctions, which amount to nearly 60 percent of all extinctions. Engineering invasive rats out of existence has become a primary goal of global conservation policy.[57]

However, in contrast to the sophisticated science surrounding lab rats, or the complex relationships among urban humans and urban rats, island eradication has seemingly been straightforward. The policy is simply to poison them all, although recently, as Christian Ross has explained in his chapter, gene drive, in multiple manifestations, has been proposed as an alternative method of control. Unlike urban rats, which, as Richter discovered in the 1940s, are nearly impossible to eliminate fully, islands offer a naturally bounded territory, and there have been hundreds of island eradications since they began in New Zealand in the late 1950s.[58]

New Zealand (Aotearoa) has no native mammals except for a few species of bats, but it has many native birds, several of them unique, such as the moa (now extinct), kiwi, and kakapo. It consists of two large islands and many smaller islands, and eradications have mainly occurred on these smaller islands. With a larger proportion of nonnative species than perhaps any other place in the world, it is not surprising that New Zealand has become the model for invasive species eradication. A 1950 study identified many introduced species of wild mammals and described their "blending into a kaleidoscope of the indigenous fauna and flora." Less than a decade later, however, the first attempt at eradication of invasive rats occurred, on two tiny islands off the east coast of New Zealand's North Island. It involved Norway rats and a warfarin-based rodenticide known as "Rid-rat," and took place between 1959 and 1961.[59]

The Māori brought the Pacific rat to New Zealand around 1,000 years ago, probably for food. Europeans brought brown and black rats, as well as mice, in the eighteenth and nineteenth centuries. The introduction of black rats to Big South Cape Island, off the southern tip of New Zealand, provides a well-documented case study of both introduction and eradication. Black rats probably reached the island in 1963 from Māori fishing boats; their presence was noted early in 1964. The effects of the rats on native birds was immediate and severe; scientists later commented, "There was absolute silence over most of the island."[60] Warfarin bait applied around the island in 1964–65 was ineffective, and attempts to relocate specimens of some of the most endangered birds to rat-free islands also mostly failed, although two species survived the transfer. Meanwhile rats ate native grasses to the ground, and a native bat also disappeared. At the time, the only islands that had seen complete eradication of rats were tiny, about one hectare in area; Big South Cape Island was over 1,000 hectares. The combination of funding, inter-agency coordination, and eradication technology did not coalesce until 2006, when aerial distribution of brodifacoum killed all the black rats on the island, which was declared rat-free in 2009. Specimens of the two species of birds that had been successfully relocated were returned to the island.[61]

The rat invasion of Big South Cape Island has been called "one of the worst ecological disasters in New Zealand history," with four species rendered extinct. However, the spectacle of species disappearing before the eyes of scientists and the public helped to galvanize opinion in favor of extermination campaigns against invasive species, convincing scientists that such extermination could be effective. Rats, mostly black rats, have now been eliminated from several of the smaller islands surrounding New Zealand.[62]

As island eradications increased in the early 2000s, joined by a widespread eradication campaign against all nonnative species across New Zealand, animal rights activists and others expressed growing concerns, particularly about the use of the poison known as 1080. This poison of choice on the mainland is a synthetic version of sodium fluoroacetate, which US government scientists developed during World War II around the same time as ANTU. It was so toxic that its manufacturer, Monsanto, originally restricted its sale to professional exterminators; *Time* magazine in 1945 likened it to "atomic energy . . . almost too hot to handle."[63] Unlike the anticoagulants used on the islands, 1080 causes vital system failure, and death from it is painful and protracted. The web page of the New Zealand government notes that 1080 is biodegradable but does not state how long it takes to degrade, noting only that it "dilutes rapidly . . . in waterways."[64] Ground and aerial application of bait laced with 1080 is directed at Australian possums, feral cats, rabbits, rats, and stoats. The US bans all applications of 1080 except for "toxic collars" employed on sheep and goats to kill predatory coyotes.[65]

In response to this growing concern about widespread use of toxic rodenticides, New Zealand ecologist David Towns and two colleagues undertook a comprehensive study of the adverse effects of invasive rats, employing evidence from around the world, including Big South Cape Island. They concluded that rats are especially damaging to seabirds: they eat eggs, chicks, and even some small adults. "We conclude that the strong evidence of pervasive negative effects from introduced rats is sufficient to justify the high costs of eradications."[66] Notably, Towns and his colleagues confined their study to islands where rats were the only introduced mammal. However, a case study of Norway rat eradication on a small island in the Indian Ocean complicates the picture they drew, and their definition of "natural," because rats were not in this case the only introduced rodent.

Norway rats may have been introduced to Tromelin Island after a shipwreck in 1761, or during a visit by an English ship in 1856. By the time humans settled there in 1954, both rats and mice were well established, and at least four species of previously recorded seabirds had disappeared. By the mid-1990s, two more species of seabirds had disappeared, leaving only two breeding species on the island. Over a month from December 2005 to January 2006, ecologists employed chocolate-flavored pellets of brodifacoum to eradicate both rats and mice. Although the rats all died, the mice did not. Because rats had played a role in controlling the mice population, the number of mice increased, but so did the number and variety of birds, since mice did not prey on birds (in fact, little is known about the impact of mice on tropical islands). While the authors of the study concluded that the eradication had achieved its goals, they did acknowledge that the continued presence of mice was a complicating factor.[67]

While the effects of elimination of a single species appear to be relatively straightforward, eliminating multiple species multiplies the impacts of eradication, often in unforeseen ways. It complicates the simple stories of invasion-eradication-recovery that ecologists have employed to justify mass poisoning of invasive species while underplaying the collateral impacts. What seems on the surface to be a simple case of cause and effect is in fact a complex engineering project whose effects are not yet fully understood. Many studies have indeed concluded that eliminating some species may have unforeseen effects on others. Invasive plants, in particular, may expand their range when predators such as rats disappear.[68]

Conclusion

These case studies demonstrate that as humans have gained knowledge of rats' lives, they have subjected rats to numerous technologies of control, engineering them to fit human needs. Laboratory rats have been standardized to conform to the demands of biological science for well over a century, and they count as one of the great success stories of modern science. On the other hand, technological mastery of populations and habitats rather than individual organisms has led to the development over the same period of techniques of mass slaughter and environmental controls, which have been applied to urban rats with indifferent success. Similar techniques when applied to rat eradication outside of urban areas, where they have shown themselves to be the cause of environmental havoc, have been more successful, especially on islands, but their long-term effects remain incompletely understood.

None of these attempts to engineer rats have been without controversy, and the varied attempts as well as the varied responses to them tell a historical story that is neither triumphant nor defeatist. While laboratory animal use has been regulated for decades, activists continue to press for ever more restrictive uses of animals, or the elimination of animal research altogether. However, in the priority list of animal concern, rats are far lower than primates, dogs, or cats, even though scientists use many more rodents than all of these animals combined. Where invasive rats are prominent, such as in New Zealand, the rhetoric of animal rights applied to laboratory rats is entirely missing from discussions of invasive rats.[69] I have found no assertion of the possible rights of urban rats.

The human valuation of few other animals is as dependent on context as is the rat's. Humans universally revile urban rats, while scientists afford high value to laboratory rats as tools for human benefit. Invasive rats have no value in comparison to the species they prey upon or displace. In each case, science and society have concluded that rats are somehow out of place without the mechanisms of human control. Yet humans have not been entirely successful in restricting rat populations to laboratories, and until scientists spend more time studying rats as rats rather than as means to human ends, we will not know what the consequences might be if the only remaining rats are in laboratories. The resilience rats have shown as a species probably means that they will always be with us, whether we like it or not.

3 * *Cloning as Orientalism: Reproducing Citrus in Mandatory Palestine*

TIAGO SARAIVA

This chapter details the importance of citrus reproduction practices in projects for engineering Palestine. It displaces the history of cloning from dystopic sci-fi scenarios produced in closed biological laboratories, into expansive political vistas like the ones normally found in histories of engineering in which state, capital, and technology are obligatory presence.[1] By cloning budwood and rootstocks, Jaffa oranges became mass produced commodities ready to be shipped long distances for international markets, the main export from Palestine in the first half of the twentieth century. This is a story of obsessive reproduction of the same at industrial scale, not of endless innovation of new forms of life at laboratory scale; a story in which productivity, efficiency, and control, all typical engineering values, have center stage. The deployment of such values, historians of engineering like to remind us, meant new forms not only of producing material objects but also of producing new humans such as white male engineers and unskilled workers of color.[2] Here, these human effects of cloning will become apparent by exploring how being Jewish or Arab became intertwined with specific practices of citrus reproduction.

The close attention to reproduction practices more characteristic of historians of biology and science studies scholars will contribute to unveil the connected histories of Palestine and California.[3] It is in fact through the correspondence of Jewish scientists at the Agriculture Experiment Station in Rehovot and American scientists at the University of California Citrus Experiment Station in Riverside that we learn much of the challenges of keeping the commercial qualities of the Jaffa orange. As Israeli historians have already demonstrated, the Zionist project during the Mandate period (1920–1948) gained much inspiration from the Californian citrusscape.[4] It has been less noticed that while other regions of the globe emulated Californian orchards by reproducing the same varieties of oranges—Washington Navels and Valencias—these varieties did not travel to Mandatory Palestine, which mostly stuck to its Jaffa oranges. But if actual oranges did not travel between the two regions, cloning practices did, remaking Palestinian Jaffa oranges in the image of Californian Navels. Maybe more unexpectedly, these travels were not a Jewish exclusive and they also sustained Arab visions for engineering Palestine that demand questioning simplistic dichotomies of modern Jews and traditional Arabs.

Biology and New Orientalism

In August 1910, the United States Department of Agriculture published Aaron Aaronsohn's "Agricultural and Botanical Explorations in Palestine." Knowledge of plants from faraway lands was central for a federal agency tasked with introducing new crops to the repertoire of American farmers and supporting the settlement of the different regions of the country. Aaronsohn followed the rules of the genre and offered an extensive list of "economic plants worthy of introduction into the United States."[5] Less common was to not trust such task to one of the celebrated plant hunters of the USDA such as Walter T. Swingle or David Fairchild, and rely instead on reporting from a foreign expert, in this case, the head of the Jewish Agricultural Experiment Station in Athlit, some ten miles south of Haifa, in Palestine.[6] Despite the service it provided to the USDA, Aaronsohn failed in his intentions of having the American federal government directly funding his experiment station in Palestine. He did succeed nevertheless in convincing wealthy American Jews to provide the funds necessary for his institution, which was chartered under New York state law in 1911.[7]

Aaronsohn's acclaimed discovery of the wild ancestor of wheat at Mount Carmel, very close to the site of his experiment station, not only guaranteed his status among American botanists and plant breeders, but also caught the attention of Jewish communities in America. The wild wheat relative confirmed the Fertile Crescent and the Syria/Palestine area as no less than one of the cradles of civilization built on domesticating grains. Aaronsohn, who grew up in Palestine and whose parents were Romanian Jews of the first Aliyah (1882–1903), challenged with his major scientific accomplishment negative views about Zionism as a project undertaken by unsophisticated Jews from Eastern Europe.[8] In fact, he had already made part of the Commission for the Exploration of Palestine (1903–1907) put together by the World Zionist Organization. Under the leadership of the botanist Otto Warburg, the Commission transformed Zionism into a technocratic endeavor at the image of the German Colonial Service.[9] While German botanists had gained recognition in the late nineteenth century for their role in promoting investments in tropical plantations (cocoa, cotton, peanuts, palm oil, . . .) across the German empire, the scientists of the Commission now surveyed Palestine, identifying resources and business opportunities to attract the capitals necessary to make Jewish settlement viable.[10] The success of Zionism was identified with the transformation of Palestine from unexplored land into a modern landscape of mines, cash crops, and industry. In other words, the Commission presented Zionism as an engineering project.

Capital and celebrity were not the only things that took Aaronsohn to the United States. He presented his experiment station as a source of plant materials for American plant breeders in search for exotic species and varieties from the Middle East, but reciprocally he expected to have access to American plant collections amassed in Washington, DC, to undertake his own acclimatization work in Palestine. Aaronsohn offered "drought-resistant stocks and dry land grains" of special interest for farmers in the southwest of the US; in exchange, he brought to Palestine American date palm trees and "24,000 specimens of fungi." Equipped also with the whole collection of USDA bulletins and reports, the Jewish Agricultural Experiment Station hoped to advise Jewish farmers in Palestine on productive agricultural methods and new plants the same way American experiment stations offered expertise to settlers farming in newly occupied lands. The Athlit station thus combined in a single institution the American experience of settling the West and German colonialism in the tropics.

In his 1910 report to the USDA, Aaronsohn characterized his work as a new form of orientalism: "Economic exploration rather than scholarly research is needed to make the countries of the Orient known and appreciated abroad."[11] The natural scientist replaced the legal and cultural expert of the nineteenth century in transforming the Orient into a valuable resource for Western development. This new orientalist made the plants of Palestine into resources expanding American economy and in the same step transformed Palestine into a territory suitable for Jewish settlement. As Edward Said famously sustained in *Orientalism*, Western authors constituted "the Orient as a decrepit canvas awaiting his restorative efforts. The Oriental Arab was 'civilized man fallen again into a savage state.'"[12] Revisionist readings of Said's work have criticized how his mentioning of "restorative efforts" referred mostly to literary work by Western men of letters, with Said showing little concern for the actual historical role of orientalist texts as a "blueprint or a road map for an effective, and not just textual, re-creation of the Orient."[13] In this interpretation, orientalism was an issue not only of Western (mis)representations of the East, but of engineering the Orient.[14] The historical figure of Aaronsohn fully belongs to this genealogy of orientalists as engineers, not only portraying Arabs, their lands, and agricultural practices as belonging to a fallen civilization, but advancing as well concrete projects to remake the Orient.

In good orientalist tradition, care and knowledge for things oriental was entrenched with contempt—as well as ignorance—of the culture under scrutiny. The diversity of plants which Aaronsohn tapped into was attributed as much to climate and geologic conditions of Palestine as to social backwardness. More than ten centuries of cultural stagnation under Arab influence "finally rapidly disappearing," allegedly determined the isolation in the territory of a "number of hostile tribes, each one living in a territory of quite definite natural boundaries."[15] The abundance of varieties of cultivated plants was described as the direct result of the inexistence of commercial relations between different tribes fated to "cultivate for centuries on the same soils without outside introduction."[16] Historical arguments became biological ones asserting the value of Palestine: cultural inertia had made Palestine, according to Aaronsohn, into a reservoir of old cultivars of very low productivity but holding important traits such as drought or disease resistance, perfect raw materials for plant breeders in search of interesting traits to introduce in their new varieties. The wild wheat found by Aaronsohn in Mount Carmel was exemplary in this respect: It could be used by American breeders in crossings to produce new hybrids sustaining farmers in semiarid regions of the West of the United States and it enabled as well dreams of new fields of grain sustaining Jewish settlers in Palestine.

Aaronsohn was especially interested in establishing equivalences between Jews farming in Palestine and Americans settling California. Palestine is about a tenth of the size of California, but Aaronsohn reminded that the flora catalog of the two regions not only had a similar size—around 3,000 plant species—but also had important qualitative communalities: "evergreen shrubs predominate. The same forms of vegetation, often the same genera, are found on Mount Tamalpais, California, and on Mount Carmel, Palestine; the maqui formation of Palestine is to be compared to the chaparral and chamiso of California, and the forms of vegetation of the Lebanon and the Hermon mountains are much the same as those of the western slope of the Sierras."[17] Aaronsohn was following here the path of botanists like Candolle, who since the late nineteenth century had made plant geography a key feature of acclimatization efforts instead of trying to introduce plants from regions with different environmental conditions.[18] Identifying similarities in climate and topography made exchanges of crops between Palestine and California much more plausible than between Palestine and other areas of the United States, not to mention other areas of the globe.

Citrus cultivation being at the turn of the century the backbone of the economy of Southern California, it is not surprising that Aaronsohn also described for his American readers the citrus orchards in coastal areas of Palestine, namely around the port city of Jaffa. The presence of citrus in both California and Palestine confirmed the accuracy of the analogies between the two regions. He praised the qualities of the Jaffa orange, its thick skin enabling long distance shipping without damage, its large size and juicy pulp, as well as the absence of seeds, earning the favor of consumers. The long harvest season from late September to early March employed a large proportion of the Arab population of the region in picking, packing, and shipping the fruit that arrived in the busy harbor of Jaffa on the back of camels.

To reconcile the patent commercial success of the Jaffa orange, a cash crop commanding high prices in Constantinople, London, and Hamburg with the orientalist portrait of backward Palestinian peasantry, Aaronsohn emphasized poor cultivation practices: "It is a truly miserable picture, the one offered by an Arab orange orchard to the eyes of an observer."[19] To compose his orientalist "decrepit canvas," he referred to the common use of camels for elevating water for irrigation through traditional norias, poor tree training, sick trees next to healthy ones, vegetable crops between orange trees, and, above all, the high density of trees, "unique in the entire world."[20] While close cultivation of orange trees increased the productivity per hectare, it allegedly prevented both the proper growth of individual trees and the mechanization of operations in the orchard. The messy dense Arab orchard suggested how easily settlers would make big profits by changing local practices using modern cultivation methods.

It is important to take stock of the orientalist nature of Aaronsohn's project, which would define orange cultivation by Jewish settlers in Palestine in the following decades. More than ignoring what already existed in the land, Aaronsohn insisted on the importance of knowing natives' practices for coping with local environmental challenges. As a son of the first Aliyah, knowledgeable of the dire conditions of Jewish settlements that relied exclusively on European agricultural practices, he built the whole project of his agriculture experiment station in Palestine on knowledge of local expertise and local plants.[21] But as a good orientalist, Aaronsohn also portrayed Palestinian society as a stagnant reality frozen in time, ignoring the major historical changes it had undergone in the previous decades. There was no mentioning in his writings of the modern character of the Jaffa orange cultivation business and how much it was propelled by Ottoman empire reforms of the nineteenth century.[22] When Aaronsohn published in *Der Tropenpflanzer*, the German journal of tropical agriculture edited by Otto Warburg, an article exclusively dedicated to Jaffa oranges, he did notice the relatively recent history of Jaffa orange exports to major European cities as well as to urban centers of the Ottoman empire. Nevertheless, this did not lead, for example, into discussions of reforms such as the Ottoman Land Code of 1858 that required individual registration of land titles, enabling a major land grab and the formation of a new class of large landowners producing cash crops such as citrus. The patronizing description of Arab orchards as belonging to a decadent civilization waiting to be restored by Jewish settlers supported by Western science promised nice returns on investment for capitalists willing to support Zionist endeavors.

California Cloning in Palestine

World War I would mean an abrupt end to the work undertaken at the Jewish Agriculture Station, its premises destroyed in 1917. Aaronsohn's death in an airplane crash in 1919 seemed to confirm the ephemeral nature of the project. Nevertheless, it does not take much to unveil the relevance of his new form of orientalism for sustaining Jewish settlements during the years of British Mandate in Palestine (1920–1948). The materialization of the famous Balfour Declaration (1917), in which the British Empire committed to support the establishment of a "national home for the Jewish people," would be achieved in large measure through the cultivation of oranges.[23] Petah Tikvah, the settlement funded by the Jewish Colonial Association (ICA) and known as the "mother of moshavot (privately owned agricultural villages)," would steadily grow to become the most populous Jewish agricultural village in Palestine in the 1910s.[24] Its success was based on abandoning obsessions with transplanting European agriculture and mimicking instead the Arab orange orchards surrounding the neighbor city of Jaffa.[25]

In 1914, Jewish-owned orchards occupied some 2,000 acres and, due both to World War I and water constraints, this area would significantly expand again only from the mid-1920s onwards, attaining 11,000 acres by 1929.[26] In the following decade, increase would be pronounced, reaching 38,000 acres in 1936, the year of the Arab Rebellion. With growing difficulties of acquiring land from Arabs afterwards, the area under citrus controlled by Jewish owners would start decreasing thereafter. Meanwhile, Arab citrus cultivation experienced a no less significant expansion from 5,000 acres in the aftermath of World War I to 35,000 in 1936, the early 1930s being also the years of more intense cultivation of new orchards. In 1945, three years before the creation of the State of Israel, Arabs owned 32,000 acres of citrus orchards and Jews 30,000. During the Mandate years, Palestine had become the second largest citrus exporter in the world, and citrus the largest single export item from the territory. No less significant, citrus cultivation was also the largest component of capital investment by Jews in Palestine.[27]

Most Jewish capitalists investing in citrus would not live in the new settlements, preferring to establish themselves in the recently built Tel

Aviv neighborhood of Jaffa, or even remain in England or the United States, hiring local expertise to manage their orchards. The London egg merchant Moshe Gredinger, for example, hired in 1926 the agronomist Baruch Ben-Ezer to design and handle his Qalmaniya estate on the Central Coastal Plain. Ben-Ezer was not only the son of one of the first farmers of the Petah Tikvah *moshava* as he had worked as well with Aaronsohn in establishing the Jewish Agricultural Experiment Station in Athlit. Ben-Ezer explicitly accepted the job at Qalmaniya with the declared intention of reproducing Aaronsohn's experiment first tried in the now ruined Athlit in the new citrus estates sustaining Jewish settlement.[28]

As already highlighted by other scholars, the Californian model became a recurrent trope for the expansion of Jewish orchards in Palestine.[29] There are multiple accounts of Jewish orchardists and their descendants following the steps of Aaransohn and traveling to California to learn methods of citrus cultivation. Also, American experts were a regular presence in Palestine advocating for the adoption of Californian practices to guarantee returns on investment of Zionist projects. Not surprisingly, such trips were in many cases paid by groups of American Jewish capitalists such as the Palestine Economic Corporation, formed by Justice Louis D. Brandeis to supply capital and credit for agricultural and industrial endeavors in Palestine.[30] In other occasions, it was the British colonial government that promoted such exchanges, as exemplified by the visit of H. Clark Powell in 1928.[31] Powell had been trained at the Citrus Experiment Station of the University of California in Riverside, the heartland of the Californian citrus industry, and he was the son of the general manager of the Californian Fruit Exchange, the cooperative organization that brought together thousands of Californian citrus growers. In 1923 Powell moved to South Africa with the hope of reproducing the Californian model in the Transvaal and the Eastern Cape. When British Mandate authorities brought him to Palestine, he arrived there as professor of horticulture of the Transvaal University College (Pretoria) and technical advisor of the South-African Co-operative Citrus Exchange.[32]

While irrigation, mechanization, or chemical fertilizers have been at the heart of descriptions of the global travels of American agriculture, I want to emphasize here the centrality of reproduction practices in such travels. One of the distinguished features of citrus cultivation in California was indeed the obsession over controlling reproduction of budding material.[33] Ever since a gummosis plague devastated orange orchards in the Azores islands in the middle of the nineteenth century, trees in commercial citrus areas around the world were propagated by budding wood of the desired variety onto a resistant rootstock instead of growing them from seed. California was no exception and branches of the most successful commercial variety, the Washington Navel, were budded onto Sour Orange rootstocks.

But American horticultural scientists employed by the USDA went a step further when they put in place a bureaucratic structure certifying selected buds to be used by Californian citrus growers. Considering how the Washington Navel variety was the result of a bud sport, scientists had identified how branches easily degenerated into less valuable strains of lower productivity bearing fruits of lesser quality (more seeds, less juice, more acidic). Only buds from trees with a proven record of producing good yields of high quality Washington Navels should be cut and sold as bud wood to avoid the risk of propagating undesired strains. As a recognition of the importance of asexual forms of reproduction such as budding and grafting in modern agriculture for the multiplication of identical standardized organisms, Herbert J. Webber in an article for *Science* in 1903 coined the term "clone." In 1912, Webber would become the head of the University of California Citrus Experiment Station in Riverside, where those American experts coming to Palestine got their training.[34] That was the case of Powell, who arrived in Palestine via South Africa, where he had already advocated for the importance of certified buds, or clones, in citrus cultivation.

In describing the "Citrus Industry in Palestine" in 1928, Powell was in fact convening a list of what he believed were Californian best practices already adopted in South Africa and that should also be embraced in Palestine. He did not mince his words when pointing the poor reproduction methods he had come across in Palestinian orchards: "Very few growers practice careful 'bud selection' and the result is seen in the number of inferior trees in the groves of Palestine."[35] The way to identify suitable parent trees had been already established by Californian scientists whose writings Powell recommended: "The best trees in a grove should be studied and records should be kept of the quantity and quality of the fruit produced by each."[36] To make things worse, branches in Palestine were usually budded onto stocks in the orchard one year after these were planted, instead of budding stocks in the nursery before these became too thick as it was done in California. Thick stocks demanded thick buds from older parts of the tree, which, as Powell explained, had higher probability than small buds cut from fruiting brush of not reproducing true type and thus of propagating undesired strains. What the California system of certified thin buds offered was the promise of transforming orange cultivation into mass reproduction of identical organisms—clones—ready to be shipped to international markets.

The year after Powell published these remarks, J. D. Oppenheim, head of the Division of Horticultural Breeding at the Jewish Agricultural Experiment Station, started his own research on budwood selection.[37] The Jewish Agency of Palestine, the Zionist organization charged with facilitating Jewish immigration into the territory namely through land purchases, had established in 1921 a new experiment station in Rehovot, a Jewish colony some fifteen miles southeast of Jaffa. Founded in the late nineteenth century on land bought from an Ottoman notable, after expelling the Bedouins who previously cultivated the land,[38] Rehovot would become the center of Jewish citrus production and the most prosperous Jewish colony in Palestine of the 1930s.[39] Oppenheim, in cooperation with local Jewish growers, chose 200 trees from several groves, visiting them four times a year for four years. He took records of "size, growth habits and health of the tree, foliage characters, blossoming, freedom from limb variations, number, size, uniformity, shape, juiciness and aroma of fruit, number of seeds, thickness and external appearance of the peel." Oppenheim provided each tree with a numbered label and after four years of observations granted a certificate to forty of them from which budwood could be taken. Such certificate guaranteed the "clonal purity" of newly cultivated Jewish orange groves, whose area would rapidly increase in the 1930s, reaching the same acreage of Arab-owned citrus orchards.

The justification for such concerns with clonal purity were the same learned from the Californian experience. As detailed by Oppenheim (together with his colleague Oppenheimer) in his correspondence with H. J. Webber, the head of the Californian Citrus Experiment Station, instead of Californian fears of degeneration of the Washington Navel, in Palestine the anxiety grew from the spreading of lower strains of the Shamouti, the local name of the fruit known in European markets as Jaffa orange: "Mr. Smilansky at Rehovot found in his grove of 8ha 70 trees which had given rise to limbs of the 'Baladi' [*sic*] variety. Yedidja observed that a high percentage of Shamouti trees in commercial groves changed to 'Baladi' [*sic*] . . . Distribution of inferior eyes may be reduced to a minimum only by constant observation of the mother trees."[40] To avoid the reproduction of less valuable Belladi oranges, exemplar orchardists should thus keep detailed records of their trees, as was the case of Shmuel Tolkowsky who as manager of orange groves in Rehovot got the rare praise by Powell of producing the "finest and most carefully kept records" he had ever seen.[41]

As Powell revealingly reminded in his account of the citrus industry in Palestine, the discussion about certified budding made sense only if trees were cultivated at least five meters apart from one another.[42] The tree records needed for bud selection could be secured only when the individual tree became the basic unit of measurement, replacing traditional practices of cultivating trees closer to each other and discussing productivity in terms of acres. Each tree should be given two numbers, one indicating the row and the other the place of the tree in each row. Or has described by another Californian expert who also visited Palestine, "the first step . . . is the establishing of an identity for each tree in the orchard. . . . Numbers not only serve to establish a permanent identity but also to designate the location."[43] This individualization of trees was allegedly impossible in the densely cultivated orchards of Palestine as the ones Arabs had been planting since the mid-nineteenth century and from which initially Jews took their clues. The program of bud certification put in place by Oppenheim at the Jewish Experiment Station to guarantee the clonal purity of the Shamouti orange demanded new well-ordered spaced orchards. Jewish exemplary orchards, as the ones managed by Tolkowski in Rehovot, thus materialized in the landscape the orientalist division suggested above by Aaronsohn between stagnant Arabs and modern Jews. Or, in an alternative formulation, Jews distinguished themselves from Arabs by cloning oranges the Californian way.

Local Belladi and Cosmopolitan Shamouti

The relation between valuable Shamouti oranges and undesired Belladi is crucial to understand how orientalism was performed through the cultivation of oranges. This insistence with orientalism is suggested by the historical actors themselves. Tolkowski, the exemplary orchardist who kept perfect tree records, was a recognized leader among citrus cultivators who made him the director of the Jaffa Citrus Exchange, the umbrella organization that marketed the fruit of Jewish growers.[44] He was also an influential political figure who served under Chaim Weizmann during the negotiations of the Balfour Declaration in 1917 and was secretary of the Zionist delegation in the Versailles peace conference two years later. To justify the famous wording of "a national home for the Jewish people" included in the declaration, the English Zionist Federation had published that same year Tolkowsky's "Achievements and Prospects in Palestine," a short history of Jewish settlement in the territory. Having left behind his family diamond business in Antwerp for a life in Palestine as an agronomist, Tolkowsky was an enthusiast of the virtues of working the land for a Jewish revival. Like Aaronsohn, he was also convinced that "the only way open to the Jewish settlers was to take a lead from the surrounding Arab population and to try to imitate as best they could the methods used by them."[45] And, like Aaronsohn, mimicry came together with patronizing: "the fellaheen, with their typical oriental lack of foresight . . . have no other principle than to try to make their fields yield as much as they can with their very primitive methods."[46]

Tolkowski's orientalism was nowhere better expressed than in his history of the city of Jaffa starting in c. 4,000 BC and ending in the British Mandate years.[47] An erudite work built on German, French, and British traditions of scholarly orientalism, it also included a brief discussion on the origin of the Jaffa orange.[48] References to cultivation of the citron in Palestine in the fifth century BC and of bitter oranges and lemons during the tenth century AD were complemented with philology considerations of the name orange and its relation to the Persian narenj and the Arabic naranj.[49] As for the Shamouti variety that gained Jaffa oranges the favor of consumers in international markets, Tolkowsky could only point at a version told by a certain Mr. Frederick Murad of Jaffa that it "was brought back from China, about two hundred years ago, by an Armenian priest whom the Armenian Patriarch had sent on a mission to Persia, India, and the Far East." Acknowledging the insufficient reliability of his sources, Tolkowsky concluded that the "question as to where and whence this fruit was brought to Jaffa may possibly ever remain unsolved."

Instead of looking for faraway origins, Oppenheim suggested that the observations of variations of Shamouti branches pointed at a local origin. As described in his correspondence with Webber, Belladi branches were consistently found in Shamouti trees, indicating the relation between both varieties. In addition, Oppenheim also reported less frequent cases of Belladi trees "bearing a branch with the fruits and the leaves of the Shamouti. So I got proof of the relationship of the Belladi and the Shamouti."[50] The Belladi, which is Arabic for local or native, is a variety bearing smaller round fruit, with more seeds and less juice, than the prized oval shaped and thick skinned Shamouti. The Shamouti then was no more than a bud sport of the Belladi reproduced by Arab growers in the nineteenth century. In other words, although Tolkowsky blamed local growers for their "typical oriental lack of foresight," they had been the ones who had identified and cloned a fruit whose characteristics sustained long distance commercial relations.

The difficulties in attributing the Shamouti variety to Arab reproduction practices by Jewish growers are in accordance with the orientalist blindness to historical change in the Ottoman empire. The emergence of the Shamouti in Palestine went hand in hand with the development of a new class of notables (a'yan) supported by Ottoman authorities that displaced the previous power of local rural sheikhs.[51] In the twenty years after the land reform of 1858 some 750 hectares of orange groves were being cultivated around Jaffa, the most evident element of the transformation of Palestinian agricultural landscape into cash crop production.[52] By the 1890s there were more than four hundred plantations[53] (*mazra'as*) constituted by the orchard, the landowner's big house, and *saknat*—the fellahin dwelling clusters that, differently from Palestinian traditional villages, had no land of their own, housing landless serfs paid per day or season. It was in these capitalist plantations of Jaffa that bud sports of the Belladi tree holding large oval Shamouti oranges were first reproduced.

Although we now have many informed historical accounts of the role of the new elite during the late Ottoman period in integrating Palestine in the international capitalist system, the predominant image in the historiography and in popular accounts is still one produced by orientalist tropes of traditional Arab notables living according to aristocrat habits and values.[54] Theodor Herzl's Zionist vision was indeed built on caricatures of wealthy indolent effendis willing to sell their lands to Jewish settlers and of impoverished ignorant fellahin that could be easily removed. It is not just a question of recognizing that, as we saw, Jewish settlers depended on local practices to make their projects viable. Beyond that, these local practices and the people undertaking them were not ahistorical realities frozen in time waiting for modernity brought to Palestine by Jewish settlers and European imperialism. Jaffa orchard owners and fellahin working the fields for wages were already living in a capitalist world, actively participating of its promises as well as injustices. Locals were already modern. The names of the oranges illuminate the point: The "national home" for the Jewish people in Palestine was not built on the Belladi (local or native) oranges, but on already existing Shamouti (oil lamp, in reference to the citrus distinguished oval shape), modern fruits able to be transported a long distance to faraway markets.

The Undifferentiated Orchard

Historians exploring the history of citrus in Palestine have uncritically accepted the distinction between traditional Arab orchards and modern Jewish orchards, thus reinforcing the orientalist account put forward by the likes of Aaronsohn and Tolkowsky. If one looks briefly at the Arab press of the Mandate years, it is apparent that such sharp distinction does not hold. In addition to being already capitalist when Jewish settlers arrived in Palestine, Arab elites involved in citrus cultivation seemed as interested in discussing agricultural innovations as their Jewish counterparts during the Mandate period. These elites were not static before Jewish arrival, and they did not remain static during the British Mandate. The *Arab Economic Journal* (Al-Iqtisadiyyat al-'arabiyya), published from 1935 onwards, was probably the best example of a much neglected form of *nahda* (Arab cultural renaissance) in Palestine that was not defined by its opposition to Zionism, preferring to promote instead an "Arab utopia built on the foundations of private property, investment, self-responsibility, and the accumulation of capital."[55] The front cover of the journal was an eloquent illustration of such pan Arabian utopia, depicting a new map of the Middle East dotted by industrial plants, shepherds, airplanes, camel caravans, trucks, ocean liners, oil wells, date palm oasis, railways, and cedar trees. Palestine was identified by the al-Aqsa mosque and orange orchards. This was an alternative Pan Arab vision of Palestine engineered.

The presence of citrus in Iqtisadiyyat was overwhelming: Issue 3 included "A study of the Citrus Industry"; issue 4 featured a discussion of "Jaffa Orange Position in Germany"; issue 5 included no less than five articles on citrus; issue 6 had a leading article on "The future of Citrus fruits in Palestine." It is revealing that almost every issue of the journal included some kind of technical or economic discussion of citrus production. Importantly for our argument, issue 8 dedicated an article to "Propagating Jaffa Oranges." And considering the alleged role of California in differentiating between traditional and modern orchards, between Arab and Jewish, it deserves mentioning how the journal discussed "The Importance of Advertising by the Californian Fruit Growers Exchange." When after World War II, Khalil Totah, a Quaker from Ramallah and head of the local Friends Boys' School, testified for the Anglo-American Committee of Inquiry established to deal with the "Palestinian question," he defended the capacity of self-government by Arab Palestinians by invoking as well how the American example had reached Arab orchards: "Scores of Arabs from Palestine emigrated to America and were brought into close contact with progress. . . . There is now hardly a village in Palestine, no matter how remote and lonely, that does not boast a son in America."[56] Totah identified Arab economic success in Palestine first and foremost with the transformation of citrus cultivation into "a substantial modern Arab enterprise. The Arabs own about 150,000 dunams (37,500 acres) of citrus trees, which is about half the citrus acreage of Palestine. . . . Palestine Arabs have benefited greatly from the fact that their sons have gone abroad to study. Many studied in California, Texas, and Florida and the Arabs are reaping the benefit of their experience."[57] According to Totah's account, citrus cultivation in the 1940s was not anymore "a monopoly of the effendis and the capitalists. The peasants turned their wheat fields, vineyards and olive orchards into orange groves" holding the promise of social mobility.

The descriptions offered by Totah as well as the multiple articles in Iqtisadiyyat seem to contradict persistent references to a Palestinian landscape divided between Arab dense messy traditional groves and Jewish spaced ordered modern groves. Granted, Jewish citrus expansion in the 1920s had occurred in areas beyond the initial citrus belt around the city of Jaffa, with new irrigation techniques and the detection of the water table level transforming the red sand belt of the Sharon into a major citrus district.

This new availability of water allowed Jewish growers to follow Californian guidelines for wide orchards in which machinery could be used to prepare the land while providing space and light for the even development of each individual tree. Some of these growers opted for cultivating no more than 80 trees/acre, marking in the landscape a stark contrast with the original dense orchards around Jaffa with densities of 300 trees/acre determined in large measure by the reach of the irrigation infrastructure. But significantly, in the early 1930s, the years of more intense cultivation of new orchards, densities averaged 220 trees/acre, lower than the original ones in Jaffa but higher than those of the 1920s that had been cultivated strictly following the Californian example. Growers explicitly acknowledged that the availability of wage workers (mostly from Arab origin) did not justify the Californian emphasis put into mechanizing the orchard and thus in widening the space between trees. The new orchards of 220 trees/acre of the 1930s were a compromise between the initial dense Arab orchard designed to increase productivity per acre and the Californian orchard cultivated to increase productivity of individual selected trees. In other words, in the 1930s density of cultivation did not perform any further neat distinctions in the land between Arabs and Jews.[58]

It was not only the human eye that had increasing difficulties in distinguishing Arab from Jewish orchards. Viruses were also oblivious to such distinctions. In 1928, a new disease was detected in Palestinian citrus orchards identified two years later as xyloporosis. Discolored wood and bending trees, eventually leading to branches wilting and dying, in recently cultivated orchards were worrisome enough to justify a major study by J. Reichert and J. Periberger of the division of Plant Pathology of the Rehovot Agricultural Experiment Station. The disease was found to affect trees with Shamouti budded on Palestinian Sweet Lime rootstock in both Jewish and Arab orchards. The rootstock mostly used in California, and recommended by the Californian experts visiting Palestine, was Sour Orange, which Reichert and Periberger confirmed was tolerant to the new disease.[59] The widespread presence of xyloporosis in the 1930s asserted that Jewish orchardists had kept preferring the Arab original combination of Shamouti budded on Sweet Lime instead of following the Californian example of Sour Orange as rootstock. Sour Orange in Palestine was used only in the heavier soils such as those east of Jaffa, growers wrongly attributing to soil properties the poor performance of the Sweet Lime in this new citrus cultivation area. Growers' insistence with Sweet Lime in the rest of the country was explained by the crucial advantage of its faster growing in the first years, bringing trees to profitable bearing two to three years earlier than Sour Orange.

This xyloporosis story points at more than Jewish citrus growers stick-

Figure 3.1. Arab orange groves at Bir Salem in Jaffa. American Colony. Photo Department. [Between 1934 and 1939]. Retrieved from the Library of Congress, https://www.loc.gov/item/2019709409/

ing to Arab growers' practices. It also shows how all growers, both Jewish and Arab, had changed their reproduction practices in the period in question. The increasing demand for new trees in the late 1920s and 1930s required the acceleration of the rootstock reproduction process, replacing the traditional propagation of occasional cuttings by the systematic propagation of seedlings from seed beds. As stated, Sour Orange rootstock had been used in the early 1920s to cultivate new trees into heavier soils, and when in the following years budwood was taken from these apparently healthy trees, they were nevertheless infected with severe xyloporosis–inducing viruses, since Sour Orange rootstocks were tolerant to these. For the subsequent rapid expansion of orchards of the 1930s, budwood from these infected trees was used to graft Sweet Lime rootstocks now grown from seedlings and not from cuttings. These Sweet Lime rootstocks were completely free from viruses and showed no symptoms of xyloporosis, since viruses are not seedborne. But once infested Shamouti budwood was used to graft virus-free rootstock these trees quickly showed xyloporosis symptoms. The presence of xyloporosis in all orchards, independently of being Jewish or Arab owned, reveals that all growers had changed their rootstock reproduction practices.

The dynamics of spreading xyloporosis, or the "new disease" as growers in Palestine first called it, bears obvious practical lessons such as the importance of carefully considering the consequences of changing horticultural practices of reproduction. But for the purposes of this chapter, what is important to underline is how xyloporosis revealed the features shared by both Jewish and Arab orchards: Shamouti budwood grafted on Sweet Lime rootstocks reproduced through seedlings. Jews had followed Arabs in sticking to the combination of Shamouti and Sweet Lime as key to their presence in Palestine; Arabs followed Jews in reproducing Sweet Lime from seedling and not from cuttings. The virus proved unwilling to follow the orientalist division between modern Jewish Californian orchards and traditional Arab local orchards. Historians would do well to follow the virus when discussing Arabs and Jews in Palestine.

4 * *Harvesting Hogzillas: Feral Pigs and the Engineering Ideal*

ABRAHAM GIBSON

By almost any measure, the domestication of animals is one of the most successful engineering projects in history. Many researchers cite our ability to control animal populations, our ability to steer them toward desired ends, as one of the hallmarks of behavioral modernity. Some insist that domestication heralded the beginning of the Anthropocene, a new global epoch defined by human dominion. Its importance to the history of biology is proverbial. After all, when Darwin sought to explain his theory of evolution by natural selection to the world for the first time, he did so by drawing analogy to domestication, which he called artificial selection. Meanwhile, domestication has had an equally profound influence on the animals themselves. These animals are now found on every continent (even Antarctica), they influence the terrestrial biosphere almost as much as humans, and they vastly outnumber their closest wild cousins.

Reframing domestication as one long engineering project opens new avenues of research. Like most engineering projects, it was not achieved in one fell swoop. Breeders have initiated domestication many different times in many different places, and they have sometimes failed. Like other engineering projects throughout history, domestication requires constant maintenance. If any animals leave their partnership with humans and establish residency in the wild, free from direct anthropogenic selection, they are relabeled *feral*. These animals are ubiquitous in the historical record, but they are typically overlooked.

Viewing the history of domestication through an engineering lens helps shed light on both processes. To limit what would otherwise prove an unmanageable scope, this chapter will focus on a specific population: feral pigs in the American South. Their story reveals the unintended consequences of engineering projects, the importance of sociotechnical context, and the extent to which domestication qualifies as a "wicked problem."[1] This chapter will focus on the long twentieth century, but it will also place these events within a much larger historical context, which is essential to understanding how domestication and engineering are entangled. This chapter begins with a brief discussion of the domestication of pigs, their arrival in southeastern North America, and the establishment of the open range. The second half of this chapter turns to the surprising fate of feral pigs over the past 150 years of southern history. Finally, a brief conclusion reflects on the lessons for remaking nature.

Philip Pauly remarked that "control of life is coextensive with civilization," but different civilizations have exercised different levels of control over the years.[2] Researchers once promoted the "master breeder" theory of domestication, which credits humans with something resembling godlike prescience. According to this view, humans deliberately partitioned some animals away from their closest cousins in an attempt to initiate domestication. Most researchers now reject claims that humans practiced such overt foresight. In the case of pigs, archaeologists believe that people in the Fertile Crescent spent hundreds of years manipulating free-ranging wild boars, starting around 10,500 years ago. This meant herding the animals and killing the least pliable among them. Whether these relatively modest efforts qualify as domestication, or, for that matter, engineering, depends on one's definition of each.[3]

Similarly, previous generations once assumed that all extant pigs had descended from a single progenitor population, but archaeologists and geneticists have since shown that pigs were domesticated many different times in many different places. In China, for example, pigs were far more likely to be confined in sties and fed at troughs at a very early stage in the domestication process. Chinese farmers made special efforts to enclose domestic pigs and control their reproductive output, which increased their propensity to fatten at a young age. By comparison, the domestication of European swine looked remarkably different. Europe's earliest farmers practiced a unique method of husbandry known as "pannage," which encouraged pigs to scour woodland floors in search of fallen acorns and other mast. These free-ranging pigs relied on their wits for survival and procreated beyond humanity's watchful eye. As a result, there was extensive hybridization among Europe's wild, domestic, and feral populations.[4] These radically different approaches to domestication remind us that there is more than one solution to any engineering challenge: the idealized method and the path of least resistance.

Since there were no domestic animals in the Americas prior to the Columbian exchange (save for dogs), the first pigs in southeastern North America were descended from the free-ranging, undifferentiated animals that accompanied the first generation of European explorers in the late fifteenth century. Spanish conquistador Hernando de Soto brought thirteen of these pigs with him when he and his army of 600 men disembarked on the west coast of Florida in 1539. They spent the next four years rambling across the South, as far north as Tennessee and as far west as Texas. During that time, the pigs multiplied to become several hundred. Even so, de Soto jealously protected the pigs, refusing to share the animals with his men under penalty of death. When de Soto died in 1543, his army plundered the pigs with abandon.[5]

Modern biologists often claim that the South's extant population of feral pigs are descended from de Soto's strays.[6] It is certainly possible. When de Soto's army crossed the Savannah River, numerous pigs were swept downstream and out of the historical record forever. Chronicles from the expedition also reveal that Native Americans developed a taste for pork and tried to steal pigs as often as possible. On one occasion, while backtracking across territory they had already covered, the Spaniards discovered that some of the Native Americans they had left in their wake were now living with pilfered pigs. On another occasion, Native Americans stole a canoe that was loaded with pigs.[7] Of course, none of this proves that de Soto's pigs went feral. After all, passing from one group of humans to another would not have entailed feralization, though living among people who had no experience with husbandry or fences would have certainly increased the chances. Until the genetic, archaeological, or historical records produce better evidence, we cannot say for certain whether de Soto's pigs survived in the southern wilds.

By comparison, feral pigs were ubiquitous in the English colony of Virginia. Soon after they established Jamestown in 1607, colony leaders passed laws that prohibited anyone from killing any pigs upon penalty of death. The animals were too valuable. The prohibition had its desired effect. By 1614, colonist Ralph Hamor wrote that there were "infinite hogs" in the woods around Jamestown.[8] Eager to protect their precious cornfields, colonists appointed swineherds to watch over the pigs, but this method proved ineffective as the number of pigs continued to grow. In 1629, the House of Burgesses ordered every person in the colony to plant two acres of corn, instructing that the plots should be "sufficiently tended, weeded, and preserved from birds, hogs, cattle and other inconveniences."[9] When that dictate failed to prevent animals from trespassing, lawmakers directed all settlers to build fences that would enclose their crops. This stipulation gave livestock the run of the land, and thereby codified the "open range," a social, cultural, and ecological institution that would define life in the South for centuries.[10]

Colonists devised several measures to control their wayward pigs, some more successful than others. Most notably, colonists carved distinctive notches into the ears of their respective pigs, and they registered these "earmarks" with the local courthouse.[11] This practice did little to dissuade poaching, especially since most of the colony's pigs were born in the woods and bore no identifying insignia. By the late seventeenth century, the vast majority of Virginia's pigs lived in a feral condition. "Hogs swarm like vermin upon the earth," colonist Robert Beverley acknowledged, adding that the animals typically run in "a gang" and that they "find their own support in the woods, without any care of the owner."[12] Colonists and the Crown both claimed legal dominion over these pigs, though neither party made any attempt to engineer the animals.

Up to that point, livestock production in the South was not so different from the rest of the world. Despite several thousand years of domestication, breeders still relied on extant forms that were readily available. To paraphrase Philip Pauly, the limits on biological manipulation were more notable than the achievements.[13] Things began to change in the nineteenth century, when breeders in Europe began importing Chinese pigs, who showcased a "remarkable tendency to fatten."[14] As a result, when American farmers in the Northeast and Midwest began cultivating distinctly "American" breeds like the Duroc Jersey, the Chester White, and the Poland China during the early national period, they were working with genes that hailed from China, by way of Great Britain.[15] They exercised tighter control over reproduction, and they exercised forethought. This was a radical departure from the colonial period. This was goal-oriented engineering.

Things were different in the South. While farmers in the Northeast and the Midwest increasingly adopted scientific breeding during the early national period, southern farmers made almost no effort to engineer their pigs. Why should they? The commons remained open for most of the nineteenth century. To be sure, wealthy planters tried to close the range following the end of the Civil War, but they were consistently thwarted by an unlikely voting bloc of poor whites and recently enfranchised poor blacks, who relied on the open commons to range their animals. "Any fool can see that if we adopt this new plan [closing the open range] the poor at least, and that means four-fifths of our rural population must give up utterly every vestige, every hoof, of domestic animals," one citizen explained incredulously, unable to conceptualize a scenario in which rural people lived without access to animals.[16] Tellingly, things began to change when Reconstruction ended and federal troops withdrew. This left southern legislatures more squarely within the hands of wealthy planters, who were quick to disenfranchise those who would oppose their grand designs. By 1900, the range had collapsed across large swaths of the South.[17]

Planters had promised that closing the range would transform agriculture in the South, but their forecast proved inaccurate. First, plantation slavery may have ended with the war, but the planter's ethos had not. Sharecropping allowed planters to retain debilitating control over their recently emancipated neighbors, and they were still not keen on committing vast stretches of potential cropland to rangeland. Rather than raise their own animals, many planters imported pigs in huge numbers from the Midwest. Unlike the antebellum period, when animals were marched on the hoof, pigs increasingly arrived on the nation's expanding network of railroads. Second, many planters had promised that closing the range would eliminate feral pigs altogether, but that was not the case. Feral pigs persisted throughout the South, seemingly little concerned that they had lost the legal right to do so. This necessarily undermined any attempt to "improve" southern pigs, as Perry Van Ewing explained in *Southern Pork Production* (1919). Whenever some enterprising farmer actually *did* procure purebred swine, they invariably encouraged the animals to roam at large, where they "indiscriminately mixed with other blood." As a result, Van Ewing estimated that 0.001 percent of the pigs in the region were "purebred."[18] As ever, the success of any engineering project depends on social and cultural context.

Southerners had lived among feral pigs for centuries, and thus seldom remarked on their apparent ubiquity. By comparison, visitors from other regions generally had no experience with feral pigs, and therefore marveled at the creatures. They were quick to report that ferality had endowed southern pigs with unique morphological traits. Tobe Hoge tried to explain the "razorback" nickname to his northern readers in 1887. "This species of hog takes its name from its likeness to a razor with the thin edge up, the sharpness of its vertebral column," he wrote.[19] Writing in *Outing Magazine* four years later, J. M. Murphy reported that that the razorback's legs were "long, lean and sinewy, its hocks short, its body attenuated to the verge of the ridiculous, its snout prolonged and tapering, its skull low and elongated, its neck scrawny and its back arched in the centre and sloping gradually toward the flanks."[20] Silas M. Shepard memorably described the southern pig as "long nosed, long eared, long necked, long legged, slab-sided, small hammed, coarse haired, large bristled, gaunt, restless, hard feeder, and an impudent 'cuss.'"[21]

While visitors obviously noticed the South's feral pigs, they did not originally regard them as potential hunting trophies. This is significant because hunting was very popular among the northeastern elite in the late nineteenth and early twentieth centuries.[22] These sportsmen began by hunting the areas closest to them, the undeveloped areas of the Northeast, where there were no feral pigs. When northeastern hunting grounds were denuded of their game, however, sportsmen began turning their attention southward. "The old haunts of the North have become drained," one hunter lamented in *Field and Stream*, adding that "the only really good shooting to be had is in the South."[23] By the early 1900s, newspapers in

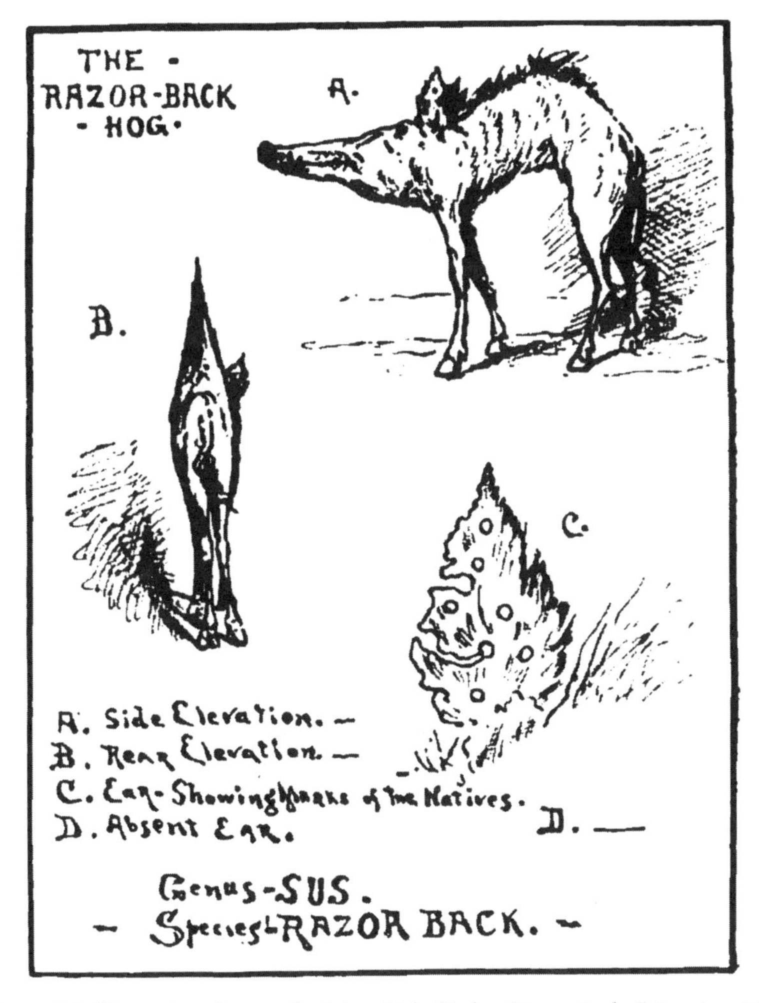

Figure 4.1. Illustration of a razor-back hog. Tobe Hodge, "Razor-Backs," *American Magazine* (December 1887): 255.

the Northeast informed their readers that a new kind of game awaited sportsmen who traveled South. "Sportsmen are not accustomed to think of wild hogs as game that may be hunted in the United States, but the fact remains that there are thousands upon thousands of the animals wandering through certain sections of the South," Leonidas Hubbard wrote in the *New York Times*. "They are as wild as deer and well-nigh as formidable as the bear."[24]

The most famous hunting club was located on Jekyll Island, just off the coast of Georgia. In 1886, a cohort of millionaires, including J. P. Morgan, William K. Vanderbilt, Joseph Pulitzer, and Marshall Field, purchased the island for $125,000. At the time of purchase, the island contained about 600 feral pigs. The millionaires originally wanted the animals exterminated, and so they contracted a hunter to accomplish the task. The pigs proved elusive, however, and the hunter met with limited success. Eventually, these millionaires discovered what many other tourist sportsmen had discovered: killing feral pigs was fun. One bemused journalist for the *Atlanta Constitution* reported that "the clubmen find no end of sport in hunting these bristly boars, with trained dogs and rifles." There was a never-ending campaign to kill the feral pigs, "affording opportunities for the Nimrods of the club to show their prowess."[25]

While these tourist sportsmen prized the novelty of feral pigs, they eventually grew dissatisfied with their relatively unimpressive carcasses. Thus, in 1909, banker J. P. Morgan arranged for the release of two wild (never domesticated) boars on Jekyll Island. These animals hailed from Europe, and they were considerably larger than their feral cousins in America. Morgan and his fellow millionaires hoped that the boars would intermix with the local pigs, and thus provide hunters with larger trophies. They wanted to re-engineer, or "breed back," these feral pigs so that they would more closely resemble their wild forebears. Three years later, additional wild boars were introduced into the mountains of western North Carolina, where a financial advisor from Detroit had established a game preserve on Hooper Bald. He released thirteen wild boars into a fenced enclosure. The animals remained unmolested for a decade, and so their numbers continued to grow. They escaped into the surrounding mountainsides during an organized hunt during the early 1920s. In short order, these wild boars began to interbreed with the untold number of feral pigs who already inhabiting the mountains.[26]

While tourist sportsmen of the early twentieth century regarded feral pigs as a quarry and sought to expand both their morphology and their range, biologists increasingly regarded the animals as an ecological nuisance. As early as 1899, biologists decried the ecological impact of feral pigs. T. S. Palmer, assistant chief of the Bureau of Biological Survey, reported that domestic pigs "may run wild and become so abundant as to be extremely injurious." Rather than referring to these wayward animals as "invasive," however, Palmer instead referred to them as "noxious."[27] In 1910, *Forest and Stream* likewise disparaged the ecological effects of feral pigs. "They travel fast and far and rake the country as with a fine-tooth comb; their sense of smell is highly developed, and the eggs and young of ground-nesting birds are never safe when they are abroad."[28] In 1938, when LeRoy Stegeman published the first scientific analysis of feral pigs, he included photographs showcasing their destructive habits.[29] Later, when ecologist Charles Elton published his famous tome, *Ecology of*

Invasions by Plants and Animals, in 1958, he referred to feral pigs as "problem animals."[30] Tom McKnight offered similar testimony six years later. "Where concentrated, the presence of feral swine is quite noticeable," the ecologist wrote. "They are persistent rooters, untidy feeders, frequent wallowers . . . and accused of causing a considerable number of problems."[31]

Contemporary biologists and management officials are even more resolute. They insist that feral pigs are highly "invasive" and that they should be eradicated with extreme prejudice.[32] One would think that recreational hunters might assist in the task. As early as 1910, contributors to *Forest and Stream* implored their fellow hunters to target feral pigs for ecological purposes. "To curtail the range of these beasts [feral pigs] wherever it may be possible is a line of work that should be taken up by sportsmen's clubs and Audubon societies in every state affected," one correspondent opined.[33] But the interests of biologists and recreational hunters do not overlap as much as one might think. Biologists want to steer ecosystems toward a specific composition they deem healthy. Recreational hunters want to shoot pigs. Their goals may overlap in practice, but their motivations could not be more different.

Many hunters have no interest in getting rid of feral pigs *altogether*. On the contrary, they actively engineer these populations. It is well documented that some hunters have deliberately released pigs into the woods for commercial and recreational purposes.[34] In May 1962, hunters released twenty-six pen-reared hogs near Crossville, Tennessee, in an attempt to establish "another huntable population," but hunters complained that the animals were too tame.[35] They desired something more. They wanted the thrill of the chase, and they wanted an impressive trophy to showcase after the kill. Pen-reared animals provided neither. To improve (or is it un-improve?) their local quarry, hunters descended on Hooper Bald, where dozens of wild boars had escaped into the wilderness decades earlier. These creatures had interbred with local pigs and, as a result, the free-ranging swine of western North Carolina were larger than their cousins elsewhere on the continent. In short order, hunters began trapping these hybridized creatures and exporting them to locations throughout the South in an attempt to engineer the local populations.[36]

Thanks to these translocations, a majority of the free-ranging swine in the South are now hybrids between wild boars and feral pigs. As a result, the region's free-roaming swine are now much larger than they were in previous centuries. This development has increased the hunter's desire to kill these animals, and feral pigs are now one of the most popular hunting trophies in North America, second only to white-tail deer in the number killed.[37] These developments have also helped invigorate the popular myth that the South is overrun with enormous, tenacious, quasi-mythical swine. Reported sightings of massive hogs have increased dramatically over the past two decades. After one image of a giant feral pig went viral in 2004, *National Geographic* sent a team of scientists to southern Georgia to exhume the animal and measure its dimensions. Post-mortem analyses confirmed that the specimen, now known as "Hogzilla," was more than 9 feet long and had probably weighed more than 800 pounds.[38]

An even more dramatic photograph surfaced in 2007. It showed a boy wearing a ball cap, holding a pistol, and kneeling behind what appeared to be an enormous feral pig. The image was reprinted in the nation's major newspapers, and the boy in the picture, Jamison Stone, was interviewed in-studio by CNN. He and his father swore that they had been hunting and came upon the animal (which they dubbed "Monster Pig"), and that young Jamison had felled the beast with eight shots from a .50-calliber revolver. People from around the world cried protest. Some objected to recreational hunting in general, but most people just doubted the veracity of the story. Many claimed that the photograph was an optical illusion, and that no pig could possibly be that large. Others insisted that the photograph had somehow been manipulated.[39] Suspicions were further raised when the boy's father began promoting a website called "monsterpig.com." Well, it turns out the story *was* a fake, but not in the way one might think. About a month after the "Monster Pig" story first broke, reports surfaced that the animal was not, in fact, feral. On the contrary, the animal had been pen-reared for more than ten years on a farm in southern Alabama. The pig's owners, Phil and Rhonda Blissitt, had even given the animal a name: "Fred." Eddy Borden, the owner of Lost Creek Plantation, purchased the animal, turned him loose on a 150-acre enclosure, and advertised the opportunity to hunt a wild boar. By the time Jamison Stone shot Fred to death, the animal had been "feral" for less than a week.[40] Scientists who analyzed Fred's skull confirmed that he had been huge, but that he had also been domesticated. In this context, domestic pigs are a failure of engineering, while feral pigs are the goal.

This mania for hunting pigs has fueled a population explosion. When John Mayer and Lehr Brisbin counted the region's feral pigs in 1991, they reported that approximately 2,000,000 feral pigs had established populations in 20 states.[41] One might think that another few decades of control measures would have shrunk the nation's population of feral pigs by now, but nothing could be further from the truth. When Mayer and Brisbin reassessed the nation's feral pigs in 2009, the biologists reported that their number and range had both increased dramatically. They estimate that 5,600,000 feral pigs have established populations in 36 states, and that the vast majority live in the southeastern United States.[42] There is no question that hunters are to blame for the nation's growing number of feral pigs.

Even hunters have admitted as much. "There's an undeniable correlation between the wild pig's spread in the past few decades and its glamorization as a big-game animal," *Field & Stream* reported in the summer of 2015, adding that "most new pig problems can be blamed on escaped animals from high-fence operations or illegal stockings by outfitters and hunters to create recreational opportunities."[43]

Finally, while most biologists want feral pigs to die, and most hunters want them alive so they can kill them, several parties now advocate for varying degrees of protection. For example, some biologists insist that the feral pigs who roam undeveloped Ossabaw Island off the coast of Georgia ought to be saved. They claim that several centuries of relative isolation have endowed the animals with unique physiological traits. The pigs contain the highest lipid reserves of any feral pigs in the world, which renders them a "biomedical treasure" for diabetes researchers.[44] Meanwhile, the Colonial Williamsburg living museum in southeast Virginia began exhibiting Ossabaw pigs in the early 2000s. Museum officials explain that Ossabaw pigs have avoided breeding initiatives for hundreds of years and are thus morphologically similar to colonial pigs, as if arrested in time.[45] In both of these examples, the pigs are valued precisely because they have *avoided* intensive engineering.

It is clear from the foregoing discussion that viewing domestication through an engineering lens sheds light on both processes, and that feral pigs are especially instructive. When machines break down, entropy takes over. Rust sets in. Organisms are not machines, though. When domestication breaks down, the animals often persist. The implications for engineering depend on one's perspective. For biologists, feral pigs signal a failure, a dereliction of duty. Their very existence implies that humans have ceded "control" back to natural selection, that we have failed to exercise proper "maintenance."[46] For hunters, feral pigs are evidence of success, of best-laid plans bearing well-earned fruit. Rather than fueling extinction, their bloodlust results in still more feral pigs. For them, natural selection is not antithetical to engineering, but an engineering tool that one can wield toward desired ends. Finally, just as context changes from one group to another, so can it change over time. The same pigs who were deliberately released by one generation may be damned as invaders by the next.

PART TWO

Knowing as Making

5 * *Design and Narrative in the History of DNA Analysis and Synthesis*

DOMINIC J. BERRY

In the early 1950s physiological chemists, organic chemists, and biochemists learned how to degrade DNA. Eventually they also learned how to synthesize two nucleic acids (dinucleotides), and in time longer lengths of DNA (oligonucleotides), possessing specific sequences of nucleic acids that could be used in a wide range of experimental investigations and medical interventions.[1] Development of methods for DNA degradation and the eventual wide availability of synthesized DNA had a profound effect on the biological and biomedical sciences. On the public stage this was because the ability to manipulate DNA, a material that at this time rapidly became *the* iconic material of life, could underwrite the molecular biologists' broader and more outlandish claims as to biological control.[2] With regard to changes in scientific practice, these came in fits and starts as new uses for DNA presented themselves. Thus far the best-known contributor, Har Gobind Khorana, has been a marginal figure, only occasionally incorporated into existing and established international histories of genetic information and code breaking. While many histories of biology have focused on the notion of DNA as encoded information, DNA is first and foremost a material. Histories of DNA as a chemical and biological material need to adopt a defiantly sober tone to avoid heaping further hyperbole on the helix. There are good reasons to consider the presence of an engineering epistemology in biology beyond rhetoric.[3] By paying attention to the matter of DNA, we can begin to tell alternative histories of molecular life as technology.[4]

In this chapter I discuss two scientific figures who came to pursue DNA as a problem of design and, in one case, engineering. In the first half I focus on the DNA degradation work of Erwin Chargaff (1905–2002), and in the second the DNA synthesis–based research of Nadrian Seeman (born 1945). My aim is to demonstrate what alternative histories of molecular life as technology look like, through researchers who pursued greater control of DNA as a manipulable material rather than as a carrier of biochemical information.[5] In both cases the pursuit of this control served a larger investigative agenda, in Chargaff's case to understand the phenomena of DNA-protein binding, in Seeman's, to understand the phenomena of DNA branched junctions.

Chargaff: Design and Molecular Architecture

Erwin Chargaff was a physiological chemist who is best known to historians of biology for producing a "rule" for the ratios of nucleic acid bases found in DNA, and as molecular biology's most vocal and articulate antagonist.[6] One of his most biting comments, that molecular biology is merely the practice of biochemistry without a license, was actually a line of dialogue in a play that he published within his *Essays on Nucleic Acids* in the early 1960s.[7] Though Chargaff investigated the constituent chemicals involved in various biological narratives, including blood coagulation, the formation of egg yolks, and the binding of lipids and proteins, it is his interest in finding DNA's general structural features that concerns us here.[8]

Physiological chemists investigated the constituent chemicals present at each stage of any given biological process, before proposing potential reasonable sequences of events that could accommodate these findings.[9] In the case of DNA-protein binding this involved learning how DNA responded to various different enzymatic, chemical, and technical treatments, the honing of protocols and experimental procedures that increased one's mastery of the molecules, improved understanding of their characteristics, and uncovered what could be done with them. Divining the "general structural features" of DNA, as Chargaff expressed it, required reasoning from the new materials produced in any given chemical reaction backwards to potential structures that could explain one's arrival at them. As Mary Jo Nye has explained with respect to the origins of structural chemistry, "theoretical explanation meant understanding what *has happened* as well as what would happen. It is this temporal emphasis of chemistry that aligns it with biology as much as with physics."[10] Indeed, Chargaff regularly referred to his method as a "post-mortem science."[11]

Chemists and biologists working in this mode should be taken seriously as occupying the position of a designer or architect, because their results did not specify any actual candidate structure, but rather only the constraints that such a structure (whatever it might be) had to operate within. The language of molecular architecture abounds in the history of chemistry and biology, both as an analytical and an actor's category, existing work from Mary Jo Nye on Linus Pauling providing a clear and helpful example.[12] However, I have yet to find a historian or philosopher of either biology or chemistry who has pushed this language beyond metaphor, such that design and engineering epistemology become relevant.[13] We should be thinking as much with Walter Vincenti as we might Pauling.[14] Knowing this helps to prevent us from seeing Chargaff as some fusty chemist; rather, he had his own cutting edge.

Chargaff: Degradation and the Core

In the conduct of his research Chargaff discovered what he termed the "core" or "limit polynucleotide" or "enzyme-resistant residue."[15] In 1949 he and a coauthor reported this unexpected finding resulting from their degradation experiments on DNA polynucleotides, which were intended to break DNA into its constituent units and subunits. Such experiments were also of direct importance for attempts at DNA synthesis in this period because they produced the nucleosides—individual nucleic acid bases without a phosphate group—which could subsequently be synthesized.

Chargaff had found that a fraction of his samples consistently failed to be degraded by the deoxyribonuclease enzyme used for these purposes; this material he called the "core." He pursued the core by comparative biochemistry, the same means that had previously revealed his celebrated DNA nucleotide ratios. Comparative biochemistry required sampling different tissues from different species and measuring the amounts of the different nucleotides contained within each sample. This he now did with the core, looking for patterns in the presence and absence of its constituent nucleotides in a range of different cellular contexts. He was prepared to conclude relatively quickly that there was something structurally important happening:

> Apart from less probable explanations, these findings could be interpreted as indicating either the presence in the nucleic acid sample of a small quantity of a second desoxypentose nucleic acid of different purine and pyrimidine composition or the existence in the desoxyribonucleic acid chain of nucleotide clusters (in the case studied, relatively richer in adenine and thymine) distinguished from the bulk of the molecule by greater resistance to enzymatic disintegration.[16]

The first suggestion entertained, that there may be further nucleic acids present, was a preoccupation of Chargaff's throughout his career, and formed part of his worry that molecular biology was practicing biochemistry "without a license." Without suitably sensitive experimental arrangements he feared that unexpected chemical constituents might be systematically overlooked. But we are particularly concerned with his second suggestion. Having found "cores," this observation became a design constraint for Chargaff's interpretation of DNA-enzyme structure and interaction. Whatever those structures and interactions might be, some could resist degradation.

The core next appears in a single-author review essay of 1950, based on a lecture series he had given to the Chemical Societies of Zürich and Basel in June 1949. His findings concerning the enzyme resistance of particular lengths of DNA were placed in relation to the question of the chemical specificity of enzyme action on DNA. His interpretation relied on an analogy with reactions studied in immunology, a field whose overlapping interconnections with the origins of molecular biology have been historicized by Lily Kay, also via Linus Pauling. The latter is a figure much more at home in the traditional historiography of molecular biology than Chargaff is, so it seems that insights from immunochemistry could belong to many different schools of molecular biological thought.[17]

> It is very inviting to assume that such relations between specific inhibitor and enzyme, in some ways reminiscent of immunological reactions, are of more general biological significance, in any event, a better understanding of such systems will permit an insight into the delicate mechanisms through which the cell manages the economy of its life, through which it maintains its own continuity and protects itself against agents striving to transform it.[18]

Typically historians have cited this paper as evidence that Chargaff was a true and early supporter of the view that the hereditary material was DNA. But this commitment reads as secondary in comparison with his reasoning about DNA design and function.

After another paper in 1954 in which Chargaff and his coauthor explored the core as a "reflection of the specificity of the enzyme," making use of an enzyme found in germinating barley,[19] the core was reintroduced to an international readership in his single-authored chapter for *The Nucleic Acids*.[20] The latter was a two-volume collection edited by Chargaff and (James) Norman Davidson, published in 1955. The essays were organized so as to produce something like a functional anatomy of DNA and RNA. Chapters progressed according to scale, from molecular subunits (the sugars, nucleosides, nucleotides, etc.) all the way up to their roles in the cell. Along the way other chapters were dedicated to the kinds of technique and analysis most appropriate for the investigation of molecules at that scale. Many of these methods relied on new instruments that were contributing to a revolution in chemists' methodology at this time, one which it has been argued placed an emphasis on rule-based thinking and experimental capacities for synthesis.[21]

In his single-authored contribution, Chargaff first sets up the problem of the impossible length of polynucleotide sequences:

> It will suffice to point out that a chain consisting of 2500 nucleotides in the proportions found for the total deoxypentose nucleic acid of ox tissues . . . could exist in something like 1015000 sequential isomers. Since the human mind does not enjoy contemplating the impossible for a long time, it either forgets, neglects, or reduces it. The latter operation results in the more modest desire, not to write the entire sequence of nucleotides, but to discern certain more general structural features, if any can be found.[22]

When it came to the latter, aside from explaining how interpretations of the degradation products of thymic and apurinic acid "have played an important part in discussions of the structure of the deoxypentose nucleic acid," the only other structural insights he offered came from the core. He first defends the usefulness of this concept from critics (who are not named or referenced and whom I have not been able to trace) by reference to the Oxford English Dictionary: "CORE: 'A central part of different character from that which surrounds it.'" Presumably this reference was intended to demonstrate his confidence in the notion and also clarify that he was arguing that polynucleotides contained specific and distinct structural elements that could be distinguished from the bulk, elements that could be regularly physically produced as reaction products. The reader is then invited to enter into a difficult and demanding experimental situation, as Chargaff explains his research narrative and the narrative of nature it is intended to clarify:[23]

> When a deoxyribonuclease acts on a deoxyribonucleic acid, we are dealing with an enzyme of as yet unrecognized specificity attacking a polynucleotide chain of as yet unknown sequence. It is a reaction whose study is likely to spread more darkness than light. . . . The enzymic attack on a substrate of the complexity of a deoxypentose nucleic acid, which results in its partial cleavage, must go through an intricate pattern: every break of the original molecule produces substrates that are new and different; the enzyme must deal with kaleidoscopic substrate changes. The attempt to solve this gigantic puzzle by fitting the innumerable fragments into a plausible sequence is doomed to failure. Unless, however, deoxyribonuclease is specific only for the size, and not the quality, of the oligonucleotide fragments which it is able to cleave, in which case the first few random events would decide all subsequent ones, the order in which different pieces are detached, and

> the composition of those that are left behind, may serve as means of distinction between different nucleic acids.[24]

This was the potential power of the core, producing a cascade of implications. The fragments left behind, along with evidence for the order in which the other pieces had been depolymerized, contained a material record of what had been (and what could not be) cleaved in a particular polynucleotide when acted on by a particular enzyme. Collecting segments as they broke away, analyzing them to record the nucleotide constituents, and also collecting the cores left behind at the end, analyzing them to record the nucleotide constituents, all provided an iterative path toward accounts of DNA-enzyme interaction.

Ideally I would now provide an example of a narrative of DNA-enzyme interaction arrived at by analysis of cores, but it seems that the promise of the core was to remain just that. Chargaff did not return to the topic after 1955, and his research program, which had been pursuing many simultaneous lines of inquiry, moved in a different (though related) direction. Around 1952 his laboratory began exploring methods that would degrade DNA without so thoroughly depolymerizing it (i.e., without so thoroughly destroying the sequence of the lengths of acid) as enzymes otherwise did.[25] All of this chemistry did result in new and less damaging methods of degradation and revealed to Chargaff further design constraints. In work on calf thymus DNA he reported: "The structure of DNA that emerges from these experiments is that of a chain in which tracts of pyrimidine nucleotide alternate with stretches in which purine nucleotides predominate."[26]

Many more years would pass improving these methods, applying them to different species and different tissues, scanning for patterns in arrangements of degradation products that might reveal insights into DNA structure and its interactivity, until—in the mid-1960s—a new term was introduced. His newer methods of chemical degradation produced lengths of pyrimidine and purine nucleotides respectively. Any collection of these lengths that possessed an equal number of nucleotides he now proposed to call "isostichs":

> The term *isostich* is formed, similarly to "distich", from the Greek *isos*, equal, and *stichos*, line. To give an illustration, the dinucleotides pTpTp and pCpCp are isostichs, but not isomers; whereas pTpCp and pCpTp are both isomers and isostichs.[27]

What was the significance of the isostich? Well, just as with the nucleotides analyzed in different samples of different tissues in different species that led to his singularly famous ratios, and just as with the cores recovered from enzymatic degradation of different DNA samples of different tissues in different species, the constituent nucleotides making up isostichs recovered from samples of different tissues in different species might contain patterns as to DNA structure and polynucleotide sequence.

"That DNA molecules of different cellular origin can often be shown to exhibit vast differences in chemical composition is well known. But do they possess common features, and is there evidence of unifying principles, beyond the base-pairing regularities?"[28] By the point of publishing this paper Chargaff had been working on this question for more than 15 years. Leaving DNA-protein interaction aside, he instead offered speculations on how his findings suggested a common "proto-DNA" ancestor for both microbial and mammalian species that then diverged by mutation through replacement of either purines (for one group) or pyrimidines (for the other), with greater frequency.[29] Entering his findings into this kind of evolutionary narrative may have been a good way to save them (and isostichs) from obscurity, but was of course nothing like the kind of biochemical narrative he had embarked upon. I have not yet found a completed narrative of DNA-protein binding that relied on his patterns in nucleotide constituents, cores, or isostichs. This did not stop the list of potential complex phenomena that he believed his work could illuminate from continuing to grow, including, by 1968, DNA replication and translation.[30] But there is no need to interpret this as failure.

Setting oneself an impossibly ambitious challenge is a perfectly productive way in which to organize the research work underway in one's laboratory, and Chargaff was exceptionally productive. His strategy of collecting constraints for DNA (no matter how obscure their significance might have been) by filling in the partial elements of an overall narrative (no matter how small) could have continued indefinitely. In the process he also learned to control and manipulate DNA more precisely, producing nucleosides that other chemists would learn how to synthesize into their own new designs, in some cases as a project for engineering.

Seeman: Immobile Junctions and Auto-Footprinting

In the later twentieth century the history of DNA synthesis also becomes somewhat intertwined with that of nanotechnology.[31] Successes in the design, control, manipulability, and automation of DNA synthesis were ideal evidence for those promoting a new molecular dawn, and were appealed to by actors pursuing a broader and more comprehensive molecular engineering.[32] In the 1980s and 1990s machines capable of synthesizing specifiable lengths of DNA entered the marketplace.[33] Nadrian Seeman, best recognized as a founding figure in the field of DNA nanotechnology, readily admits the importance of the availability of synthetic DNA and DNA synthesis machines for his work. Seeman's construction of immobile junctions to investigate the nature of DNA recombination would not have been possible without the availability of synthetic DNA.

As with Chargaff, Seeman was also interested in learning about the many conformations that DNA might be capable of, rather than understanding only that of the helix. But in contrast to Chargaff, Seeman was also interested in what DNA conformations might be possible *regardless* of whether or not they occur in nature. In so doing he stepped beyond understanding of design in nature into engineering new natures. Also, as with Chargaff, Seeman was dealing with a complex set of chemical interactions that made DNA structures and mechanisms almost impenetrable—in his case, those that DNA enters into during recombination. And, as Chargaff had done, he sought a foothold into that complexity. Chargaff's approach to understanding the narrative of DNA-enzyme systems always relied on evidence collected after the phenomena had finished. Seeman instead constructed "immobile branched junctions" that trapped an otherwise unstable phenomena in place, and made its actions subject to analysis by instruments measuring its physical states. One could use these synthetic analogs to shed light on corresponding phenomena occurring in cells.[34] Aside from making these immobile branched junctions to improve understanding of DNA recombination, he also envisioned them being grown into large extensive networks of complex architectures (see figure 5.1). In Seeman we have a case where design thinking was an explicit feature of his research program. He developed a number of his own DNA design rules into FORTRAN code that could then automatically produce the correct sequences for the DNA construct desired, and he also expanded this approach into an entire workflow.[35]

Seeman's route to possible DNA was arrived at after initial training in crystallography at the University of Pittsburgh, completed in 1970.[36] One of his first postdoc appointments included time in the laboratory of Alexander Rich, famous for completing the first experiments leading to polynucleotide hybridization. Seeman eventually became dissatisfied with crystallography as a primary investigative method, in large part because of an inability to regularly acquire suitable crystals. In 1977 he joined State University of New York at Albany (SUNY) as an assistant professor, and it was there that he conceived of a new method for investigating DNA recombination. This focused on branched junctions in DNA, and a research object first proposed by Robin Holliday in the 1960s, the eponymous "Holliday junction."[37] It is necessary to explain this object in more detail in order to understand Seeman's own research program.

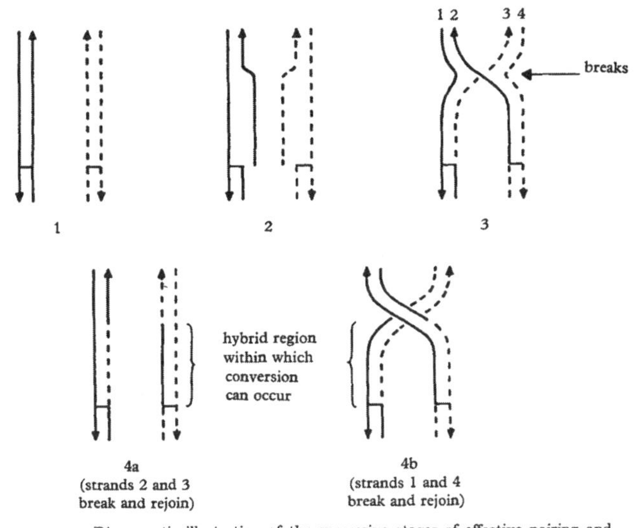
1 2
3 4
breaks
1
2
3
hybrid region
within which
conversion
can occur
4a
(strands 2 and 3
break and rejoin)
4b
(strands 1 and 4
break and rejoin)
Diagramatic illustration of the successive stages of effective pairing and recombination of homologous chromatids. Solid lines represent the DNA strands of one chromatid and broken lines those of the other chromatid. The polarity of the strands is indicated by the arrows, and the short horizontal line gives the position of the linker or genetic discontinuity.

Figure 5.1. Figure and description taken from Robin Holliday, "A Mechanism for Gene Conversion in Funghi," *Genetics Research* 5 (1964): 282–304.

In his 1964 paper, "A Mechanism for Gene Conversion in Funghi," Holliday laid out his interpretation of how recombination must happen. The conventional explanation at this time, the "copy choice hypothesis,"[38] he first analyzed according to how well it could accommodate available evidence:

> One advantage of the copy choice hypothesis is that it makes several quite specific predictions: (1) that the genetic material replicates conservatively; (2) that genetic pairing must take place prior to or at the same time as genetic replication; (3) that if the hypothesis is used to explain crossing-over as well as conversion, and sister strand exchange by breakage and reunion is ruled out, then successive cross-overs along the length of the chromosome should involve the same two chromatids; (4) that at a given heterozygous site conversion from mutant to wild allele and from wild to mutant allele should occur with equal frequency; and (5) that in crosses between different mutants within the same gene, the frequency of conversion to the wildtype allele should be proportional to the distance between the mutants, thus allowing the construction of linear maps with additive recombination frequencies.[39]

Reasoning from the whole copy choice narrative was a way to produce counterfactuals that would confirm or undermine its status. Holliday reported that predictions 1, 2, and 3 had been disproven, and that 4 was unlikely to be true, while 5 seemed well supported. Holliday now offered a competing narrative, one that did not require replication but instead required only the breakage of strands and their reunion.[40] Some of the key features of his alternative narrative can be found in figure 5.1.

By the time Seeman began his research, it was well known that branched junctions occurred naturally as part of genetic recombination, and they were therefore thought to be a good focal point for those attempting to understand these processes as a whole. He was directed to look closely at the phenomena after being approached by Bruce H. Robinson, then a postdoc at SUNY using electron paramagnetic resonance spectroscopy to study how much DNA "liberates" (vibrates in a rotational direction). Robinson asked Seeman for assistance making a model of branched junctions.[41]

The model Seeman and his collaborators proposed was not much use without experimental means to explore it. "So the whole thing about DNA branched junctions was that . . . naturally-occurring branched junctions don't sit still. They move around."[42] Inspired by a conversation with a protein crystallographer about how to get hold of stable intermediate protein structures (which are not unlike branched DNA), and in conversation with an undergraduate working in his lab, Seeman landed on a potential strategy: to stop them moving once they had formed so that they could be studied in situ.[43]

> And suddenly I realized, *if* you could make the DNA synthetically—and in 1979, that was a large if, not a small if—then you could put together something like this, where you'd have four different [base pairs] flanking the junction. And, Jesus, I mean, I was higher than a kite when I realized that. And then I [. . .] built a model, because now I'm starting to think about [branched] DNA as something [. . .] we could approach experimentally.[44]

He knew that sections of DNA could be ligated together thanks in particular to his being based in a biology department:

> I said, "Gee, you know . . . if I could organize these guys this way. . . ." In the preceding three years, I had had to go to listen to other people's students talk about how they made their miserable little constructs. . . . And every one of them had done the same damned thing, and I was bored to tears. What they'd done is they'd taken their plasmid, they restricted it [with] either one or two enzymes . . . and then they'd taken their gene and they'd put it in there, and they had some way of checking that it was there. And

> they'd stuck them in there with sticky ends. Sticky ends, you know. I was not unfamiliar with sticky ends.[45]

This experimental approach was not possible without specifiable lengths of synthesized DNA that could then be ligated together using sticky ends. It was absolutely embroiled in a design and engineering approach, as we can see in Seeman's immediate publications on this research.

His first publication describing some of the key ideas was published in 1981. His chapter, "Nucleic Acid Junctions: Building Blocks for Genetic Engineering in Three Dimensions," also included a passing statement regarding genetic engineering: "In so far as genetic engineering consists of constructing specific structures of genetic material (nucleic acids) by these techniques, this procedure can be termed genetic engineering in two or three dimensions."[46] But these comments and the title were dropped when a more fully formed version of the paper was published in 1982 in the *Journal of Theoretical Biology*.[47]

The bulk of these papers was dedicated to the design rules that Seeman had come up with for creating DNA sequences that would generate immobile junctions surrounded by double helices, each of which was resistant to migration (i.e., the point where the Holliday junction had formed would not change). Confirmation that the desired branched structures had indeed been made was achievable thanks to their method of "auto-footprinting." Here we can draw another direct line between Seeman and Chargaff because the origins of the original hydroxyl radical footprinting method that Seeman adapted came directly out of research on enzymic degradation of DNA.[48] It is also interesting to note that the key design feature that Seeman had to adopt in order to accomplish his goal was to reduce the amount of symmetry present in the nucleotide sequences used for making immobile junctions. Chargaff had spent the greater part of his career searching for meaningful patterns in DNA at this scale. In these respects, Chargaff and Seeman can be understood as contributors to a longer and broader material history of DNA, its investigation, manipulation, and design. This is true despite their very different training, methods, and goals.

The auto-footprinting method was used to characterize branched junctions and ensure that intended designs had definitely been produced.

> [W]e were involved a lot in sort of working out the structure of the Holliday junction in solution. There was a collaboration there with Tom [Thomas D.] Tullius, and we had a couple of papers with him using his hydroxyl radical attack method, and we kind of made a variation on the theme. He was doing hydroxyl radical footprinting. So a footprint is you have a piece of

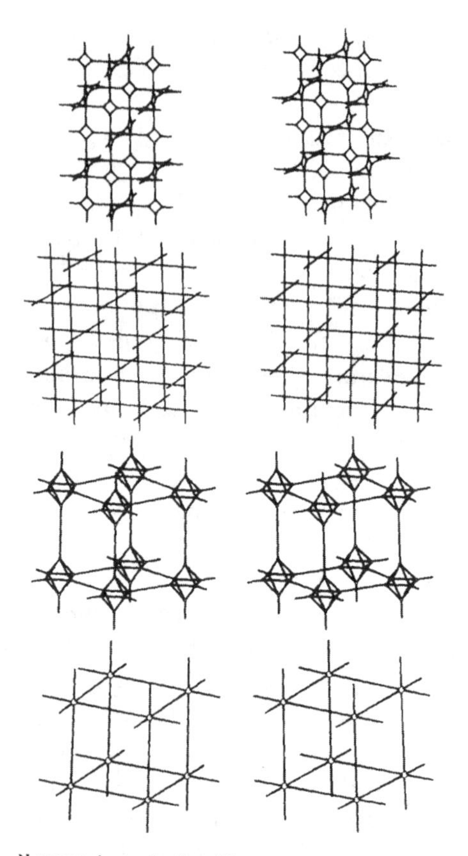

N-connected networks. Four different N-connected networks are shown in stereoscopic projection. These are all indicated as forming cubic lattices, although this is certainly not necessary. The dark lines represent double helical stretches of nucleic acid. The large circles represent junction regions of the appropriate rank. The short lines on the periphery of each figure represent unsatisfied valences. From the top, these are respectively units of 3, 4, 5 and 6 connected three-dimensional networks.

Figure 5.2. Figure and description taken from Nadrian C. Seeman, "Nucleic Acid Junctions and Lattices," *Journal of Theoretical Biology* 99 (1982): 237–47. This article also includes diagrams corresponding to nucleic acid sequences making immobile and semimobile junctions developed according to his design rules.

DNA and you want to know where a protein binds. So you bind a protein to it and you throw hydroxyl radicals [at it; they] are probably the best way to do it. You put iron complexed with EDTA [ethylenediaminetetraacetic acid] in there, [in the presence of] hydrogen peroxide, and that in turn generates hydroxyl radicals. . . . And then that attacks the DNA, and you can see where is it broken and where is it not broken . . . I mean, it's single-hit kinetics . . . so it cuts here, and then you've got a label up here, so you run it out on a gel and you say, okay, we cut here . . . so this is a fragment this long. That's that position. And then there's that position, that position. But in this position, so where my wristwatch is, well, it couldn't attack because it's protected there. . . . So that's a straight-up footprint. So we made a variation on that theme, sort of auto footprint, where you take a motif like the Holliday junction and you hit it with hydroxyl radicals, and then you take each of those same strands and you put them up against the Watson-Crick complement and you see what's different.[49]

Where Chargaff used enzymes to break DNA, finding cores, Seeman used hydroxyl radicals, finding where DNA had or had not formed Holliday crossovers. Both methods create sections of DNA following application of a degrading factor, sections that can then be used to reason backwards to the structural features that most likely caused them. In Chargaff's case, one could make inferences from distinctions between polynucleotides; in Seeman's, one could make inferences from the presence or absence of crossovers. Where Chargaff found a use of the core to make more fine-grained narratives of the phenomena, Seeman used crossover formation to test competing narratives.[50] Seeman's strategy mirrors Chargaff's in other ways, concerning the complexity of the phenomena.

Seeman described the epistemic challenge that he faced and his ambitions for immobile branched junctions:

> To date, it has not been possible to study the structural and dynamic properties of these [Holliday] junctions in oligonucleotide model systems, where the junction will contribute a significant signal. This is due to the sequence symmetry . . . the strands there are unlikely to form junction structures in preference to double helices; if they did occasionally form such structures, the process of branch point migration . . . will result in the rapid resolution of the junction structures into double helices. Conversely, abandonment of this sequence symmetry results in the possibility of generating junctions which will form preferentially, and which are immobile with respect to the migratory process.[51]

Here Seeman and Robinson laid out the groundwork for an exploratory research program in which the use of junctions designed to resist migration (unlike a typical length of DNA) could be used to probe what was happening when recombination occurred. Due to the difficulty of explaining the molecular model that they were proposing, they worked through an analogy with a staircase:

> The difficulty in understanding a complete molecular model of the junction has led us to represent it both schematically and by comparison with

> a familiar staircase. In the staircase, the bannisters correspond to the backbone structures, and the steps correspond to the base pairs. The junction region is suggested by the plateau in the middle. . . . From this figure, it can readily be understood how each backbone (bannister) is part of two separate double helices.[52]
>
> The staircase . . . illustrates another point of importance. . . . Two individuals symmetrically walking down the two staircases above the plateau would face each other; however, if they continued below the plateau, down opposite staircases, they would face away from each other. Thus, although two-fold rotational symmetry exists between helices attached to the opposite sides of the rhombus none exists between helices attached to adjacent sides of the rhombus.[53]

Hoping that readers now understood what an immobile branched junction was, Seeman and Robinson explained their intention to use these synthetic junctions to test candidate models of recombination.[54] It would not be until the early 1990s that such assessments began to be published. Here investigating the characteristics of the immobile junctions themselves—to demonstrate their adequacy as a research tool—was as important as informing models of recombination.[55] In addition, and in ways that are again similar to Chargaff's case, Seeman had set himself the challenge of providing a narrative of a phenomenon that remained beyond his grasp. I have not yet found a complete narrative of recombination produced with and through Seeman's research using immobile junctions, and he is himself ambivalent about the status of this work. "We did a ton of solution physical chemistry on branched DNA. I don't know the ultimate value of some of that work. . . . [I]t'll ultimately be decided sometime after I'm dead."[56] Instead, as with Chargaff, Seeman found the enterprise rewarding on its own terms, in his case moving on to explore the design space afforded by these immobile junctions, and other structures like them. Engineering and working out designs of DNA offered pleasures and challenges all of its own, regardless of any light it might potentially shed on phenomena occurring in the cell.

Conclusion

For researchers attempting to create plausible narratives of complex interactions, the adoption of experimental systems that produced molecular artifacts (be they cores or immobile junctions) was a particularly fruitful strategy. This has been so in cases where scientists did not understand the complex chemical processes that produced their artifacts, though they could nevertheless treat these artifacts as evidencing some structure and specificity (as with Chargaff and the cores), and also when scientists engineered a design of their own, which could then be sent—like a probe—into chemical complexity (as with Seeman and immobile branched junctions). Future studies at the intersection of biology, technology, design, and engineering, will expand the cast of characters that we might think with. Each case can be analyzed so as to learn its constituents, before the overall collection of historical cases is then studied comparatively for patterns. If the examples of Chargaff and Seeman that I have laid out seem superficial or incomplete, consider them instead as useful synthetic analogs that can be sent, like probes, into historical complexity.

Throughout the above I have referred to the *narratives* of DNA-protein binding and DNA branched junctions. I have a particular purpose in doing so, as part of an effort to expand the forms of analysis that scientific work is thought to invite in the history and philosophy of science.[57] Narrative resides in science as a means of conceiving of phenomena and reasoning about them. As a way of knowing, narrative provides essential tools for scientists working on complex interactions. In the particular context of molecular design explored here, this argument builds on insights from a range of historians of chemistry, including Mary Jo Nye, Katherine Jackson, Evan Helper Smith, and M. Norton Wise.[58] Design and narrative are complementary analytical frames for the scientist and engineer because where design thinking provides access to, and a means of conceiving of, possible structural elements in the world, narrative provides access to, and a means of reasoning about, the interrelations between them. These starting points open up a wide field of investigation for intersections of science and engineering.

Chargaff, in contrast to Seeman, never transitioned from design thinking into engineering. Instead he explored nature exhaustively, pursuing innumerable analyses of different tissue samples taken from a wide range of different species. Why do researchers begin to engineer biology? Unlike Seeman, Chargaff either believed a) he had a sufficient method for exploring his phenomena as it occurred in nature, or b) there was no possible way to make DNA-protein binding easier to study. Seeman by contrast was directed toward an engineering solution because he a) did not believe he had a sufficient method to study the phenomena and b) believed engineering would indeed make it easier to study. The decision to engineer biology is dependent on having means to do so and calculating whether those means are worth the investment. When grinding out the actual ceases to pay, dipping into the possible can begin to look more attractive, provided one has means and method.

Acknowledgments

I am deeply grateful to Patrick McCray for suggesting that I look at the case of Seeman in the first place, and for all his assistance and willingness to share his extensive archive of primary and secondary research materials. I am likewise grateful to Nadrian Seeman for extending me the same access to these materials that he originally granted McCray, and also for his help clarifying certain aspects of his research by email. My considerable thanks to Luis Campos, Robert Meunier, Mary S. Morgan, Alok Srivastava, and Zachary Pirtle, who all provided extensive feedback on numerous drafts, and to Konstantin Kiprijanov and Hanna Lucia Worliczek for help finding relevant resources in the historiography of chemistry and biology respectively. I am of course very grateful to the editors of *Nature Remade* for inviting me to contribute to this collection. A version of this paper was presented in the University of Leeds senior seminar series, and I am grateful to that audience for their insightful feedback and criticisms. This paper was written while employed as Research Fellow on the Narrative Science project, supported by the ERC under the European Union's Horizon 2020 research and innovation programme (grant agreement no. 694,732). More about that project can be found at www.narrative-science.org. A network of scholars is now being built around integrative and interdisciplinary questions concerning biological engineering, about which more can be learned at www.bioengcoll.org.

6 * *Behavioral Engineering and the Problems of Animal Misbehavior*

EDMUND RAMSDEN

"Fully automatic" was how a magazine article of 1954 described an exhibit designed by psychologists for the food manufacturer General Mills.[1] A hen, "Casey," tugged at a rubber ring, and an electrically operated bat knocked a small ball toward a wire screen outfield. If it cleared the mechanical defensive players, the hen would run frantically around the bases to her reward, a few grains of wheat that had fallen into her trough. At one time, General Mills had forty-five of these shows on the road and Casey even made it onto prime-time television. The "base-ball playing hens" were one of a series of acts that included tap-dancing chickens, rabbits that played piano, raccoons playing basketball, and turkeys that won at pinball. For Marian and Keller Breland, the creators of the exhibit, behind the frivolity lay a serious program of applied research dedicated to the expansion of B. F. Skinner's methods of operant conditioning into commerce, industry, the military, and wider society. Commercial animal training, based upon efficient and consistent laboratory methods, could transform animals into useful and dependable helpers in a wide range of human activities.

Skinner was initially supportive of this new field of "applied animal psychology."[2] As Alexandra Rutherford has argued, Skinner's most enduring legacy is not his philosophy of radical behaviorism but "his *technology* of behavior."[3] Skinner has been described as a "psychologist with the soul of an engineer," and he saw his distinctly modern project of remaking the organism in terms of "behavioral engineering."[4] An organism's behavior was malleable, it was a material that could be shaped, manufactured, and improved with "surprising ease."[5] The natural and the mechanical were woven together into a powerful system of behavior control. However, with the expansion into the mass production of animal behavior exhibits, this system began to break down. In 1961, the Brelands coined the term "misbehavior" to describe their experiences with recalcitrant and unpredictable animals. While historical studies of behaviorism have tended to focus on external criticisms of the paradigm, such as those emanating

Figure 6.1. An automated "Dancing Chicken" exhibit from the Animal Behavior Enterprises collection, V115. Image courtesy of the Drs. Nicholas and Dorothy Cummings Center for the History of Psychology, University of Akron.

from biology, linguistics, and cognitive psychology, this chapter will focus instead on the problems generated within the field by the material conditions of animal training. It is informed by Hans-Jörg Rheinberger's analytical framework of the "experimental system," within which there exists a dynamic relationship between "technical objects," which are stable and predictable elements characterized by standards of purity and precision, and "epistemic things," the unpredictable and vague subjects of investigation that we do not yet know.[6] This very productive tension maps readily onto the perception of the organism as a predictable artifact or tool for laboratory work and as a "sample of nature" that is never fully understood or controlled and thus can continually, and very usefully, generate new questions or epistemic things.[7] As the Brelands took operant conditioning beyond the confines of the laboratory, Skinner's tidy system began to fracture, and the "nature" of the organism began to override the machine-like predictability of conditioned behavior.

While misbehavior proved costly, it also presented an opportunity for those who had a broader conception of experimental psychology as a discipline that appreciated nature, living diversity, and the processes of evolution, or those who sought to counter the broader social implications of behavioral technologies of control through reference to a biologically unique and autonomous organism that maintained its own meaningful behavior in a given environment. Many of these criticisms were reminiscent of the philosopher Georges Canguilhem's objections to behaviorist psychology and what he saw as its overly deterministic and mechanistic understanding of the environment acting on the organism. This he sought to counter by privileging the animal as a biologically unique, autonomous, and active being, adapting and maintaining its own "vital norms" in a given environment.[8] The practice of conditioning, he argued, was an inherently pathological process. It was abnormal to the organism and was resisted by "a being on which not everything can be imposed, because its existence as organism consists in its proposing itself to things on the basis of certain orientations that are proper to it."[9] And yet, in the very distortions and disruptions that ensued from the "autopoietic character" of an organism in the context of an unwelcome milieu, there were opportunities for understanding or new epistemic things for the sciences of psychology and biology.[10] For Canguilhem, like Rheinberger, failure or error has generative and creative value, whether for the organism facing the challenge of adapting to its changing environment, or the scientist, seeking to understand the truth of these relations through experimentation. The variability among test subjects, the unexpected consequences of interactions between natural organisms and technical objects, are "not obstacles but stimulants to invention."[11] The history and philosophy of biology here complements recent work in the history of technology, particularly that of Edward Jones-Imhotep.[12] In his study of defense science in the Canadian north in the 1960s, attempts to control, master, and exploit the natural order through a series of "machinic orders" failed as nature proved antagonistic to telecommunications instruments. However, as Jones-Imhotep argues perceptively, the narrative of technological failure need not place nature and machine in opposition but can also prove useful in crafting alternative and equally productive relations between them. Further, the idea of "recalcitrant natures and failing machines" can also be used to shape an identity in late modernity that countered the view of the organism as an inert and malleable mass suitable for industrial production, and instead celebrated life, in the words of Keller Breland, as "self-organizing . . . spontaneous, purposive, self-determining."[13]

Pigeons in a Pelican: A New Technology of Behavior

Jones-Imhotep describes World War II as having "instrumentalized nature" as researchers grappled with wartime technologies. They translated natural phenomena “into languages and concepts born out of the world of machines.”[14] He then takes as an example Skinner’s attempts to make pigeons into devices to guide air to ground missiles. Skinner described the birds as “cheaper . . . compact . . . good at responding to patterns” and “readily expendable.”[15] The pigeon was harnessed in a jacket, and then placed in front of a translucent screen. It then learned to peck at a specified moving target for a food reward. By 1943, a complex machinery had been constructed in the nose of the missile, nicknamed the Pelican, in which a lens threw an image on a translucent plate within reach of three pigeons in a pressure sealed chamber. Their pecks on the target image controlled a series of air valves working to control the direction of the missile. Skinner’s team had constructed a prototype of an animal-machine hybrid in which the organic component gave the system flexibility and reliability. It was “fool proof,” Skinner declared, as while radio signals could fail and circuits misfire, the pigeon’s behavior was unfailingly predictable.[16]

For Skinner and his fellow behaviorist researchers at the University of Minnesota, Project Pigeon was designed according to the fundamental principles of instrumental conditioning and the Pelican was a glorified “Skinner box.” The box was a standard experimental preparation. Within a sound and light-proof environment, a hungry rat pressed a bar or lever and, with a click, a pellet would be released from magazine and fall into a food tray. At the same moment, a mechanism caused a vertical movement of a pen upon paper that covered the surface of a revolving kymograph drum.[17] The result was a cumulative record. With this relatively simple fusion of organism and machine, Skinner sought to remake the field of psychology. The movements of the organism strengthened if rewarded, and if not, they dissipated—behavior determined by its consequences. Once the animal learned to associate the click of the feeder with food, the experimenter could build in countless variations in procedure, the animal reinforced after every other press, or an average six presses on a random schedule, or one lever press every 30 seconds. Alternatively, the animal would learn to discriminate between variables, two different shapes or a light signal, in order to secure the reward. The animal was, on the one hand, acting voluntarily and instrumentally on the environment, as an operant; but, on the other, through carefully selecting a response, the behavioral engineer could make the organism behave in preordained and predictable ways. This technology granted them immense power over the organism as “the stimulus, the response, and the reinforcement are completely arbitrary and interchangeable. No one of them bears any biologically built-in fixed connection to the others.”[18]

While the Navy declined to continue funding Project Pigeon, it played a critical role in the development of Skinner’s work. He described in his autobiography how the behavior of organisms "appeared in a new light. It was no longer merely an experimental analysis. It had given rise to a technology."[19] Not only did his wartime work refine his techniques in shaping behavior, the bird having been transformed into an organic device or machine, but he now perceived how the total control of behavior could serve a wider purpose.[20] His work was no longer oriented around "bringing the world into the laboratory, but . . . extending the practices of an experimental science to the world at large."[21] Project Pigeon was inspiration for his novel *Walden Two*, written in 1945, in which the principles of operant conditioning and positive reinforcement are put to work in the creation of a technological utopia.[22] In order to remake the human it was necessary to abolish, he later described, the "autonomous man . . . a device used to explain what we cannot explain in any other way [. . .] A scientific view of man offers exciting possibilities. We have not yet seen what man can make of man."[23] Scholars have focused much attention on Skinner's influence beyond the laboratory, in institutions, psychiatric medicine, and the field of behavior modification. As Alexandra Rutherford argues, by treating behavior "like any other technological problem" and providing "an effective set of engineering tools," Skinner's approach "aligned itself seamlessly with the American technological imperative."[24] Technology could solve so many social problems, with the material world being manipulated to shape a more efficient, productive, and happier citizen, and it therefore embodied "a concrete ideal about the kinds of people we ought to be."[25]

Beyond the Box: The Field of Applied Animal Psychology

Two of Skinner's assistants on the pigeon project were recently married students from the University of Minnesota, Marian and Keller Breland. In 1943, they founded Animal Behavior Enterprises (ABE), convinced that such principles "could apply to all animals anywhere, in any conceivable fashion."[26] They sought to transfer the technique of training a pigeon to other species, beginning with their own pets, a dog and two cats, and then a variety of small animals. They purchased a small 10-acre farm in Minnesota and expanded into farmyard animals, the most important of which would prove to be the chicken—easy to source, tame, hardy, reliable, and easily trainable. The chicken was the animal of choice in a rapidly industrializing agricultural sector, as improved breeding techniques, the science of nutrition, and intensive confined rearing allowed for the mass production of chickens for consumption in the postwar era.[27] The Brelands took advantage of these developments, selecting popular, robust, and homogeneous commercial "broiler breeds," and transforming the chicken into an advertising icon for industrialized agriculture.

In 1947 they sent an ambitious, and successful, prospectus to General Mills, the rapidly expanding corporation based in Minnesota that produced much of America's cereals and animal feeds. They declared that any animal could be trained in almost any act: "it is now easily possible to develop behaviors which are a far cry from anything that an animal does innately. Bizarre poses, contortions, the operations of mechanical gadgets and mechanical toys are only a few examples of possible unusual performances."[28] Not only could the animal manipulate mechanical objects, but they themselves could be "controlled with a degree of precision that approximates physical mechanisms. . . . In fact, it is now possible to write the specifications for a trained animal performance in advance, just as for a mechanical device, with the assurance that they can be met."[29] This combination between the mechanical and the organic, with the animal serving as a manipulator of machines and as reliable, predictable, and replicable as a mechanical device, would be continually reworked through ABE exhibits.

Keller produced a series of acts that toured county and state fairs across the country on behalf of Larro Feeds of General Mills. The following year, they began training groups of salesmen to oversee the acts in farm feed stores and wrote the first manuals in behavioral engineering. The fact that it was possible to train trainers so successfully further advertised the reliability of the behavioral methods and the capacity for mass production. A "printed instruction manual" could cover "an entire animal show in a few pages," just like a manual for a machine, and thus, the "control of an animal . . . passed from person to person in the matter of a few minutes or hours, as opposed to the many years of apprenticeship necessary in pre-scientific times."[30]

Every year they introduced new acts. They had Barney the basketball bunny, a kissing bunny, the piggy bank, in which a pig deposited coins, a drumming duck, and a golfing hen. The level of automation was increased with a series of coin-operated features, the animal released from its home compartment when a coin was placed into a slot, and then, after completing a simple trick, received a reward on its return. These exhibits were easy to replicate and could shipped almost completely assembled. They described their new larger 260-acre farm in Hot Springs, Arkansas, as having become a "real 'assembly line' . . . a mass production system, with several trainers, batteries of equipment, and two, three, to a dozen animals being trained at one time by one person."[31] By 1970, hundreds of coin-operated displays were in action at various tourist attractions and amusement parks across the United States and abroad. As Keller Breland declared in a speech to the US Navy in 1962:

> Anyone can do it—in fact a great deal of our stuff now is absolutely automatic—you simply plug it into a wall like you plug in a display of a mechanical or electronic nature. The animals go through their paces on cycle—say every minute, every three minutes, every five minutes, and there is no necessity for personal handling. As you can see, we have taken the human element out of the thing and given the animal only the things that he needs—some stimulus support in order to fire off his behavior properly. The animals have, under these circumstances, proved to be highly reliable.[32]

This speech to the Navy was part of a pitch to secure government work and that year they began training dolphins—devising ways of communicating and controlling their behavior in the open ocean to allow them to help Navy divers in a variety of projects. Like Skinner before them, they were constructing hybrid systems that fused humans, animals, and machines into productive working units.[33] The work for the Navy had emerged, in turn, from their work developing shows for marine parks based on the acrobatics of the marine mammal, such as Marineland in Florida, Marine World, and Sea World. As the ABE director Robert Bailey later reflected, they had done rather well in expanding from training chickens for feed stores to training killer whales in facilities costing hundreds of millions of dollars.[34] And yet, as if to emphasize the universality and transferability of techniques, they began their aquatic mammal instruction with an ideal "behavioral model," as they described it, the chicken pecking a spot for a reward.

In 1951, the Brelands published a paper in the *American Psychologist* in which they presented their innovations and declared that a "new field" had been founded: "applied animal psychology."[35] Automatic training methods could now "outstrip old-time professional animal trainers in speed and economy of training. . . . We can turn out multiple units—200 'Clever Hanses' instead of one."[36] There was efficiency not only in production, but in use, and the "packaged act" could be shipped anywhere and "operated day in and day out with no more instructions than are necessary for the operation of any machine designed for such use. . . . Live animals take the place of puppets and robots."[37] They then sketched out the potential of a vast new field in which operant conditioning was applied to farm animals, pets, guide dogs, animals in the police and military, and even to children in parenting and the care professions. They saw their work as improving the lot of animals by improving their usefulness, making them easier to handle, more reliable, automated and involuntary, and thus better fitted to the needs of an advanced industrial economy; they could, in their own words, "train animals to live in man's complicated world."[38]

The Problem of Animal Misbehavior

The Brelands saw their work as validating and expanding the "new science" of behaviorism by applying it to "practical problems . . . much as the engineer makes his living by applying the principles of physics to the construction of bridges and roads."[39] They were showing scientists just what their technologies could do and that the field was "much farther advanced than anyone previously dreamed possible."[40] However, the more they advanced the technology, boasting in the early 1960s of having trained some 6,000 animals of over 40 different species, the more the organisms began to behave unpredictably. By running such a variety of organisms continually in a set of apparatuses on a scale far in excess of anything attempted in a psychological laboratory, performance problems began to emerge that upset the all-too-comfortable relationship between animal and machine and, most controversially, seemed to privilege the animal as a biological organism and an individual subject.

In 1961 they described these problems in a paper titled "The Misbehavior of Organisms," a play on Skinner's foundational text of 1938, *The Behavior of Organisms*.[41] It rapidly became one of the most influential papers in the history of behavior research. In it they described a "persistent pattern of discomforting failures."[42] One of their most famous acts, the dancing chicken, they revealed to have been "wholly unplanned"—when trying to get the animal to stand still for 15 seconds, half the birds developed a "very strong and pronounced scratch pattern."[43] In another coin-operated act where chickens were trained to collect and give a small plastic capsule containing a small toy to the observer, around 20 percent would "begin to grab the capsules and drag them backwards into the cage. Here they pound them up and down on the floor."[44] A pig that was conditioned to pick up large wooden coins and deposit them in a large "piggy bank" began, after several weeks of solid work, to repeatedly drop the coin, root it, drop it again, toss it up in the air.[45] There seemed no behavioral explanation for these "horrendous failures" as they violated the fundamental "law of least effort" and prolonged the "interval between the response and the reinforcement."[46] This represented "a clear and utter failure of conditioning theory" and undermined the assumption that all species and behaviors were interchangeable, each as easy to condition as the other.[47]

They argued for a process of "instinctive drift" whereby innate biologically determined sequences of behavior interrupted and dominated the conditioning routine. "It seems obvious," they declared, "that these animals are trapped by strong instinctive behaviors" and that these could override those recently conditioned.[48] These were instincts relating to food. Having harnessed the hunger of the animal through food reinforcements, the animals' species-specific feeding behaviors—the chicken's scratching and pecking, the pig's rooting—were coming to the fore and overwhelming the conditioning routine. This meant that there was a limit to conditioning. Understanding these limits required that the scientist turn inside the organism, to address issues of genetics and physiology, and outside, to understand the animal as environmentally situated and adapted to a certain ecological niche. They needed to find a way of looking beyond the production of predictable patterns of behavior, beyond the continuous whirring and clicking of levers, feeders, and recording instruments, and ask, "what the animals are 'up to' . . . what are animals 'doing' in a laboratory situation?"[49]

Putting Misbehavior to Work

Skinner responded aggressively and dismissively to the Breland paper. He chided them for the title, arguing that each case could be explained in a way that was consistent with his experimental system.[50] Skinner argued that accidental connections could be made between reinforcer and response, the animal associating a random behavior just prior to the appearance of food with its presentation. Such "adventitious reinforcement" could result in ritualistic stereotyped patterns as the animals attempted to influence the feeding schedule through performing this so-called "superstitious" behavior.[51] The organism did not "misbehave" as "the organism is always right. . . . If *anyone* misbehaves, it is the experimenter in making a bad prediction."[52] The animal served as a reliable technical object in the study of the operant response and the principles of instrumental conditioning held.

Marian Breland later reflected that their paper seemed to have "hurt" Skinner, and that even though relations remained amicable, "I still believe he regards us even now as highly deviant sinners, lost from the field."[53] Skinner had been further incensed by the Brelands turn to ethology as a means of explaining problems of "misbehavior" and learning how to deal with species-specific behavior in their routines. They dismissed Skinner's solution through adventitious reinforcement, pointing out that these were not arbitrary units of behavior, but patterns that showed high degrees of consistency within a species. As Marian stated: "Since the 1960's we have made a regular practice of studying an animal ethologically before we start a training program. We find out what types of behaviors the animal uses in its niche, what is easy to condition, what is hard."[54] The implications for the whole science of behavior were even greater, and they criticized the psychologists' tendency to assume "that you can pick up any organism as if it were made in the shop last night, pop it into the apparatus with no consideration of its evolutionary history, or the fit of the apparatus to the ecological niche."[55] In order to be able to interpret the results of an experiment, scientists needed to understand that not everything could be controlled, that they were not dealing with an empty organism in which stimulus mechanically generated response, but an animal with a natural history that needed to be considered.

Support for the Brelands' "apostate" position began to grow through the 1960s and 70s.[56] Skinner's student and leading behaviorist Richard Herrnstein broke ranks and argued that the "Skinnerian paradigm" did work well, but "if you zoom in on the fine texture," you discover an animal continuously emitting "new movements":

> The rat that occasionally jumps the other way or runs instead of jumps tells us in its simple way about an underlying "structure" of its behavior. . . . Hidden away in the behaviorist concept of generalization, and obscured by unwarranted assumptions of physically simple classes, is a perennial tenet of antibehaviorist psychology, which is that behavior must be described by entities with psychological reality.[57]

Herrnstein used misbehavior as a means of building stronger and more productive links with the fields of genetics and ethology as behaviorist psychologists sought to address the internal states of organisms more directly. Herrnstein argued that the organism's responses were not arbitrary, each as easy to condition as the next. Innate drives could determine response classes that demanded a much broader and more interdisciplinary study of different species in different experimental situations: "Variety, people say, is the spice of life, and they may be right."[58] Psychologists were crediting the Brelands for having identified the existence of "innate motivational dynamics" that now undercut the "simplifying assumptions" upon which behavioral engineering was based.[59]

While Herrnstein focused on species-specific behavior, others attended to another problem that had emerged from ABE training practices. The psycho-physiologist Horsley Gantt focused his attention on the immense individual variability identified in the Brelands' work as only 20 percent of the animals would make the final cut for any exhibit. The fact that not all animals within the same species could be conditioned in the same way revealed the limits of Skinner's experimental system. The psychologist had become too reliant on mechanized laboratory technologies and yielded to "the precisely working, continuously clicking, machines with which he is surrounded—and perhaps often confounded."[60] Gantt invited the Brelands to present their work, and some physical examples of animal misbehavior, to scientists as a means of encouraging psychologists and psychiatrists to engage more fully with physiology and genetics.[61]

Gantt also celebrated how problems of misbehavior undermined the brave new world of behavioral engineering.[62] Gantt was concerned that Skinner's approach was so mechanistic, deterministic, and totalizing that the individual was reduced to a mere automaton, their actions utterly predictable and stereotypical. It was reminiscent of the "French Encyclopaedists in their zeal to apply deterministic science politically and culturally. . . . In *Brave New World*, Aldous Huxley predicted theoretically the nightmare that could follow if the trend continued unabated."[63] His fears were shared by many in an era increasingly characterized by a rejection of authority and conformity. As Rutherford has argued, Skinner's suggestion to "bring human behavior within the purview of the technological world order," collided with a shift toward humanism and an antitechnocratic backlash against governmental control and authority in the late 1960s.[64] In a society deliberately engineered by planners to ensure harmony and efficiency, the organism appeared objectified and stripped of its freedom, agency, and personal control. As humanist, cognitive, and social psychologists began to counter the influence of behaviorism, the fact that two of Skinner's "staunch supporters" had, in their own words, "run afoul of a persistent pattern of discomforting failures," proved very useful to critics.[65] The problems of misbehavior revealed how Skinner's system was built on an unstable synthesis of the natural and the mechanical that had begun to break down as a consequence of its internal contradictions. As Jones-Imhotep argues, it is "actionable spaces between nature and machines" that offer "possibilities for resistance."[66] The experimental organism, just like modern humans, once trapped in an artificial and industrial environment and forced to conform, was now reasserting its independence and resurfacing as a creative and purposive individual.

Conclusion

In 1966, B. F. Skinner published a response to growing criticism of his experimental system. This was spurred to a large degree by the Brelands' presentation of the problems of misbehavior, but was also a reaction to the growing influence of ethology among psychologists.[67] He acknowledged the debt he owed to Darwin, natural selection having served as a useful analogy when he had sought to explain the selective value of behavior without recourse to mental entities. He was concerned to show that he was fully cognizant and respectful of the contribution of the biological sciences, that he recognized that nature provided the raw materials upon which the technology of conditioning acted, and that he had been one of the first to recognize the intrusion of phylogenetic behavior into the conditioning routine with his studies of "superstition." He saw his experimental system as complementing that of biology in its creation of a "naturalistic science of behavior."[68] But it was also a science that was uniquely productive. Skinner synthesized the natural and the mechanical through a technology of behavior that made a biological organism even more reliable in its actions than a man-made machine. More "builder than an experimenter," Skinner was committed to what Jennifer Alexander describes as the "central value" of engineering, that of efficiency.[69] It is through efficiency, and its required techniques of measurement, management, and control, that the natural world is made to conform to a desired end. The Skinner box with its cumulative recorder was a model of efficiency, allowing for the mastery over the functional relations between organisms and their environment that not only produced consistent data but could, he supposed, create efficient and productive citizens.

The Brelands had built upon this promise of efficiency, allying Skinner's experimental system and engineering ideal with the realities of a commercial business serving American industry. They promised customers "speed and efficiency" in an exhibit or animal trained for a role.[70] They had designed a system to mass-produce trained animals to reduce costs, meet demand, and increase profits. They described their facilities as bringing together psychologists and technicians of "high order of skill" for the design of exhibits and devices. The combination of applied science, practical engineering, and artisanal skill, allowed them to "get quickly to the heart of the controlling processes involved" and judge "what the organism is really "up to" in an experimental situation."[71] Keller Breland argued that this gave them a unique position of being "the sole validators of behavior theory at the actual point of reality. We have been in the business where the theory has to work or we don't get paid."[72] This reality had shown that when they moved beyond conceptualizing the organism as a machine and began to actually treat it as a mechanical object to be manufactured, the system began to fail. They had experienced errors in production that were inconsistent with the ideal of the automatic animal with unfailingly predictable behavior.

The problems of misbehavior generated serious problems for such a totalizing system as operant conditioning. However, this did not mean that the problems were not productive. In psychology it helped generate new research questions by transforming the experimental animal from a stable and predictable "technical object" into an "epistemic thing," encouraging scientists to examine what was going on inside experimental animals, whether cognitively or physiologically, or between animals and their wider social and physical environments beyond a mechanically programmed stimulus and response. As Keller Breland argued, they needed to stop viewing the organism as if it were a "passive mass of matter waiting for the impetus from the external world to control its behavioral processes," and accept "nature's organism for what it is, not for what we could like it to be."[73] Beyond experimental psychology, we can also see the value of a narrative of "failure" for reimaging the relationship between human, animal, technology, and society.[74] The problem of misbehavior was seized upon as revealing the limits of technologies of behavioral control. It allowed critics to read into the erratic behavior of animals, forced to perform for humans in an artificial and mechanical way, a potent alternative character of an active, rebellious, and autonomous individual.

The Brelands did not come to reject the mantel of behavioral engineering. Rather, as they moved beyond the confines of the laboratory and put the system to practical use, they came to appreciate the need for a more sophisticated understanding of the organism and its relationship with the environment. In the search for ever increasing control, speed, and efficiency, the engineer could not simply impose a machinic order on natural phenomena, but needed to understand and learn to enroll existing biological processes and behaviors: "By painful experience, it has been learned that it is best to let the animal 'tell' you its needs."[75] The Brelands saw their work as having given them a privileged epistemic position that would help unite psychology with the biological sciences, describing the increased specialization among the behavioral and life sciences as "natural for the scientist but unnatural for the subject matter."[76] Their later exhibits expressed this growing appreciation of the power, independence, and value of the natural world, in their own work and in wider society. In their two amusement park attractions in Hot Springs, the IQ Zoo and Animal Wonderland, and in other parks and zoos across America, the Brelands turned to more naturalistic exhibits, believing that the "day of the naturalistic zoo is rapidly approaching."[77] Rather than training animals to manipulate mechanical objects like miniature humans, separated and isolated in cells, they used the same technology of behavioral control to demonstrate the natural behavior of animals in their particular ecological niches, creating natural scenes in which an otter slid into the water, a duck dabbled for food in a pond, a raccoon climbed down a tree and found food that it washed carefully before eating, all once again "automated—controlled completely by timers and the sequence of events."[78] In one particularly ambitious proposal they even sought to include the scientific laboratory, showing the public just how these naturalistic behaviors were put to work to further our understanding of the natural world and the "psychological properties of mankind itself."[79] The Brelands were seeking to open up the system of behavioral engineering to the public to reveal the dynamics at play, and, most importantly, show how animal instincts were not an "annoying friction" to be isolated, dominated and overcome through technology, but "the primary and driving force in the system of behavior."[80]

7 * *Engineering Spaces for the Biological Effects of Fission*

JOSHUA MCGUFFIE

Introduction: Politics, Infrastructure, Biology

Mr. McBean: I have observed info copies regarding the Donaldson survey.
Mr. Miller: Yes—what is all of this about—do you know?
Mr. McBean: I have no idea, but I was just going to suggest that if this is going to impose any burden on the guest facilities, that we better alert them on it.[1]

Eleven biologists working for the University of Washington's Applied Fisheries Laboratory made a radiological survey of Bikini and Enewetak[2] atolls in the Marshall Islands during July and August 1949.[3] Lauren Donaldson, the lab's director, and his scientific team from Seattle had come yearly to these atolls since their participation in 1946's Operation Crossroads. Yet they remained invisible to Kelly McBean and Herb Miller. These two men worked for the Los Angeles engineering and construction firm Holmes & Narver, Inc., the Atomic Energy Commission's (AEC) contractor at the nascent Pacific Proving Grounds. As McBean was the firm's chief of operations, everything from the blueprints for buildings and the towers that hoisted test bombs aloft to personnel and logistics concerns came across his desk. Hence his confusion on 20 May 1949. The firm had few personnel in place at Enewetak and fewer places to house them.[4] McBean had no room at the inn for the unknown biologists from the Applied Fisheries Laboratory who wanted to study the biological effects of radiation. Instead, the Navy would house the team aboard a vessel retrofitted for their research trip.[5] Only four short years later, in 1953, Holmes & Narver built the biologists a lab on Parry Island in Enewetak. As they became more and more of the Proving Grounds, rather than at the Proving Grounds, the Seattle biologists internalized the site's mission. Their work after the Castle Bravo test disaster of 1 March 1954 highlights how the site's infrastructure became determinative for scientific practice. Their biology helped repatriate the Rongelapese, exiled after their atoll and their bodies suffered dangerous irradiation from Bravo's fallout.[6] But repatriation required biology attuned more to the Proving Grounds' exigencies than to the health of the Rongelapese. The Seattle biologists obliged.

This study places the Pacific Proving Grounds alongside other Cold War atomic sites at which the relationship between infrastructure, science, and politics unfolded in tragedy. Gabrielle Hecht, in her study of African uranium mining, has argued for a move away from stories of "the bomb" toward accounts of the political and cultural mechanics underlying Cold War atomic projects.[7] Another call to see the veiled dynamics at atomic sites comes from Traci Voyles as she describes how Navajo lives and land became marginal things naturally exploited for the US's national interest.[8] At the Proving Grounds, infrastructure fostered biology that aided political ends. In this story, engineers undoubtedly created a landscape useful for testing bombs. But the piers, water taxis, mess halls, and labs they built also afforded Donaldson and the Seattle biologists the ability to biologically reify American Cold War ambitions by calling dangerous places safe and sick environments healthy.

A Politically Engineered Landscape

The civilian Atomic Energy Commission longed for a permanent base of operations in the Pacific after the rough and ready atomic tests conducted by the military in the wake of World War II. Operation Crossroads' two shots in 1946 involved 42,000 personnel who visited Bikini over a period of six months. Operation Sandstone in early 1948 employed just over 10,000 personnel at Enewetak for three shots. The atolls' remoteness and the lack of permanent structures complicated both operations. To develop the atolls for such massive exercises, D. Lee Narver, J. T. Holmes, and Kelly McBean, their chief of operations, traveled from Holmes & Narver headquarters in Los Angeles to Los Alamos on 12 September 1948 to discuss the design and construction of long-term infrastructure at the Proving Grounds.[9] "It was accepted by everyone that it would be necessary for an engineering team to go to the Atoll to make a reconnaissance survey of the conditions existing there with a view to carrying out the proposed work for the development of a permanent Proving ground."[10] The meeting went well and a team under McBean left for Enewetak on 1 October 1948.

McBean and his engineers spent their visit envisioning how to transform Enewetak's landscape into test sites, camps to house scientists and military men, and workspaces to support long-term US atomic ambitions. First they addressed the three shot islands used during Operation Sandstone. Radiation levels remained dangerous at the sites, but "two feet of coral sand over the present craters should reduce present radioactivity to a sufficiently low level for all construction purposes."[11] The Caterpillar blade could renew the radioactive land. The survey team also inspected photographic towers, roads, causeways, airstrips, piers, communications antennae, and buildings left over from Sandstone. They found rust and decay. Even the markers from the Navy's wartime geodetic survey of the atoll proved useless, since the engineers thought the effort too slapdash.[12] The firm would have to conduct a new survey and design everything from the ground up. In spite of these challenges, McBean saw a landscape that could be engineered to support scientific and military personnel responsible for long-term atomic testing.

The landscape that McBean engineered for the political goal of continual atomic testing very quickly pushed the Seattle biologists to gather quantitative data about radiation's movement amid the environment. To do this, they turned their efforts away from qualitative practices. Rather than creating radioautographs, or photographs exposed by the radiation inside an animal or plant, they increasingly focused on ashing.[13] Ashing first required collecting specimens. The biologists poisoned fish in shallow water with derris root or shot rats on land. Next they took these specimens to their shipboard laboratory as quickly as possible for dissection. The blade yielded small, homogeneous samples of whatever organs they wanted to measure for radiation. The roll of the ship, the USS *Chilton* in 1947, LCI (L) 1054 in 1948, and LSI (L) 1091 in 1949, made weighing samples difficult. Donaldson complained in 1947 that "the samples were then weighed again on a torsion analytical balance accurate to .01 gram. Due to the motion of the ship, however, an accuracy of probably not more than .03 gram was obtainable."[14] Approximate weight in hand, the biologists used an oven to dry the samples and then a furnace heated to 800 degrees centigrade to render them ashen. The ash's radiation could be measured by a Geiger meter wired to a device called a scaler that created counts from meter readings over time. The process effectively turned plants and animals into radiological data points from the most sensitive sites at the Proving Grounds.

Quantitative data helped Donaldson and his team as they sorted through the flux of radiation at Bikini. They could not distinguish between cesium from a Sandstone shot and cesium from a Crossroads shot. Confronted with this problem, they began to think in terms of aggregation, space, and cycles. Ashed specimens could show how species metabolized radioactive elements and aggregated them in tissues over time. Specimens also bore the geographies of the radiation, or how radioactive elements traveled from a bomb to an organism and then traveled within the organism. Based on their position in the food chain, specimens could also show them how fission products moved through the biota as a whole. By 1950 Donaldson could talk about "translocation and concentration" or the unexpected accumulation of certain radioactive elements in particular species over time.[15] Ashing yielded useful data about radiation's biological effects even as more and more tests yielded fission products.

Interpreting quantitative data from ashing allowed the Seattle biologists to integrate themselves into the political project of the landscape that Holmes & Narver engineered. The biologists followed the physicists to the test islands. In 1948, just three months after Operation Sandstone, they visited Enjebi, Runit, and Aomon islands to collect specimens for ashing.[16] In 1949, when McBean had decided that the unfinished camp on Enewetak Island had no accommodation for the biologists, they nevertheless revisited those islands.[17] Ashed samples from the two years created a quantitative picture of how radiation behaved biologically at the shot sites. In this way they contributed to perceptions that some islands were safe for long-term habitation while others were inherently dangerous and needed Holmes & Narver to mitigate their radioactive risks. A September 1949 *New York Times* article picked up the biologists' work and reported, "One apparently can swim safely in the blue waters of Bikini lagoon."[18] The article acknowledged the problem of lingering radiation in local foodstuffs, but presented a picture of scientific rigor in service of patriotic aims. Even though they had no room at the inn, the Seattle biologists made the case that their science benefited the Proving Grounds' political mission.

A Laboratory for Disaster

In late 1953 the biologists got a room at the inn when building 218, the Eniwetok Marine Biology Laboratory, took its place among the aluminum sprawl on Parry Island. The AEC's Division of Biology and Medicine requested that Holmes & Narver construct building 218 to support the biology operations associated with 1954's Operation Castle. The lab constituted a small part of the $22,000,000 that had "been budgeted for the engineering and construction of the added base and scientific facilities."[19] Holmes & Narver had the bare bones of the building up by November 1953, an impressive feat since the AEC had also tasked them with the construction of four new camps on remote islands in Bikini and Enewetak. Still, the firm scrambled to complete the work before the operation. "The overall construction program," a bulletin from Joint Task Group 7.5 noted, "has progressed from 43% completion on 30 September 1953 to 96% on 20 January 1954."[20] The new building meant that the Seattle biologists' years of relegation to tiny shipboard laboratories had come to an end. Biology had a place amid the Proving Grounds' infrastructure.

Unfortunately, the engineers neglected important facilities for the biologists in the scramble leading up to Castle. Willis Boss, the AEC Division of Biology and Medicine point man for the project, visited the nascent lab the week before Christmas 1954 on an inspection tour with Sydney Galler from the Office of Naval Research and Robert Hiatt from the University of Hawai'i.[21] Hiatt came because the University would act as building 218's administrator for the AEC. The three men noted a number of deficiencies. A lack of climate-controlled storage for sensitive radiological instruments topped the list. Controlling humidity required electricity and Holmes & Narver's diesel power plants generated only so much of it on Parry. Boss noted another problem. He wrote to Donaldson on 29 December with the unhappy news that the building lacked a water table for dissections or space for aquariums. Boss accordingly harassed the Joint Task Force to requisition the fixes to make the building ready for Operation Castle. Donaldson responded gleefully, "After some of the patched-up and make-shift arrangements we have experienced in the past operations in the Pacific testing grounds, the present arrangement seems almost too good to be true."[22]

Building 218 was not ready in advance of Operation Castle. When Castle Bravo's dangerously radioactive fallout blanketed Rongelap Atoll on 1 March, the lab building still needed work inside and out. Two advance members of Donaldson's team, the botanist Ralph Palumbo and the marine zoologist Ed Held, waited impatiently in the half-finished lab for the installation of major appliances and modifications to the structure. On 15 March, Palumbo requested "repair of 6 faucets which do not set properly; installatn [*sic*] of 2 deep freezers, drying oven and muffle oven; louvring of porch to protect water table from sun."[23] The following day, electricians from Holmes & Narver arrived to install two International Harvester Deep Freezers, a drying oven from Precision Scientific in Chicago, and a Lindberg muffle furnace. Unfortunately for Palumbo and Held, the two furnaces required "additional wiring . . . plus a crew of seven men."[24] The electricians had the furnaces basically installed by the 18th, when Donaldson arrived with Paul Olson, a younger member of the Seattle lab, and Charles Barnes, a biologist from the Air Force.

The Seattle biologists expected to use the equipment in building 218 to streamline the preparation of field specimens for ashing during and after Operation Castle. On Parry they would number and catalog specimens on punch cards.[25] After weighing and dissecting each specimen, they placed the resulting tissue samples in small pliofilm bags. A plastic-like rubber substance, pliofilm could withstand the heat of the drying oven. The new process yielded biotic samples individually sealed. These they dropped into numbered envelopes for quick shipment by air to the mainland along with their punch cards. No matter who was in the territory, Donaldson kept a senior biologist at the home lab to ensure careful operation of the expensive and delicate radiation counting machine housed there. This dedicated machine produced much more thorough radioactivity counts than the shipboard Geiger meters ever had. By assigning some of the steps to the lab on Parry and some to the lab in Seattle, Donaldson made the most of the Proving Grounds' infrastructure. His biologists could turn animals and plants that metabolized fission products in the Pacific into ashes and finally into data in Seattle in a matter of weeks.

The biologists found that the building they so longed for would push their research to serve political ends in ways they had not anticipated. On 21 March, Paul Pearson, from the Division of Biology and Medicine, sent a message "with a request for some opinion of the feasibility of a trip to Rongerich [*sic*] to have a look at the food supply that was possibly contaminated by the fallout."[26] The Seattle biologists learned that the Marshallese populations on Rongelap, rather than Rongerik, and Utirik atolls had been exposed to dangerous fallout. So had American sailors aboard the USS *Bairoko*. In response, Barnes, Olson, Held, Palumbo, and Donaldson left Parry on 24 March and arrived at Rongelap two days later. They had only one day to collect and take measurements with their Juno Survey Meter, which could measure α, β, and γ radiations. In a grove of trees on Kabelle Island, the Juno read 2.8 Roentgens-per-hour, a dangerously unhealthy rate over time.[27] They collected coconuts and other plants, "algae, invertebrates, birds, and fish."[28] They froze the specimens for transport to Parry and slept as their ship steamed throughout the night for Bikini in order to witness the Castle Romeo test the next morning.

The following day they put the Romeo shot behind them and set to offloading the Rongelap specimens that bore the scars of Bravo. Building 218 became a disaster response center. Once in the lab, "it was decided because of the press for time to only dissect and freeze the tissues for return rather than dry as originally scheduled."[29] The biologists wanted to get back on schedule for the work they had planned to do for Operation Castle at Enewetak. At the end of the long work day Donaldson mused, "Packaging tissue etc seems like old times although our environment for working seems pretty lush by contrast of the stern of an LSI or back by the garbage shoots of the *Chilton*."[30] Their hurried efforts on 28 March shed light on the proper purpose of that lush working environment. Though ostensibly designed for basic biological research, the lab had a political purpose built into its frame, its wiring, its plumbing. Holmes & Narver had not constructed it to expand the frontiers of ecology or marine biology. Like the rest of the infrastructure at the Proving Grounds, it served American dreams of atomic hegemony. Investigating Castle Bravo's biological effects, the Seattle biologists did the work the lab was engineered to support. They did biology designed for atomic testing.

Determinative Infrastructure

The bureaucratic power brokers at the Proving Grounds wanted to ameliorate the Castle Bravo disaster even as the last of the snow-like fallout fell on Rongelap. This meant repatriating the Rongelapese after they had been evacuated from their home atoll by the Navy. On Parry Island, Colonel William Cowart from the Joint Task Force spoke with Ed Held on 13 April because he had "been plagued with questions from Trust Territory [*sic*] regarding when natives can be returned + what foods will be safe to eat."[31] Cowart felt enormous pressure because every day the Rongelapese remained refugees reflected poorly on the test site. The Proving Ground's administrators believed they had spent public funds and political capital to build a well-managed, permanent base where they could fulfill their military and scientific ambitions. They wanted the shiny, rust-proof aluminum buildings, laboratories, airstrips, causeways, and towers that Holmes & Narver built to continue their Cold War purpose. The Rongelapese had to go home to protect the site.

The plan to repatriate the Rongelapese rested on the infrastructure that already existed at the Proving Grounds. The Navy had transported the Rongelapese, who suffered from radiation-induced nausea, vomiting, and diarrhea, to ramshackle short-term quarters on Ebeye Island in Kwajelein Atoll in early March.[32] Around the time in April that Cowart spoke to Held, the AEC directed Holmes & Narver to build a long-term village for the displaced islanders. By June, the eighty-two men, women, and children removed from their home atoll settled in their new quarters on Ejit Island in Majuro Atoll.[33]

On 6 July 1954, Joint Task Force Seven in Washington, DC, assured the physicists at Los Alamos that "parties of medical and radsafe personnel under the direction of Division of Biology and Medicine, AEC, will visit the natives and atolls concerned periodically in order to observe the medical progress of the natives and to ascertain the earliest possible time for the return of the Rongelap natives to their homes."[34] Dr. Robert Conard led the medical team from Brookhaven National Laboratory. The radsafe work relied heavily on the Seattle biologists' efforts in building 218 alongside the work of the Naval Radiological Defense Laboratory.[35]

Accordingly, the Seattle biologists turned their focus to the problem of radiation in foodstuffs at Rongelap. They collected samples at the atoll in July 1954 and in January, October, and November 1955. In each case:

> Collections in the field were retained on ice or frozen until they could be returned to the Division of Biology and Medicine field laboratory on Parry Island. There the organisms were identified, selected tissues were dissected, weighed, and then dried. The packaged dried samples, together with the data cards, were sent by airmail to the Applied Fisheries Laboratory . . . for further processing.[36]

Ashing became properly political as the Seattle biologists transformed animals and plants into data designed to answer the question of when the Rongelapese could return. Building 218 became a clearing house for Castle's fission products that could be detected only in the lab on the mainland. By 30 December 1955, the biologists had produced two comprehensive reports of their findings. In both, they showed that radiation levels generally decreased in the foodstuffs they collected, ashed, and counted.[37]

But the curves they created for the foodstuffs told a nuanced story. Some would show declining levels of radioactivity only to be followed by small spikes in activity levels. While the biologists could not fully account for the mechanics of the curves they did warn that "radioactive material is being redistributed throughout the atoll"[38]—or at least throughout the atoll's biota.

Based on the reporting of the Brookhaven doctors and the decreases in radioactivity reported by the Seattle biologists, the AEC directed Holmes & Narver on 16 June 1956 to rebuild the settlements on Rongelap.[39] On 9 July of that year, engineers from the firm met with representatives from the US Trust Territory and with five members of the Rongelapese council on Rongelap Island.[40] The firm considered customary land-use practices as they rebuilt the atoll. But the spirit of the age prevailed and Western-style buildings replaced traditional ones. Rongelapese involvement largely comprised rubber-stamping decisions made on the mainland and posing for photos produced to reassure the distant US public that repatriation was going well. The process certainly went quickly. In just under a year, Holmes & Narver built three settlements. On 29 June 1957, the firm orchestrated the return of 250 Rongelapese and their livestock. "Before debarking," the press release explained, "all the Rongelapese gathered beneath the deck awning. There they offered prayers and hymns of thanksgiving to God for their safe return to their native land."[41] Engineering, founded on biology, seemed to overcome atomic dangers to produce an ideal American neighborhood on Rongelap.

But the biology that supported the engineering had become so integrated into the Proving Grounds' political infrastructure that it could not effectively raise alarms about repatriation. Just over a month before the Rongelapese returned home, Ed Held published a report concerning radio-

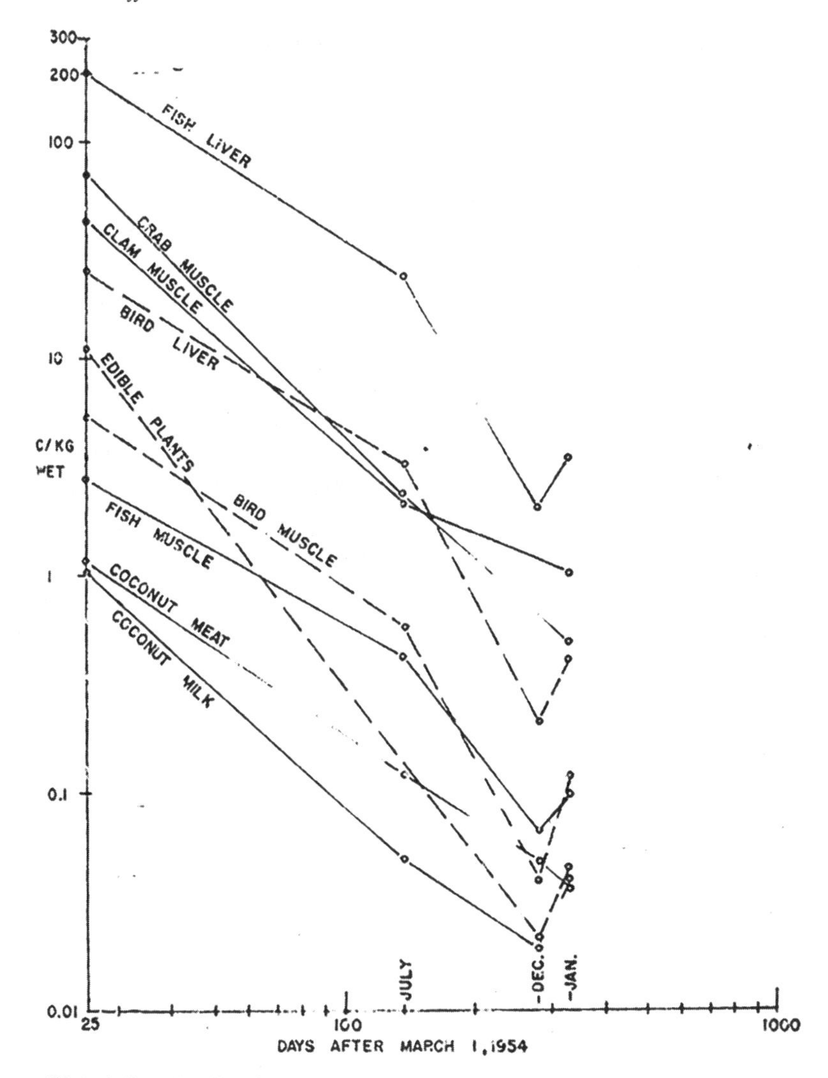

Figure 7.1. Curves of radiation from Castle Bravo in Rongelap food stuffs, 1954.

activity in land crabs on Bokombako Island in Enewetak. He found low levels of radiation in crab specimens that he prepared in building 218 and ashed in Seattle. Counts from the omnivorous crabs showed that "in so far as the long-lived radioactive fission products strontium, cesium, and cerium are concerned there is what might be called a strontium, cesium food cycle on land and a cerium food cycle in the lagoon."[42] This 1957 observation moved beyond the vague concern in 1955 about the redistribution of radioactive material. Some danger remained from local

Figure 7.2. The Seattle biologists at ease in building 218 in 1957.

foodstuffs. Fission products did not simply wash out in the vastness of the Pacific from Enewetak or, by proxy, Rongelap. But by this point, the infrastructure ran the game.[43]

Held's findings moved up the administrative chain of command from the Division of Biology and Medicine to the bureaucrats in the Department of the Interior who ran the Trust Territory Administration. They found a political use for Held's work in the press release they sent out on 24 May 1957 heralding the repatriation of the Rongelapese:

> The results of the latest survey indicate the presence of residual radioactivity at a level that is acceptable from a health point of view . . . with the possible exception of land crabs . . . land crabs are not a major item of their [the Rongelapese] diet.[44]

By the time the Rongelapese had been repatriated, the Seattle biologists' privileged position in building 218 determined that their science would be rigorous but never critical. When Held returned to the atoll in the late summer 1957, he attempted to collect representative samples of an average daily Rongelapese ration to show that their diet was safe. He shipped what he was given back to Seattle for ashing and counting. He also wrote a letter to Robert Conard, head of the Brookhaven doctors assigned to study the Rongelapese. Held alerted the doctor that "the Rongelapese are dubious about the necessity for continued medical examinations . . . about being 'stuck by needles.'"[45] He then conveyed that he had assured the Rongelapese "that you are sincere in your efforts and that you are the logical person to go to for a true evaluation of the medical aspects."[46] Held and Conard had spent three weeks together at Rongelap earlier in the year and had built a good rapport. Both men had become functionaries guided by the place and its mission. When Held published a 1958 report about the land crabs at Rongelap that were sometimes part of the daily ration, he noted no direct connection between them and the radioactive cesium and potassium in the Rongelap food cycle.[47]

Conclusion: Tragic Infrastructure

The Seattle biologists and the Los Angeles engineers went to the Pacific Proving Grounds because of fission's brute force and its subtle ability to insinuate radioactive products into the environment. Their paths crossed since fission awoke technological and political ambitions within the US that required not just scientific expertise, but a landscape engineered to accommodate and even direct that expertise. Federal science in the Cold War had to simultaneously produce technical successes and display total American scientific hegemony. Politics and atomic competition with the Soviets drove the development of the Pacific Proving Grounds. The engineers who poured the concrete and erected the built environment created a landscape designed for risky, large-scale extraction of scientific knowledge, a particularly Cold War space. Building 218 facilitated biological research. It also embodied the holistic scientific grasp that Holmes & Narver engineered for the AEC at Enewetak and Bikini. Ancillary to atomic physics and bomb design as they were, the Seattle biologists played a key role in the Cold War project that found form in the site's highly curated landscape.

Infrastructure at the Proving Grounds' curated landscape could maintain this very particular biology only for a season. As a mechanism for supporting the site's mission, ashing and counting radiation in biotic samples worked marvelously. These practices quantified the danger from testing to species important to the local environment. Ashed fish, land animals, and plants offered assurances that the US could safely play its high-stakes Cold War game. Donaldson and the Seattle biologists not only affected this mechanism, they interpreted its results and shared them with an apprehensive public back on the mainland. By the time of the Castle Bravo disaster, when the biologists had to ash and count foodstuffs for a human population, the practice's bona fides were without blemish. But the biological data existed as artifacts of the site's infrastructure. When atmospheric testing ended in 1963 after the Partial Test Ban Treaty, the AEC and Holmes & Narver withdrew from the atolls. The concrete crumbled and the aluminum corroded. The technoscientific detritus that came to characterize so many terminated Cold War projects littered Bikini, Enewetak, and Rongelap. The infrastructure abandoned, Donaldson and the Seattle biologists no longer worked to maintain the biological truths that once kept the atolls safe, or at least kept the public back in the metropole believing they were safe. Their yearly trips stopped and they studied landscapes back at home.[48]

But the radiation failed to follow the program. It lingered and the Rongelapese got sicker and sicker.[49] The body of each man, woman, and child exposed to fallout from Castle Bravo and the bodies of their children continued to create data about the bomb. But with Holmes & Narver gone, with the infrastructure dead rather that living, the biologists and doctors of the AEC found only marginal academic reasons to care. Like so many other boom and bust landscapes on the fringes of empire, the infrastructure at the Proving Grounds left a long-term legacy of sickness and tragedy for the local population and toxicity for the land and the sea.

8 * *A Matter of Taste: Making Artificial Silkworm Food in Twentieth-Century Japan*

LISA ONAGA

I first encountered silkworms in 2006. In a ten-gallon terrarium tank displayed in the Yokohama Silk Museum, they were eating a block of what looked like a dull green–colored paté. I felt excited to finally see silkworms after initiating my research on the history of silkworm genetics in Japan. At the same time, it puzzled me that I did not get to see them eat what every historical sericultural text that I had read promised they would with finicky dedication: mulberry leaves. In sericultural treatises of Japan such as *Yōsan Hiroku* of 1803, curious illustrations depict men and women harvesting leaves from mulberry trees by precariously climbing up into the arbor, cutting branches, and returning with bales of fresh fodder to the house where the mulberry leaves are then chopped and sprinkled over newly hatched larvae (fig. 8.1).[1]

The process of making silkworm fodder noted by the revered early nineteenth-century treatise required attention so as to match the hatching of young silkworms with the budding of new foliage that had to later be fed to the larvae, but before the formation of flowers. The expansion of mulberry acreage took the form of collecting and planting mulberry seeds, or a layering process known as *toriki* that trained low-lying long tender mulberry branches radiating from the trunk of an existing plant beneath the earth to serve as runners that could send down roots and establish a new plant clone.[2] During the second half of the century, the distribution of land dedicated to mulberry fields reflected the growing importance of the plant for silk production. Mulberry fields occupied about 14 percent of all agriculturally available land in 1904 (231,400 *chō*), and grew to around 24 percent by 1933 (640,178 *chō*).[3]

This green substance the silkworms were chewing in 2006 is called *jinkō shiryō* (人工飼料) in Japanese, which translates as "artificial feed" in English. This reconstructed mulberry is man-made and nowadays a "black-boxed" technology dispensed to classrooms, laboratories, and hobbyists where mulberry is hard to come by. It comes in different forms, from refrigerated, bologna-style packages sold on Yahoo! Japan for roughly US$20, to

Figure 8.1. Harvesting mulberry leaves, in *Yōsan Hiroku* (1803). Digital image courtesy National Institute of Japanese Literature.

powders. The packaged feed is sliced like cheese, thinly or into blocks, and fed to silkworms (fig. 8.2). The dehydrated powder mixes, which require no refrigeration, just require some water stirred into a "silkworm chow" (the name by which it is often marketed in the US).

The popular use of silkworms in primary schools usually focuses on the insect itself in order to teach young people early lessons about biological life. The use of artificial silkworm feed, often the only feed option available in the absence of mulberry, invites further questions about the human cultivation of biological life. This essay suggests that reflecting upon the food that silkworms eat and how that food has changed to allow new uses for silkworms—including in classrooms around the world where mulberry has been introduced—can lead to productive discussions about the relationship between humans and land use, plants and microbes, and animals that far exceed the insect itself. In some parts of Australia like Queensland, for example, *Morus alba* spreads easily and has earned an unwanted invasive status, although in Melbourne, the plants are integral for cultivating *Bombyx mori* sold to classrooms by "Australia's most trusted Silkworm supplier."[4] Once introduced to create shade, planting new mulberry (and olive) trees has been banned since 1984 in Pima County, Arizona, as a means to curb pollen counts.[5] Aside from highlighting changes in human

Figure 8.2. Silkworms on display, eating *jinkō-shiryō*. Image courtesy Yokohama Silk Museum.

land use, delving into the history of making artificial silkworm food may also prove a useful means for understanding the history of industrialized food manufacture more thoroughly. Plant-based silkworm feed may not involve the same combination of animal, grain, and vegetable ingredients used to make feed for goldfish, chickens, dogs, or cats; but highlighting how a concoction of sciences surrounding nutrition enabled the making of processed animal feed shows how key elements of industrial, rational engineering could enter the foreground of biology.

Artificial silkworm food directs attention to new aspects of the scientific history of silkworms in Japan situated in the world, as well as the history of biology more generally. Whereas previous research on the history of silkworm science has centered mainly upon the history of silkworm breeding and experiments leading up to what Japanese scientists have called *keishitsu idengaku*, or "morphological genetics," the stuff of artificial silkworm food adds a twist to assumptions about the parameters of biological inquiry in Japan, and how inquiries about regimes of care in sericulture have persisted over time.[6] Moreover, this history provides a counterbalance to the domination of breeding and heredity in the history of biology be they studies of domestication reaching back into archaeological records, or those of genetic research on plants or animals.[7] The issue of "feeding" located at the agricultural crosshairs of the intellectual field highlights nutrition and its relationship to political developments, a hitherto overlooked factor in the history of biological experimentation. Switches away from habituated food choices for livestock often indicate some deeper reason underlying justifications, be they food shortages or desires for food sovereignty. The switch to artificial food in the case of silkworms was not a leap but a process that meandered through soya as a substitute for mulberry that raised questions as to why soya, and how food rationing or searches for human food alternatives, followed the military development of "Total National Mobilization" in Japan.[8] The scientific relevance of the adage "you are what you eat" thus dovetails with historical questions about the materiality of the infrastructures of silkworm care and the seasonality of sericulture.[9] Considering the observable shift in the feeding habits of silkworms from exclusively mulberry leaves to "artificial silkworm food" or "semi-synthetic food," this brief history of making artificial silkworm food takes a step toward a deeper understanding of the twentieth-century molecular relationship between humans and insects, especially after World War II.

By asking what artificial silkworm feed is, means, and signifies, one can understand how the changing materiality of silkworm diet—its molecularization—resulted from a series of curiosity-driven and economically driven choices that occurred more disjointedly than seamlessly. Historical works on animal feed have gained sharper attention in the twenty-first century in the wake of outbreaks of bovine spongiform encephalitis, globalization, and concerns about genetically modified grains and demarcations between food for animals or for humans.[10] These contemporary concerns built upon earlier developments of the twentieth century to scientifically understand and optimize nutrition for livestock. By adding minerals, vitamins, or other chemical substances such as antibiotics or hormones, for instance, chickens could be cured of rickets, and cattle and sheep could mature faster and increase productivity.[11] In addition, many of the concerns about the efficacy of raising animals for human food have been assessed and reassessed in view of the energy costs required to produce such food and intentions to determine how these externalities undermine environmental sustainability goals.[12] These environmental concerns have been magnified even further by numerous environmental histories that show how the foods consumed by animals of agricultural value are at times tainted, with devastating effects, including extinction, due to a combination of toxic materials and metaphorically toxic politics that grant capitalist interests undue power.[13]

The molecularly described and reconstituted silkworm food qua chow opens up an inquiry into the history of silkworm nutrition and biochemistry that genetics of the early twentieth century could only incompletely explain. At least two other questions are necessary to pose in order to deepen an understanding of the history of silkworm science in relation to molecularized understandings of the insect's nutritional needs. First, what sociocultural, economic, and political conditions in Japan remade the conditions leading to scientific interests in making artificial food or granting silkworms with a new nature? Second, how should those intents to re-engineer the mulberry-silkworm interface be understood, if not only as purified efforts to define silkworm nutrition or to understand silkworm feeding behaviors? These questions facilitate an outline that maps upon two distinct eras in Japan, first in the aforementioned rise of total war mobilization in the 1910s through the onset of the Pacific War (scholars have referred to this period alternatingly as eras of fascism, militarism, and developmentally exceptional feudalism in modern clothing, owing to Emperor-worship wrapped in a mantle of nationalist ideologies), and the postwar occupation period shortly followed by the over-valorized "economic miracle" era.[14] An overview of mulberry use and the situation of mulberry acreage in Japan spans these two eras in the mid-twentieth century; moreover, the focus on mulberry itself sets the stage for analyzing the consequences spurred by the then-emergent prospect of colonially available soya. A reconstruction of the different approaches to researching silkworm feeding behaviors by Japanese chemists and geneticists operates in a foreground lush with mulberry, against a much more distant background of soya. Experiments reflected questions about why silkworms seem to eat mulberry exclusively, and they also motivated the development of silkworm nutritional physiology. These and other scientific engagements leading to "artificial" feeding practices were a byproduct of dealing with the space in between the imperial expansionist agenda and nation-building projects. A discussion addresses the postwar use of artificial silkworm food for re-engineering the macro-level strategy of sericulture for the nation. The essay ends with some contemplation as to what silkworms have become useful "for" and to whom.

Mulberry at a Glance

Over twenty species of mulberry are found in Japan, and among these, the most widely cultivated is *maguwa* (マグワ), or *Morus alba*. Commonly known as white mulberry for the color of its young fruiting body (that later turns red-purple), teas of dried mulberry leaves are used to treat high blood sugar and weight loss, among other things.[15] Since the ancient period, however, the main purpose of moriculture has remained to provide fodder for silkworm larvae.

The nutritional content of mulberry leaves depends on several things:

the variety, season of harvest, locations on a branch, geography of cultivation, soil and fertilizer, method of cultivation, and sunlight. These things explain the range of crude protein content of mulberry leaves, for nature does not produce a standard mulberry leaf. Leaves grown under shaded conditions produce less protein compared to those of sun-exposed conditions.[16] Common sense understandings of when and how to collect and dispense mulberry leaves belied the complications of developing artificial food suitable for silkworms of different developmental stages.

The variability of mulberry leaf compositions posed a sufficient challenge to scientists historically in pursuit of understanding the nutritional requirements of silkworms. On the one hand, this was a familiar issue, exemplified by efforts in the late nineteenth century to standardize silkworm rearing practices and to determine the weight of leaves necessary to feed a given number of silkworms and generate a certain volume of cocoons. On the other hand, we can understand the nutritional requirements of silkworms as a more recent problem that is less about making a table of conversions and more about developing substitutions for mulberry leaves. This twentieth-century look at sericultural science falls upon mulberry and efforts to disturb the exclusive nutritional partnership it has with *Bombyx mori.* Silkworm nutrition formed a major counterpart to the genetics of the silkworm that began to define a growing field of twentieth-century sericultural science research.

Before delving into the key experiments connected to the problems of silkworm nutrition and how molecular sense of mulberry was made, let us review some of the problems that faced the Japanese raw silk industry in the twentieth century. About 2.2 million farming families in Japan engaged with sericulture at the height of Japan's silk production and export in 1930 before the effects of the stock market crash were felt fully. A look at agricultural land use dedicated to mulberry farming according to Ministry of Agriculture reports shows that the 707,550 hectares of land then allocated to mulberry cultivation yielded about 400,000 tons of cocoons. For key silk-producing areas like Gunma Prefecture, mulberry occupied as much as 47 percent of the farmed area.[17] After World War II, the price of raw silk collapsed even further as the USA stopped importing raw silk from Japan, coupled with the development and popularization of synthetic fibers. Yet, the GHQ (General Headquarters, Supreme Commander for the Allied Powers in Japan) and the Japanese government agreed upon a Five Year Emergency Rehabilitation Plan for the raw silk industry. This plan, which went into effect on August 13, 1946, consisted of a forceful government policy to maintain the acreage of mulberry (170,000 *chobu*, roughly 170,000 hectares) in Japan. Already, farmers were digging up their mulberry trees in favor of growing other crops that were less laborious and involved less risk. Even if they did not destroy their trees, the possible neglect of mulberry added to concerns of Japanese abandonment of sericulture altogether. The five-year plan laid out an ambition to recover the silk industry enough to export at least 10,000 bales of raw silk per month, if not more. Farmers were guided to follow the policy under the expectations that the prices of cocoons would increase.[18] The dwindling mulberry acreage following the war may seem like a logical explanation for the initial development of artificial silkworm food. Although mulberry does regularly figure in postwar Japan, the actual story of making artificial silkworm food reflects a patchier development that bridges economic concerns of pre- and postwar sericulture with scientific inquiries into the biology of silkworms.

The prewar investigations circling around what silkworms eat, why, and how, would call into question not just what was appropriate nutrition for silkworms, but what was mulberry in a molecular sense. These intertwined with matters of Japanese colonialism, desires to avoid microbial diseases transmitted to larvae by consuming infected leaves, hopes to industrialize production, and intellectual questions like why silkworms eat mulberry. At the most pressing level, the problem of completing a fall season of sericulture successfully depended on overcoming problems caused by feeding silkworms yellowing leaves. Seasonality features prominently in sericultural materiality, and this challenge of taking advantage of autumn leaves with their diminishing nutritional content propelled feeding studies to understand how to supplement the silkworms.[19]

What's Soy Got to Do with It?

"Silkworm chow" is a compound of powdered mulberry, starch binders, and nutritional substances such as soybean extract or purified amino acids, among other things. In some cases, especially in Japan, it is possible to purchase the food in a raw form, but otherwise, a human must reconstitute the powder with water. Although information about some European experiment reports of adding hormones and amino acids to silkworm diets reached Japan through translation of abstracts, these results reported negligible effects in the cocoons. Despite various feeding experiments that also continued in Japan involving carbohydrates, vitamins, inorganic salts, lipids, and other substances like thyroid powder from animals, these for the most part did not lead to increased rates of development nor did they yield larger cocoons.[20] Instead, the process of arriving at an ideal formula interestingly seems to have begun with the proteins found in soybeans.

The precipitous fall of silk cocoon prices after 1929 served as an indicator of economic health in Japan, which led to manifold responses.

Manchuria, the northeastern-most tip of present-day China that Imperial Japan invaded in 1931 and subsequently opened up for settlement, offered Japan a range of alternative autarkic hopes and promises.[21] By 1932, the Rural (Farm, Mountain, and Fishing Village) Economic Revitalization Campaign particularly encouraged farmers to seek out their own solutions for economic recovery from economic crisis. By encouraging education, farm management, and village culture and social life, the vast reforms made possible by the Ministry of Agriculture and Forestry's 80 percent budget increase ultimately served to prop up the small villages in Japan over the span of three years.[22] This movement differed from the late 1930s German effort to create a peasantry by establishing hereditary homesteads and enlisting young men in the *Arbeitdienst* (Reich Labor Service). Yet they share some similar motivations, such as the recovery of farm debt and curtailing food shortages. In Germany, the desire to eliminate dependence on imports such as animal fodder (e.g., fish and meat meal, bran, other grains) had motivated the Nazi policies that sought to control the production and distribution of agricultural products, and to carry out settlement and reclamation activities.[23] Meanwhile, public works projects in Japan developed at the same time as the Economic Revitalization Campaign in order to increase wage-earning jobs. The tasks that the Home Ministry could fund included the redevelopment of mulberry fields, which notably included the labor of some women farmers as well.[24]

As historian Kerry Smith has observed, the Economic Revitalization Campaign occurred on a highly localized scale, in which village reforms were directed inward, making several thousands of villages more uniquely operable, although the structure of procedures, like submitting proposals to the Economic Revitalization Committee in order to strategize plans for earning income, did lead to trends that could be observed later on a national scale. In Smith's case study of the village Sekishiba in Fukushima Prefecture, sericulture had slowly become edged out of strategic focal view in order to free up acreage for food production. More generally, not only in Fukushima where mulberry cultivation was crucial fodder grown for silkworms, farmers and planners were expected to produce as much mulberry on their remaining fields as they had before the campaign took effect.[25]

As the radical and utopian arm of economic revitalization, Manchuria symbolized an economic panacea. Manchuria began as an economic security strategy that had to handle the rural population and poverty in the home islands. Most of the Japanese emigrants hailed from northeast Japan's silk-producing region, the Tōhoku region. Sericulture was Tōhoku Japan's main revenue source until the Manchurian Incident in 1931, which prompted consumer boycotts in the US and a crash in Japanese silk prices. It was not motives to purify Japan of the ills of overpopulation, but as historian Louise Young has explained, these price drops, compounded by rice crop failures in 1933 and 1934, particularly catalyzed the disproportionate migration of 66,522 individuals from Tōhoku; and Manchukuo, as the client-state was called after 1934, reached a population of 1 million in 1940. In the context of a stagnant silk trade, the steadily growing soybean industry must have appealed to Japanese farmers, many of whom had worked originally in the heart of Japan's silk country, Nagano Prefecture.[26] The land in Manchuria was viewed by planners as a space that could accommodate a military reserve and food. Its inhabitants also absorbed Japanese exports, including yarn and cloth and textile piece goods. The idea of Manchukuo had morphed into that of a sovereign destination yet its affordance of fused technocratic, bureaucratic, and military control was also understood as a means to overcome the shortcomings of Japan's original natural resources. Such juxtaposition marked Japanese fascism. It is in this imperial and extractive context that technological endeavors and ideology coupled with national political and economic decisions and ideologies about multiethnic authenticity.[27]

Japanese reliance on Manchuria's resources existed long before Manchukuo did, as demonstrated by the import of cakes of soybean lees, the byproducts of the extraction of plant oil, in the previous century. The nitrogen-rich lees were used mainly as a source of fertilizer until faster-acting synthetic fertilizers came into use in the 1930s. In addition, the lees have been used, including to this day, as animal feed and media for microorganisms.[28] Soya's demand did not diminish, as it easily fulfilled demands as livestock fodder. The settlers' attraction to Manchukuo thus included tantalizing promises of livelihood that included soybean cultivation. From Japan's perspective, Manchuria was an important granary. Of the world's soybean-producing regions, the projected output of Manchuria's soya was an incredible 208 million bushels compared to Japan's roughly 15 million.[29] The settlement and control of Manchuria not only meant a supply of food, but also promised a ready source of fertilizer resulting from the byproducts of soybean oil extraction. While the oil was used for cooking, lubrication, painting, and illumination, the 64-pound cakes of soybean lees presented an alternative to fish fertilizer used for enhancing the growth of mulberry plants (and rice) on the home islands of Japan. Although the export of soybeans was widely appreciated, the plant was not a total panacea. In the "soy frontier," historian Sakura Christmas has examined, the cultivation of soya in Manchuria also led to depletions of selenium that led to Keshan disease, but this did little to change land use. Instead of remediating the environment, the ministration of mineral supplements to people salvaged only human health.[30]

Still, soybeans commanded an appealing interpretive flexibility. Encouragement of soybean cultivation among the emigrants may be especially understood as part of Japan's imperial quest for economic autarky. The scientific soya experiments that took place during the 1930s and the aura of the critique of capitalism must be understood in the light of the strategies of Manchukuo's planners to bolster the soybean industry in the 1930s. Experiments of the plant itself underscored a multitude of uses alongside the energetic promises embodied in soybeans. These experiments focused on questions about how to microbially enhance its delivery as a fertilizer, the optimal growth requirements of the plant (including disease control), potential industrial applications, how water content relates to soy sauce production, and the protein content and nutritional value of the beans for humans and animals.[31] Uses for soybeans were also explored in Manchuria. In 1933, the journal *Shōyu to Miso* [Soy Sauce and Miso] published a short piece that reported on the successful feeding of a mixture of soybean and mulberry leaves to silkworms. The experimenter, identified separately in a sericultural journal as a certain Mr. Naruse of the port city of Dairen, had immersed and boiled mulberry leaves into a soup of soybean lees and fed them to larvae up to the third instar.[32] After that, the lees alone were said to be used as feed. Although the exact recipe was not made clear, the article heralded the reality of lowering the overall cost of raw silk production by half, and boasted that this innovation would undoubtedly impress the founding fathers of Japan's raw silk export trade.[33] Three months later, the journal reported that soybean lees were used as silkworm feed for the first time in Japan, in a village of Kagawa Prefecture, but the larvae reached their cocoon-spinning stages at a slower rate.[34] Despite the various competing ideologies about self-sufficiency, the highly particular activities of individuals residing in Manchuria connected back to a paradoxical substrate of Japanese activities dependent upon Western capitalist markets. Manchuria's transformed landscape, both environmentally and figuratively, provided a ground for imagining and prototyping new applications of soybean lees for use on the home islands.

The soy experiments that unfolded represented inquiries into how to maximize plant resources. These interests entailed scientific questions engendered by the particular constellation of networks and concerns that constituted the geographic space of the client-state of Manchuria. The groundcover of soya fields converted from boreal forest point to an embodiment of autarkic priorities and institutional, formal, and informal networks in the making of particular crops in fascist regimes.[35] On an institutional level, the Japanese colonies had their own agricultural experiment stations that extended from a network of agricultural and sericultural experiment stations on the home islands. In colonial settings in Taiwan and Korea, agricultural experiment station scientists surveyed and prospected indigenous agricultural resources and facilitated crop development, including silk, in hopes of encouraging local inhabitants to undertake new cultivation objectives and enterprises. The experiment station that served the client-state of Manchukuo had extended this network and facilitated the expansion of agricultural land while occasionally carrying out studies that would have been better suited for a sericultural station. The preexisting Agricultural Experiment Station of the Kwantung Agency (of Kwantung Leased Territory, acquired by Japan after the Russo-Japanese War), which was understood to be part of the sovereign body of Manchukuo after its establishment, carried out soybean feeding experiments that led only to the death of the tested silkworms. The experiments were short-lived and limited to the 1930s. Like many other agricultural experiment stations in Japan, the work of this experiment station was dedicated primarily to the promotion of food production, both to support the inhabitants and to send grains and pulses (i.e., seeds) back to Japan or other destinations. Without a justifiable *Bombyx mori* industry, one can understand that silkworm feeding experiments were not intended for large-scale field tests; rather, knowledge of the experiment, too, circled back to Japan.[36]

Toward Molecular Understandings: Silkworm Feeding Behavior

The enthusiasm for feeding experiments was more practicable on the home islands. In the late 1930s, studies that extracted carbohydrates and the protein glycinin from soybeans developed, taking the research onto a molecular scale. One aspect of research focused upon the biomolecular content of the mulberry plant, which paved the way to test supplements for silkworm diets. Feeding tests by Itō Toshio, a scientist at the Sericultural Experiment Station, sought to develop feed to compensate for the loss of proteins and sugars from autumnal mulberry leaves.[37] In these feeding tests, the scientist developed several versions of enhanced feeds including things like *kinako* (parched soybean powder), potatoes, and regular soy protein powder that could be sprinkled over top of mulberry to deliver glycinin as well as other nutrients into silkworms' bodies.[38]

Aside from the analysis of mulberry leaves, interests in the physiological chemistry of silkworms had also emerged as another facet of the nutritional science of silkworms. In the 1930s, the behavior of the silkworm had attracted the attention of chemist Hamamura Yasuji (1901–1985). He had neither training nor background in entomology, but when he took a post at the Kyoto Institute of Textile Fibers in 1935, he explored the question of silkworm monophagy. Something about the fresh, leafy aroma of mulberry leaves must have attracted silkworms to the plant, he thought. So, he placed fresh young mulberry leaves in chilled ethanol to create an extract, which he then separated using paper chromatography. When he placed this filter paper near silkworms, they gathered toward it as they would gather toward mulberry as he expected, but none of them tried to bite the paper. Hamamura continued several other experiments such as soaking the leaves of sakura or poplar trees with the extract, but these did not yield behaviors of actual eating. It would not be until after the World War II that Hamamura would isolate the other chemicals critical for understanding the sequence of things necessary for a silkworm to ingest food.[39]

Hamamura's efforts to understand silkworm feeding behaviors from a molecular vantage can be appreciated in hindsight as joining up with earlier studies in dietarily enhanced soybean powder in a vein of research on polyphagy. Additional research framed around the question of why silkworms eat mulberry exclusively added to this. Around 1941, the silkworm geneticist Tanaka Yoshimaro (1884–1972) carried out rudimentary experiments of sprinkling dry mulberry leaf powder over top of lettuce leaves in order to entice them to eat.[40] The method, which allowed silkworms to enter the cocoon-spinning stage, lent promise to the rearing of silkworms in midwinter.[41] Testing silkworm feeding behaviors carried on after the war. In a 1945 experiment published in 1948 by Torii Ichio and Morii Kensuke, removal of one of the two maxillae of the silkworm inhibited its ability to differentiate between mulberry and other leaves such as sakura and cabbage.[42] Whereas Torii was quite confident that the biomechanics of the mandible determined feeding choice, subsequent research by Itō Toshio showed that the maxillectomized larvae did not continuously eat these sakura and cabbage leaves as they did eat mulberry. A follow-up mandible experiment in 1959 with Yasuhiro Horie and insect nutrition scientist Gottfried Fraenkel of the University of Illinois (who also published a now-classic paper in *Science* in 1959 that opened up the field of insect-plant coevolution) concluded that taste organs other than those on the maxillae must exist.[43]

A new generation of scientists using genetic approaches simultaneously joined the band of chemistry and microsurgical experiments investigating silkworm polyphagy. In the 1950s, silkworm geneticists like Tazima Yataro (1913–2009), a protégé of Tanaka, developed a mutant strain of silkworm without feeding inhibitions, allowing it to eat apples and chard, among other things.[44] Tazima believed that food preference in silkworms had to do with the sensory organs in the maxillae. Since 1952, he had begun experiments with X-rays to induce mutations in silkworms with his assistant Kobayashi Kazuo. Using chard to test the responsiveness of resulting silkworms to detect any such mutations was a trial-and-error process. First, the bite marks of first and second instar silkworms were too small to detect. Second, it turned out that chard was toxic for silkworms, so as soon as the scientists noticed bite marks, they had to remove the silkworms and place them onto mulberry. Even though they were able to count 114 larvae out of 37,900, the silkworms fared so poorly that only one male matured to produce offspring. In a second round of experiments, they produced 314 mutants that could eat soft leaves such as those of chard, sakura, persimmons, satsuma oranges, and soybeans. Tazima and Kobayashi published the results of breeding these resulting silkworms for ten more generations in 1954. From this genealogy of silkworms, they eventually identified the chromosome location of the gene for this nonselective eating behavior.[45]

Understanding that similar mutations could have occurred naturally, Yokoyama Tadao suggested the need for the Sericultural Experiment Station to survey existing strains for similar mutations. Yokoyama indeed found individual silkworms among Japanese and Chinese strains that responded positively to cabbage and these were interbred, forming a line known as Sj (沢J). The strain was also known to eat other things: leaves of apricot, cherry, and persimmon. The study of this Sj line was shown to have genetic congruence with the mutant found by Tazima.[46] That said, breeders did not use these silkworms, due to their inferior cocoon quality. Additional researchers would go on to study the physiology and biochemistry of silkworm feeding itself, constantly blurring the lines of where practical application and basic research began and ended.[47]

In the 1950s, Hamamura also resumed his experiments, which provided a clearer picture as to what chemical substances governed the exact feeding behaviors of silkworms. He placed silkworms onto the remains of a mulberry leaf that had undergone extraction. The silkworms ate the veiny remnants of the leaves, which gave him a clue about the molecules that were necessary to induce silkworms to perform the action of biting. Hamamura carried out other extraction experiments with acetone and methanol to pinpoint the substances that were necessary to make the silkworms bite. A third substance was also identified as necessary for the subsequent action of swallowing by the larvae.[48] Since then, Hamamura's biochemistry laboratory gained fame in the silkworm science community for developing an agar jelly based diet containing the three categories of factors necessary for sustained eating by the silkworm: Attractants, which included citral (also found in mandarin orange [*Citrus reticulata*] and trifoliate orange [*Citrus trifoliata*] tree leaves), tepinyl acetate, linalyl acetate, linalol, and ß-γ-hexenol; biting factors (beta-sitosterol and isoquercitrin or morin); and swallowing factors (cellulose), as well as co-factors (sucrose, inositol, inorganic phosphate, and silica).

The infusion of the agar medium with chemical agents was spurred by work in the laboratories of Robert Koch and his disciples in Japan, including bacteriologist Kitasato Shibasaburō (1852–1931).[49] While agar was used increasingly to encourage the growth of microbial life, it is also curious to consider the underexplored historical relationship between Japanese culinary consumption of *kanten* (agar jelly cubes) and experimental biology practices in Japan. As early as 1917, *Drosophila* scientist J. P. Baumberger had developed a banana agar medium that he argued made it possible to observe behaviors such as oviposition that would have been difficult to do with conventional medium of fermented banana. Following this, geneticist Komai Taku formulated a recipe for agar-based fruit fly medium by 1927.[50] While Hamamura's citation choices do not suggest that the *Drosophila* system was a direct inspiration, his work shares a connection with the need to clearly observe insect feeding behaviors and the usefulness of agar-based media. The discussions in Hamamura's papers do not project an intention to develop a new diet for silkworms. That said, the subsequent synthetic silkworm diets developed by other teams built upon Hamamura's key work.

Artificial Silkworm Diets

The development of technologies for growing pure bacterial cultures in the nineteenth century presaged broader inquiries concerning the growth of living things in laboratory conditions.[51] Endeavors to create artificial silkworm diets in the 1950s and 1960s, largely practically motivated by interests to enhance cocoon yields, dovetailed with broader scientific questions that were surfacing about the nutritional requirements of insects, not only in relation to the development of feeding media for research insects like *Drosophila* and *Ephestia*, or fish kept in captivity.[52] Research into the phytosterols, lipids, proteins, and growth factors necessary for the silkworm diet continued into the later decades as scientists sorted out what would guarantee stable and ideal growth.

The relationship between silkworm diet and silk quality and yield highlights another dimension of the research that went into making *jinkō shiryō*. By the late 1950s, the scientific thought prevailed that reducing the percentage of crude ingredients with purified compounds would lead to better larval growth and thus larger cocoons and greater egg production. The processes of "improving" the artificial silkworm diet thus also prompted the analysis of the insect's nutritional requirements. For example, the diet developed by Itō reduced the crude mulberry content to ten percent. Bound with potato starch, sucrose, defatted soybean casein, Wesson's salt mixture, and cellulose powder, this diet tested the effects of the presence and absence of ether extracts from soybean powder. Soybean oil had previously been shown to help prevent outbreaks of flacherie disease in larvae, but its nutritive value was yet unknown. This 1960 experiment by Itō helped pave the way to eventually show that the soybean oil's sterol and fatty acids together functioned integrally as a growth factor that facilitated the normal molting and growth of the larvae.[53]

Scientists continued to try to develop diets consisting of simple nutrients without dry leaf powder. In 1962, a team at the Sericultural Experiment Station led by Fukuda Toshifumi tested three types of diets consisting of different ratios of potato starch, sucrose, amino acid mixture, vitamin mixture, mineral substance mixture, cellulose powder, ß-sisterol, alcohol extract of mulberry leaves, and water. Their results, published in *Nature*, showed a clear possibility that silkworms could persist on artificial food. Interestingly, Fukuda highlighted that different "races" of silkworms (Japanese, Chinese, European) perform differently per diet. This was but a small comment, but it is mentioned here in order to remind that the vestiges of earlier decades of genetic thought intermingled with the newer biochemical analyses of silkworm nutrition. Above all, the qualification reminds that the concern that propelled this research was not about simple viability of the food that would permit the breeding of a next generation of insects, but that the nutritional content would also allow the larvae to spin cocoons of a certain decent quality.[54] The overall rate of growth and development of larvae raised on artificial diets would become more reliable but could not replace the efficacy of mulberry leaves.

The pursuit of a semi-synthetic diet for silkworms continued. In 1966, Itō developed an amino acid diet that did away with powdered mulberry leaves and soybean casein and used amino acids as the sole source of nitrogen. On a practical level, semi-synthetic diets allowed minute adjustments to their compositions in order to meet the needs of one of three qualitative stages of larval growth (starter, grower, producer). Although larvae on semi-synthetic diets can spin cocoons, the feed is used primarily for the sole purpose of nutritional investigations into the insect.[55]

Re-engineering Sericulture with Artificial Food

The history of scientific investigations into silkworm feeding behaviors and the development of a viable artificial medium for growing silkworm larvae have illustrated the continuity of research across wartime. Attention to Japan's postwar development with respect to the commercialization of silkworm feed sheds additional light on the continuity of this research. Plans for the wider-scale use of artificial silkworm food involved revisiting sericultural planning at the national, prefectural, and corporate levels of governance. The rationalization of sericulture in the postwar period would not only strike a familiar tune of macroscale strategizing. It also meant that the regulation of artificial silkworm food itself would be used to re-engineer Japanese sericulture for its survival in the postwar period, especially after 1952 following the Allied Occupation of Japan.

For sericulture industry stakeholders, artificial silkworm feed served as a technological answer to the dilemma of how to revive the silk industry of postwar Japan, namely by intervening in workers' wage distribution. The specialization of sericultural work that had supported the Japanese silk industry since the Tokugawa period had continued to encourage integration of diverse rural and cottage-industry economies into a capitalist system as silk production and trade intensified through the 1920s and 1930s, to sell high volumes at low cost. Many of the underlying reasons for female laborers' unrest in the textile sector, such as low wages and long working hours, carried over to the 1950s. The silk (and cotton) spinning industry was compelled to elevate its labor standards, which included retreating from infringements on personal freedoms and increasing wages.[56] Among those idealized interventions was the notion of economizing the amount of mulberry fed to silkworms. The concern was not so much how to offset agricultural labor and cost required to cultivate mulberry to generate silk, but how to increase the output of silk without a commensurate acreage of mulberry that would have been normally grown as supplemental income by farmers.

The idea of artificial silkworm food, seeded in the imagined potential of Manchukuo's soy fields, gestures to a turn in which the driver of self-sufficiency gradually manifested in the economic aspirations of a reintegrated postwar Japan. Although most discussions of reinvigorated postwar managerial practices refer to assembly line manufacture linked to Japan's expansion into new markets and the diversification of goods, a commitment to high growth and production optimization was shared across the sectors. One may especially understand the excitement surrounding the potential of artificial food in the light of the growing costs of female labor in textile factories and efforts to move forward from the era of management by coerced obedience with data-centric scientific labor management.[57] In contrast to the time it takes to cultivate mulberry plants from seedlings to maturation and the labor costs not just to pluck the leaves (at the height of the fourth instars of the voracious caterpillars), but also to cut entire branches and haul them on trucks and distribute them to farms, the continued interest in using soybeans to make portable artificial food during the postwar era symbolized the growing appeal of emergent scientific management priorities to economize and gratify the compulsion to produce silkworms again on a scale as before.[58] The regularity of anticipatory news headlines in the 1970s such as 「蚕もマスプロ時代　人工飼料の予算認む　桑園なくても"養蚕工場"」 ("Even silkworms in the age of mass production: Budget for artificial feed set—'Sericulture factories' without mulberry fields"), which appeared in the *Asahi Shimbun* on December 27, 1973, must be understood in the scope of the developmentalist discourse marked by the 1960 Income Doubling Plan. The plan focused on creating economic growth through enhanced productivity facilitated by improved management—high wages ensured by low costs. The vibrant clashes between workers and managers that surfaced throughout the 1960s are just as much a testament to that period of "modernization" of Japanese enterprises as were the training of scientists and technicians, and the development of new industries, encouraged by the government.[59]

Although artificial silkworm food was no icon of Japan's so-called economic miracle, it was certainly a product of its time. Its development followed the era of efficiency-driven adaptation of ideals weighed by the heavy industries involved in making power looms and other machinery for textile factories in the early 1910s and 1920s, which helped implement time-consciousness in factories that employed great numbers of young during the interwar period. Apart from the layoff of factory workers, the devalued mental and manual skills of farmers never fully recovered after the global recession and the onset of the Pacific War. The introduction of artificial silk (rayon/viscose) by this period has often also explained the demise of Japan's silk trade. Interestingly, this period saw a plenitude of Taylorite ideals that were not actually implementable on the floors of the numerous yet small textile factories of Japan; scientific management rather centered upon the rationalization of government policies concerning industry.[60] The new project of making artificial silkworm food embodied the reckoning with tensions as Japan shifted the investment of its resources from low-return agricultural sectors to higher-return nonagricultural sectors. It promised cost-saving enabled by reducing dependence on land for mulberry that would in turn minimize labor costs. Above all, artificial silkworm food was seen as a strategy to prevent the eclipse of the relevance of silk for Japan—or of silkworms, at least, by bringing the focus of concern to neither the factory floor nor to government policies alone, but to the interface between plants, insects, and capital once again.

The history of the research surrounding artificial silkworm food helps show the creative ways by which Japanese scientists, technologists, government, and sericulture endeavored to maintain the salience of silk. The restriction on the commercial release of artificial silkworm feed until 1977 represents the value of coordination between various government and private entities. Deliberate time buffers encouraged the co-development of new products and plans while reorganizing existing silkworm nurseries. At a glance, the release date may seem arbitrarily set by the government in a top-down fashion, but it indicates a close public-private sector relationship that took into account mutual practical understandings of the state of research and its potential, alongside shared anticipations about how the silk industry fit into overall anti-agricultural trade liberalization concerns. As a response to the demands for an open market dominated by the US, these discussions bore a reminder of earlier autarkic calls for self-sufficiency.[61] Above all, they generated an aura of competition both within Japan and amongst new corporate entities in agriculture and reconfigured silk industries. For example, in 1975, Japan Fertilizer and the Saitama Prefectural Sericulture Farming Cooperative Association collaborated to manufacture 30,000 liters of "Mayugen," a nutritive additive feed made from rice and chlorella (single-celled green algae) designed to be sprinkled over mulberry for silkworms. Its anticipated price of 1400 yen per liter claimed to help reduce the actual volume of mulberry leaves distributed to larvae by twenty percent and elevate the production of cocoons.[62] The animal feed company Nihon Nōsan Kōgyō Corporation (NOSAN) entered the fray in 1976 when it announced its plan to develop and commercialize artificial feed for sericulture, with the aim of reducing costs even more by cultivating their own mulberry fields and finding alternatives to expensive agar.[63] The Ministry of Agriculture and Forestry regulated silkworm feed quality by inviting companies to submit their artificial feeds to trial in advance of 1977. Five companies participated: NOSAN along with Takeda Pharmaceutical, Katakura and Gunze Corporation (two famous silk reeling companies), and the food company Ajinomoto, known for manufacturing a flavor enhancer for human food, monosodium glutamate.[64]

While some observers opined that farmers could perhaps use their freed acreage for growing other food crops, the trial runs, model manufacturing plants, and infrastructural redesigns that aspired to fit artificial feed into actual sericultural practices concerned those on the ground in different ways. Silkworm rearers who participated in the trials reported that it took silkworms several extra days to spin cocoons. Moreover, among the smallholders left in the postwar period, worries abounded that the silk industry would change too much and prioritize the larger organizations that were committed to the ideals of mass production.[65] For scientists, the project of artificial feed both diversified the scope of silkworm science and opened up new research inquiries; it also ensured the relevance of silk research, for instance, through biochemical analyses of the two silk proteins, fibroin and sericin, into their amino acid components.[66] In this sense, I suggest that artificial silkworm food helped re-engineer the silk industry for the postwar era, and the work of developing synthetic food was part of a longer process of thinking about ways to reverse-engineer the making of silk itself in a way that clearly justified ongoing research of silkworm.

Discussion

Artificial silkworm food would not completely replace mulberry in twentieth-century sericulture. Nonetheless, the time it did take to gain more widespread use widens our historical perspective for understanding its application. From its development in laboratories in the early 1960s to farms in the late 1970s, the remaking of mulberry—the leaves of which provided a perfect food for silkworms already—into an even more portable and time-defiant food represents how biological questions went hand in hand with the technocratic apparatus of postwar Japan. The development of artificial silkworm food is an intermediary for understanding the connection between the science of taste reception and palatability and the industrial fermentation of substances like MSG.[67] Although intellectually motivated by questions about insect behavior, insect nutrition scientists ran a race against time toward the development of new artificial foods for silkworms. This sense of time was not just about corporate competition. The seasonal time of the mulberry plant in the temperate zone of Japan, along with changing land use in the postwar period, and dropping prices of cocoons that marked the seemingly fragile viability of Japan's staying power in the silk market, all mediated the practical deployment of this special silkworm food. While much of what is popularly known of the uses of artificial silkworm food comes from a postwar context, a more precise way to consider the temporality of this research is as something that has taken place in a post-mulberry time—a dual notion of the mulberry's seasonality and that of the uprooting of mulberry following the 1930s when falling cocoon prices spurred farmers to alter their strategies.

The research, policy, and commercial endeavors that went into formulating artificial silkworm food in Japan show that although the food was far from the silver bullet solution for making low-cost, high-yield silk, it holds importance for situating histories of biological research in their cultural contexts. Mulberry is still cultivated in Japan and elsewhere to maintain silkworm genebanks and university stocks, and to cultivate artisanal silk. Mulberry tea is neither common nor uncommon. Different types of artificial silkworm foods exist on the market today, some of which have soybean powder, others with crude mulberry. The story of artificial silkworm food and its odd coexistence with mulberry plants does not feed into a narrative of clear success or failure. Most important, the development of artificial food has diversified how people think about the purposes of silkworms.

Artificial silkworm food helped scientists conceptualize ways to rear silkworms out of season, or any time of year in a climate-controlled building. The biochemist Hayashiya Kenzo, who worked on developing antiseptic artificial food in order to rear "germless" silkworms in laboratory conditions, dreamed of "*birudingu* (building) sericulture," a riff on the borrowed word *biru* for Western-style brick-and-mortar or concrete multi-story structures.[68] In addition to paving the way for the rearing of silkworms en masse in clean room laboratories where biomedical or veterinary substances are developed, and beyond the production of silk itself, silkworms are increasingly seen as a source of protein. The process of mail ordering "silkworm chow" in Europe, Japan, or the US is not at all extraordinary today. The main purposes of this food include feeding silkworms that are themselves fed as wiggly treats to pet reptiles. Artificial silkworm food has also been used to cultivate silkworms in space—not just to study silkworm behaviors in zero gravity, but also to develop a way to reliably supplement astronaut diets with a complete protein in the form of silkworm pupae, consumed in many parts of Asia.[69] Artificial silkworm food must not be thought of as only a formula, based on a series of experiments attempting to determine equivalent, purified, substitutes for mulberry. By situating this aspect of silkworm science in both molecular and cultural contexts, we can understand how this thing—artificial food—is simultaneously a result of experimentation and engineering that was intended to re-engineer sericultural strategies in postwar Japan, and how it ultimately helped generate new uses for silkworms altogether.

9 * *Cybernetics without the Cyborg: Biological Modernism(s) in Biomimetics and Biomimicry*

RICHARD FADOK

What is cybernetics without the cyborg?[1]

This chapter opens with a question that strains the limits of enunciation.[2] Which is to say: the figure of the cyborg has insinuated itself so thoroughly within our popular and scholarly discourses of cybernetics that it has proven nearly impossible to conceive of the sciences of control and communication without the amalgam of organism and machine that Manfred Clynes termed "cyborg"—the word itself a verbal amalgam, or portmanteau, of "cybernetic organism."[3] Early cyborgs aspired toward the artificial evolution of the human, organic bodies mechanically augmented for anthropo-phobic environments, like outer space. The cyborg was nature remade.

Nearly impossible, because for the past six years, I have been conducting ethnographic research on *biomimetics* and *biomimicry*, two contemporary design paradigms in architecture and engineering that owe their existence to cybernetics, in the sense that they are its derivative or legacy discourses, but that traffic in non-cyborgian figures, ones which pose a different set of biopolitical questions about the relations between life, power, and knowledge in the twenty-first century.[4]

"A cyborg world is about the imposition of a grid of control on the planet," wrote Donna Haraway in "A Cyborg Manifesto," a laser-sighted work of feminist theory, which, like its subject matter of command-and-control, came to impose its own grid of intelligibility on the historiography of cybernetics.[5] In the reticulated space she described, cybernetics, as represented by the "image of the cyborg," couples a scientific epistemology in which biological form appears as a control mechanism, with an engineering norm to control living organisms. The concept of control thus operates doubly, first to reduce organisms and machines to servomechanisms, then to plug them into ever-heightening systems of order. Life and technology, interior and exterior, form and norm—all become functions subordinate to the totalizing machinations of control. For Haraway, the cyborg promised feminists an ontology and politics. So too for the cyberneticians, those communication engineers who articulated more than mere matter and metal. In the cyborg, they articulated nature with artifice, life forms with a form of life geared to their manipulation.[6]

Cyborgs, however, were not the only artifacts manufactured across the better part of a century of cybernetic research. In fact, the cybernetic project to remake life through technology comprised only one-half of its original ambitions, which also encompassed the use of biological models to remake technology in their like-ness, their life-ness.[7] This chapter unearths this lesser known lineage—its figures, materials, and rhetorics—to reassess the centrality of the cyborg, and thus of control, in our understandings of cybernetics. This lineage focuses on the transformation, not of the living through the technological, but of the technological through the living, that is, on the biological analogies that communications engineers, and then other engineers and architects, would draw upon to imagine and construct new modes of artifice. Beginning with *biomimetics*, known also as *bionics*, and later joined by *biomimicry*, this alternative tradition sprang from second-order cybernetics and acquired a life of its own, with laboratories and centers dedicated to the discovery, and application, of technical design principles embodied in living organisms.

Like the cyborg itself, my approach to this lineage is a methodological amalgam, soldering history and anthropology. It begins with a conceptual departure, with the genesis and development of the cyborg as a figure of thought, which the remainder of the chapter decenters. The argument next alternates between history of science and ethnographic data, the past and the (near-)present, to chronicle the evolution of biomimetics and then biomimicry from cybernetics. I weave comparative ethnographic analysis of cybernetic design at two sites—a biomimetics laboratory at MIT, and an itinerant biomimicry workshop in Mexico—with longer institutional and intellectual histories in order to discern how discourses and practices of making technology, of technopoiesis, have been remade in ways irreducible to the cyborg and its telos of control. This shift in perspective both builds on, and breaks down, the edifice of the cyborg that has been a monument and a touchstone for cultural theories of technoscience, and, in so doing, opens up new possibilities for examining alter-configurations of life and technology obscured by the unitary period of "biological modernism" with which Philip Pauly categorizes biology today.[8]

The Nature/Culture of Control: Norms and Forms of the Cyborg

Cybernetics, famously defined by one of its founders, Norbert Wiener, as the "scientific study of control and communication in the animal and the machine," brought together an interdisciplinary cast of mathematicians, engineers, biologists, and social scientists in a postwar program organized to investigate the phenomena of feedback in communications engineering.[9]

In Geoffrey Bowker's words, cybernetics was a "meta-science" or "universal discipline" that offered tools or terms applicable across disciplinary lines and, indeed, boldly crossed those lines without trepidation, expanding the domain of communications engineering from telephones and other traditional media of communication to organic life.[10] For the cyberneticians who gathered at the germinal Macy Conferences from 1946 to 1953, the concept uniting their diverse areas of expertise was Claude Shannon's idea of "information," a ratio quantifying signal to noise.[11]

These cyberneticians found, in information, a bridge to liken organism to machine, and vice versa; information analogized both as means of communication, their embodiment made secondary to their information-feedback behavior.[12] By the application of a communication engineering framework to the ontology of life, biological organisms became paragons of control, a transmogrification that made their interdigitation with other systems of control all the easier.

For some cyberneticians, information therefore permitted more than analogy: physical hybrids. Such cybernetic artifacts took diverse form, but the most salient one today remains the cyborg, a human mechanically outfitted to improve his survivability in hostile environments. In a short paper from 1960, Manfred Clynes and Nathan Kline coined the word "cyborg" to characterize products of non-hereditary, machine-assisted evolution.[13] Initially conceived for extraterrestrial travel, exploration, and, ultimately, colonization, the first cybernetic organism was, in actuality, a white rat, technologically retrofitted with a homeostatic pump for regulated hormone secretion.[14]

For Americans in the mid-century, the cyborg became the most "evocative object" to come from cybernetics, its synecdochic representative in popular literary and televisual culture.[15]

In Martin Caidin's 1972 work in science fiction, the esteemed pilot Steve Austin suffers a violent and near-fatal accident that disfigures him, only to have his limbs and organs surgically replaced by mechanical analogs "better than the original" but "that could not be distinguished by the observer as artificial."[16] After his operation, he becomes a Soviet-fighting cyborg spy and "superagent"—better known, in the spinoff television series, as *The Six Million Dollar Man*.[17] While the cyborg persists as an enduring icon in popular media, it reached the apotheosis of its cultural stature in the decades-long span of time that stretches from the early 1940s to its denouement in the 1970s, an era that Robert Kline labels the "cybernetics moment," and that, in a tellingly pejorative register, James Baldwin dubbed the "cybernetic craze."[18]

Even after the height of its popularity, cybernetics, and in particular the cyborg, exerts a perennial influence on scholarly debates about the "posthuman" and the "Information Age."[19] Throughout such conversations runs a pervasive association between cybernetics and control, both in the predicate assumption that biological form *is* control and in the engineering norm that therefore biology can, and should, be controlled. This is perhaps not surprising given the centrality of control in Wiener's definition, and given his and Shannon's wartime involvement in projects funded by the National Defense Research Committee, the former working on antiaircraft weaponry, the latter on cryptography.[20] This view of cybernetics as control is further encoded in the etymology of the term, which Wiener loaned from the Greek *kybernetes*: a marine vessel's governor, its steersman, he who guides the ship's turbulent passage through water and wind.[21] In the shadow of the cyborg, scholarship on cybernetics and its legacy discourses has assumed an identity between the grammar of life and its pragmatics—between an ontology of life-as-control and the cultural propensity, or ideal, to control and re-engineer a nature always already artificial.

Nowhere is cybernetics welded more tightly to these thematics of control than in the historiographical tradition initiated by Donna Haraway's acclaimed essay "A Cyborg Manifesto," which reified these themes of domination in her exploration of the cyborg as a figure of thought. Therein, the cyborg refers, not strictly to the new hybrid species imagined by Clynes, but more generally to any fusion of nature and technology enabled by the cybernetic ontology of control. Although she admits and advocates an alternate possibility, that the cyborg might signify—as well, a relation of addition, not substitution—the possibility of intimacy, proximity, and boundary transgression, scholars almost invariably route the cyborg and its legacies back to control.[22] In his historical critique of the "cyborg," Peter Galison contends that this bellicose figure cannot be so superficially dissected from the "Manichean sciences" of war and their emphasis on power.[23]

In contrast to Haraway and Galison, Andrew Pickering has shown that in cybernetics rests another story that passes, not through a militaristic preoccupation with cyborg control, but through psychiatric investigations into the brain. His counter-genealogy frames cybernetics, not as the prediction of an already preexisting enemy, in other words, as a technique for anticipating and outmaneuvering tactics and ballistics in conditions of siege, as its wartime birth suggests, but as a performance, an epistemology tuned to the fundamental "unknowability" of the world.[24]

Pickering's work opens up a clearing for reappraising the cybernetic legacy without the cyborg, for reevaluating the articulations of biology with technology in our so-called cyborg society.[25]

If, for Tim Choy, articulations, any "contingent unification across distance," pose questions of the un-articulated, the forgotten or left behind, they raise, to me, the specter of dis-articulation, of being broken apart or asunder.[26] In the ethnographic vignettes juxtaposed below, I pay attention to the dis-integration of the cyborg, a hybrid that never had much integrity anyway, as the materials engineers and architects with whom I have conducted ethnographic fieldwork variously uphold, amend, and even reject the inherency of control to the legacy of cybernetics. Through a focus on the design histories of biomimetics and biomimicry, two cybernetic discourses that have been largely forsaken by scholars in their emphasis on life's manipulation, I intend to contest the hegemony of the cyborg as a figure of thought. Foregrounded instead are the figural substitutions, or absences, that designers produce as they recreate artifice in the shape of nature. Disfiguring and refiguring the cyborg, they repurpose control in cybernetics. "We inhabit and are inhabited by such figures that map universes of knowledge, practice and power," according to Haraway.[27] Onto new universes then, without the cyborg, space-suited as it was.

Control Transposed: From Information to Materiality in Biomimetics[28]

In a dimly lit, blue-carpeted room off one of the many labyrinthine concrete hallways that constitute the circulatory system of MIT, a graduate student silently manipulated a rather mundane object: a white, palm-sized square of four smaller squares interwoven together. With imperturbable focus, she subjected it to a tacit test of plasticity, quietly bending it in all manner of directions. A curly-haired woman, seated at the end of the table and partially obscured by a small mountain of papers, broke her silent reverie, "How is the Tango Plus performing?" The woman was Christine Ortiz, a professor in materials science and engineering (MSE) who joined MIT in the fall of 2000.[29] Having received an apparently satisfactory answer, she cross-examined the student about the surprisingly evocative object: how it had responded to mechanical "stress" and "fracture" measures and whether they should try materials other than the rubber-like Tango Plus. The object, I learned during my fieldwork, was a 3-D printed prototype of body armor for soldiers that Ortiz and her students made from a digital model they derived from *Polypterus senegalus*, a long, eel-like fish that inhabits the artery and branches of the Senegal River.[30] I wondered: How and why did this fish become analyzed, plasticized, and militarized at MIT?

When I later asked Ortiz what this fishy artifact offered over conventional material technologies, she replied, "Biological materials [are] so much more precisely designed and [have] so much more control over structure and properties [. . .] I just found it much more well-designed than anything we could."[31] Through what it calls "biomimetics," the Ortiz lab aspired to master the control that nature had already achieved—a mimetic method grounded in a cybernetic ontology of life-as-control, transposed, not as the control of information, but of materiality itself.

The origins of this method may be traced back to cybernetics through its filiations into bionics and its twin, biomimetics. The word "bionics" was coined by Colonel Jack E. Steele, then a major with the US Air Force's Aerospace Medical Division. Drawing from the famed cyberneticians Heinz von Foerster and Warren McCulloch's research at MIT on "biological computers," a project which sought to mimic how frog eyes process visual stimuli, Steele created the term by merging *bios* (the Greek term for life that Steele chose for its orientation to function) with electronics, as "the discipline of using principles derived from living systems in the solution of design problems."[32] Over the 1960s, Steele and others passionately explored the potential of this "biological-engineering science" to revolutionize military technologies, particularly in the areas of data processing and power efficiency, through a series of Bionics Symposia held in Dayton, Ohio, at the Wright Patterson Air Force Base.[33] The meaning of bionics as a discipline later shifted toward biomechatronics, medical prosthetics, and other cyborgian studies, thanks in no small part to the tremendous success that Caidin's *Cyborg* and its TV spin-off *The Six Million Dollar Man* had in capturing the public's imagination of the "Bionic Man," which directed attention from nature as a model technology to mechanisms replacing once-anatomical parts.[34]

"Biomimetics" encompassed what was formerly bionics for the rest of the twentieth century.

While biomimetics didn't receive its first published naming until 1969, it was initially preconceived in 1957 by the American polymath Otto Schmitt as a response to what he saw as the limited exploration of the vast possibilities of biophysics, which had matured in his time as an interdisciplinary subfield of biological research and which he had pursued in his own doctoral research in a squid-inspired feedback circuit: the titular Schmitt trigger. "Biophysics is not so much a subject matter as it is a point of view," he wrote. "It is an approach to problems of biological science utilizing the theory and technology of the physical sciences. Conversely, biophysics is also a biologist's approach to problems of physical science and engineering, although this aspect has largely been neglected."[35] During the 1970s and 1980s, biomimetics became an expansive signifier for a diverse research program that recuperated the then-flagging meta-discipline of cybernetics by fusing the latter's practical interest in the applicability of biological knowledge with extant lines of inquiry in mechanics, chemistry, and materials science, not just electronics.[36] Spurred by the Revolution in Military Affairs (RMA), which increasingly drew on the biological sciences to enhance operations research, the United States armed forces again visited the potential of the cyborg in the 1990s, in programs for research on biomimetics, not just from the army, air force, and navy, but also, and more heavily, from the Defense Advanced Research Projects Agency (DARPA).[37] These programs have commanded large sums of money for university-based projects related to military-defined biomimetic goals.

The Ortiz Laboratory for Structural Biological and Biomimetic Materials, housed in the MIT Department of Materials Science and Engineering, is one such recipient.[38] From the outset, Ortiz has received funding from DARPA and the Institute for Soldier Nanotechnologies (ISN), an on-campus collaborative venture set up in 2002 between the school, the US Army, and various corporate entities.[39] With over $100 million to "dramatically improve the survivability of the soldier" through material innovation at the nano-scale, the ISN offers funding schemes to research that will enhance the fighting and defense capabilities of US infantry.[40] One of the ISN's founding faculty members, Ortiz has received financial backing from the organization since the outset of her research program into *P. senegalus*, the fish that inspired the artifact before me. "I was one of the original faculty in the grant," she later disclosed to me. "We had been interacting with the ISN for a number of years. We got a chance to do all kinds of engagements with the military and the Department of Defense, so we knew pretty well what all the needs were."[41] Clarifying to me exactly what these needs were, Ortiz mentioned that "the students knew that the needs of the soldier were flexible armor." In other words, the Ortiz laboratory needed to develop a material with precise control over flexibility and durability.

This exact combination of material properties serendipitously presented itself to Ortiz when one of her graduate students, Benjamin Bruet, happened upon an exhibit on *P. senegalus* at a natural history museum in France in 2007.[42] Given his advisor's interest in rare biological materials, Bruet at once appreciated the fish's unique scale patterns and asked the curator for a sample. Back at MIT, Ortiz immediately recognized in the skeleton of *P. senegalus* the potential to satisfy the design criteria given to her by the ISN. The exact physical configuration of the fish's scales evoked to her both flexibility and protection, a certain expression of "multi-functionality" that Ortiz viewed through an epistemic lens provided by her training in MSE. Since its origins in the 1960s, MSE has been concerned with the military-industrial utility of materials, having been spawned during the Cold War, matured in step with the Japanese electronics industry, and progressed through a "materials by design" phase linked to an interest in composites: mixed materials with numerous functions outfitted for specific uses.[43] Ortiz's graduate training in particular coincided with the disciplinary recognition that the design principles of "nature itself" offered countless examples of this multi-functionality.[44] As one of Ortiz's students explained, "Nature has somehow engineered this process to be able to create structures, and orders, and hierarchies on so many different levels, and that creates a lot of different mechanisms by which it can improve its strength, toughness, versatility, and multi-functionality."[45] In this teleological misreading of natural selection, biological nature acts like the ideal engineer, devising new dimensions of material control through evolutionary tinkering—a misreading that recalls theological questions of nature's divine provenance.[46] Life is a composite par excellence, distinguished from the synthetic by its capacity for hierarchical and multi-functional control that Ortiz, via biomimetics, worked to emulate in life-like artifacts.

For over a decade, *P. senegalus* has inspired an ongoing multimedia research program into the structural and functional characteristics of its scale array: in its skeletal form, in a once-living specimen, in a digital model, and in 3D printed prototypes.[47] The Ortiz lab has investigated the scales' characteristics across three research phases. First, they used microscopy to visualize the indentations a diamond tip made when forced into the scales; each scale of *P. senegalus* possessed a quad-layered structure composed of various materials, from "hard" to "soft," that help dissipate impact. Second, the team scanned the scales using micro-computed tomography to obtain a series of two-dimensional renderings. Using computer software to trace the scales, they removed the original images and left only the manual lines, which generated the geometric shapes that formed the basis of their model. From the digital fish, they made two discoveries: first, that the scales of *P. senegalus* connected to one another vis-à-vis peg-and-socket joints that permit a high degree of rotation and movement; second, that there was significant variation in the geometry of the fish's scales across its body. Subsequent experiments revealed this structural variation clustered into "functional zones," which, according to their hypothesis, perform a specific function in the living fish: The clusters near the head of the creature protect its vital organs whereas the clusters near the tail enable its flexibility. Third, these functional zones, structure-function relationships embedded within geometrical forms, are the fish's "design principles," which inform a software program, MetaMesh, that permits them to extrapolate these principles into 3-D printed prototypes to fit figures other than the fish, namely, the human body. It is this bodiless abstraction that the group patented in 2014 and used to generate not just prototypes but also evolutionary knowledge of life.[48]

Natural selection has generated an "untapped encyclopedia," as Ortiz calls it, of design principles that researchers read from the book of life.[49] *P. senegalus*, Ortiz told me, is an "ancient" fish, a living species whose contemporary individuals are presumed to resemble their ancestors, owing to evolutionary isolation. "It was around here 500 million years ago," Ortiz explained. "At that time, everything was in the ocean, and everything had armor. When evolution took place, these fish started to lose their armor because they wanted to choose mobility over perfection, and things started to come out of the water and turned into amphibians." *P. senegalus*, however, stood alone, for its design principles equipped it with both protection *and* mobility. "This fish got isolated in estuaries in Africa, so it didn't evolve," she continued. "Basically, it has no predators because its armor is so tough." Through mimicry, the Ortiz laboratory can extract the knowledge embedded in life's diverse embodiments—time in condensation—and bypass the drudgery of conventional synthetic approaches to material design. In a loop from engineering practice to knowledge of life, the team has concluded that their prototypes actually reveal new findings on evolution: The diamond tip from their previous studies, as a proxy for a "toothed biting attack," suggested to Ortiz and her team that the scales are designed to resist such forces and have contributed to an "evolutionary arms race" in which predators and prey evolve improved offensive and defensive capabilities in step with one another.[50] As these biomimetics researchers reach toward greater control over nature by imitating how the design principles of life enact organizational control, they recursively affirm a cybernetic ontology of life-as-control, reframed as material, as well as the ideal *to control*.

Against Control: Toward Unintentionality in Biomimicry

> What will make the Biomimicry Revolution any different from the Industrial Revolution? Who's to say we won't simply steal nature's thunder and use it in the ongoing campaign against life?[51]

Exigent and forthright, these questions opened the training handbook I was handed upon my arrival at Casa Barragán, home and studio of the Pritzker Prize–winning Mexican architect Luis Barragán and the site of an itinerant "biomimicry workshop" I attended during fieldwork, which moved from the Miguel Hidalgo *alcaldía* of Mexico City to the rainforests of Veracruz. They implicated biomimicry designers, like the ones I was there to observe, in a collective crisis of the self, an existential interrogation of the ethical conditions that separate biomimicry from the Promethean. The author, I noticed, was Janine Benyus, an ecologist who, in 1997, penned *Biomimicry: Innovation Inspired by Nature*, in which she presented biomimicry, the "emulation of life's genius," as a design-based remedy for all kinds of ecological ills, from global warming to biodiversity loss.[52] To promote this message, Benyus and her colleagues established in 1998 the Biomimicry Guild (now Biomimicry 3.8), a for-profit design consultancy, and in 2006 the Biomimicry Institute, her nonprofit pedagogical arm. Through the Institute, Benyus offers several week-long workshops a year, often in remote and "wild" destinations, where aspiring biomimicry designers learn its major methodological tenets through their "immersion in nature."

"This is not an idle worry," the manual continued. "The last really famous biomimetic invention was the airplane (the Wright brothers watched vultures to learn the nuances of drag and lift). We flew like a bird for the first time in 1903, and by 1914, we were dropping bombs from the sky. Perhaps in the end, it will not be a change in technology that will bring us to the biomimetic future, but a change of heart, a humbling that allows us to be attentive to nature's lessons."[53] Having assumed biomimicry was identical to biomimetics, given their identity in popular media, I was surprised to hear Benyus engage in this anti-militaristic "boundary work," distinguishing her approach to nature imitation from its predecessors, particularly after my exposure to the militarization of nature in the Ortiz lab.[54] What I soon confronted at this workshop was an explicit concern with the designerly self, the moral obligations of designers.[55]

Manual-in-lap, I sat with thirteen designers, among them landscape architects, industrial designers, and environmental engineers from the United States, Mexico, and South Africa, in a wood-beamed, white stucco room that Barragán used as his drafting studio. A single beam of light illuminated the front of the room, where Raúl, a pioneering architect of bamboo structures, was recounting an anthropological "history" he borrowed from Benyus.[56] "*Homo industrialis* thrived on a logic of *control*," he said, an excess of the self's desire to remake the world.[57] Unlike the "industrial" or "modern" approach to design, biomimicry renounces "intent." With biomimicry, he finished, "the locus of control is inside of us," a growth in our "state of being." In this tale, biomimics can aspire to "grow" their "being" by limiting the self. Designers should repress their "Western" fantasies of technological control and "domination" to open up a relationship to the environment Benyus calls "respectful imitation," a shift from a dynamic of "extraction" to "learning."[58] This is accomplished by "quieting your cleverness," a vernacular discourse of intent by which Benyus targets an aspect of selfhood that designers must reform.[59] This reflexive relationship of the self to the self amounts to what Michel Foucault calls "problematization," that which "allows one to step back from this way of acting or reacting, to present it to oneself as an object of thought and to question it as to its meaning, its conditions, and its goals."[60] By denouncing intent to restrict its extent, such narrative fictions offer blueprints for a new designerly self that adheres to an cultural ideal of anti-control—the cyborg's inverse.

Studying instead how "life's genius" accomplishes "technical" feats, like movement and thermoregulation, "biomimics" hope for a new "technological culture," free of anthropocentrism, militarism, and other pathologies of modern design.[61] For Benyus, Raul, and other practitioners, biomimicry differentiates itself from biomimetics through its subjective orientation to nature—an orientation premised on an attitude of intimate "humility," not detached control. To Benyus and her followers, the ideal of controlling nature, born from a desire to bend life to human will, has motivated designers to construct technologies that ignore context. For Benyus, control is, in her symbolism, at the heart of their ecological crisis, and a change of heart, its transcendence.

In this sense, biomimicry departs from the engineering ideal to master and control nature that structured the ambitions of the cyborg. Benyus develops this argument most explicitly in her discipline-defining book *Biomimicry*. First, she advances the belief that biological life considered in totality has acquired physiological adaptations to the environment through natural selection. Second, she compares the relationship between organism and environment to that between technology and context; both depend on adaptive ecological fit. If, as she reasons, "Western" technologies have ignored the limits of context, with the implication that such breaches have caused our current environmental predicaments, then the diversity of life affords a sourcebook of well-adapted "design principles" for reforming material culture.[62] By copying, or "echoing," these principles, Benyus believes that biomimicry can harness the power of nature in designs that "emulate" the functions of life.[63] Biomimics register this power in qualities they find missing in extant technologies: multi-functionality, efficiency, wastelessness, and self-reproduction. Biomimics promise to revitalize design culture by imitating biological life in non-biological technologies—a "sympathetic magic" for conjuring a sustainable form of life.[64] For Benyus, "life creates conditions conducive to life"; by copying nature, they can create these conditions too.[65]

One might think that this refrain—bio-inspiration will transform artifice—reiterates the underlying premise of biomimetics. Indeed, Benyus draws much of her reference literature from biomimetics literatures.[66] Insofar as the value of biological nature, the reason for imitating it, lies in its organizational complexity, its exquisite material control, biomimicry is a successor science to biomimetics and, before it, cybernetics. Yet Benyus also cites a number of other intellectual influences that offer the building blocks for re-envisioning the cybernetic project: environmental activists (Rachel Carson, Bill McKibben, Stewart Brand, Paul Hawken, Wendell Berry), sustainable agriculturalists (Wes Jackson, Michael Ableman, Masanobu Fukuoka, and Bill Mollison), and ecological designers (Ken Yeang and John and Nancy Todd).[67] Biomimicry thus fuses the language of control in biomimetics with the discourse of sustainability. Early iterations of this strategy can be found in the 1974 writings of the German cybernetician Felix Paturi, who drew upon evolutionary biology to critique industrial energy consumption and advance an "economy" of natural materials.[68] Benyus takes this argument to its logical conclusion by coupling the goals of sustainable development, defined by the famous UN Brundtland Commission of 1987 as "development that meets the needs of the present without compromising the ability of future generations to meet their own needs," with the Living Systems Theory approaches of James Grier Miller and Fritjof Capra.[69] Through such theoretical affiliations, Benyus presents biomimicry as a paradigm of design that pulls the ambitions of sustainability into the ambit of the cybernetic view of life-as-control but away from the ideal of controlling it.

On the second day of the workshop, we departed Casa Barragán and, in a caravan of Volkswagen vans, stuttered through standstill traffic to make our escape away from the deafening volume of Mexico City and into Veracruz, an ecotourism destination. Raul remarked that "all that noise and electricity" would dampen our attempts to "listen" to nature. After four hours in the agave-flecked deserts of Puebla, we stopped at Cantona, a three-thousand-year-old Mesoamerican ruin that Raul reverently labeled "the most spectacular archaeological site." As we ascended the abandoned avenues that abutted domiciles then leveled and overrun with opportunistic species of grass, we at last reached one of the highest points, where we recessed to initiate in what Raul called a "Be Here" activity. He sat on the igneous soil and enjoined us to "close [our] eyes and open [our] minds, souls, and senses to the experience" of this "sacred space." Instructing us to "quiet our cleverness," Raul beckoned us, his windswept audience, to suppress our active "thought-ing" and focus on our senses of hearing and feeling. "You must slow down," he advised, "before you can focus on your surroundings." After a period of quiet meditation, Raul prompted us to attend to the humidity of the air, its heat, which added to the warmth of the overhead sun and the gusts of wind that billowed and crashed against our skin.

Through "being here," Raul problematized a profusion of thought, identifying an active inner life with an intrusion of intent, an excess that distracts the self from sensing place and "nature's lessons," as if thought could be surgically excised out of sensation.[70] "Slowness" of mind, as opposed to the "speed" associated with industrial modernity, was the ideal qualitative experience of biomimicry, the absence of distraction and cleverness. Braking thought's velocity became an ideal, a precondition to the redirection of a designer's senses toward the genius of life. In his history of Western subjectivity, Foucault calls such corporeal acts "techniques of the self," inward-facing modulations of inner states, like intent, the expression of a desire to control.[71] Biomimics accomplished this self-work through contemplative acts that took their bodily dispositions as objects of re-design and as vehicles for reforming their intents.[72] The embodied acts of seeing, hearing, and touching anew provide a renewed phenomenology of design for biomimics, not merely a mode of subjectivity but a new vista of objectivity, of design principles. Before these architectural designers could become sufficiently attentive to "nature's lessons," they first dedicated themselves to denying control as an ideal.[73] In biomimicry, an ideal of un-intentionality, based in the virtue of humility, precedes the knowledge and work of emulation.

Coda: Ethnographic Notes on the Inversion of Biological Modernism

In the iron-and-flesh grip of the cyborg, cybernetics has appeared as little more than a scientific veneer for the extension of belligerent logics of control into the nature of the organic. From the vantage point of the cybernetic application of biology to technology, the presence of the cyborg recedes, its hold loosened. Within biomimetics, the cyborg is displaced as the goal of design, to be substituted with nature-inspired artifice, only to return once more with its military objectives, in the securitization of soldiers through biologically inspired armor. These engineers seek to control nature, circuitously, by imitating it, extracting the "design principles" of organic life, exemplars of material control, and operationalizing them for human ends. In this universe of design, nature is materials engineering incarnate—a masterful engineer, whose skillfulness they affirm and reify with their own technological creations. Control commands both form and norm. In contrast, biomimics admire the control life exhibits, without wishing to control it themselves. Within their register of apprenticeship to nature, a new politics of life unfolds.[74] Such contrasting legacies of cybernetics suggest an alternative nexus of biology, technology, and power not easily captured by the dominant and simplistic figure of the cyborg and the "cyborg metaphor."[75] Within the practice of biomimicry, one finds, not a manifesto for the cyborg, but its send-off.

The introduction to this collection presents Philip Pauly's concept of "biological modernism" as a compass for thinking through the twentieth-century shift toward life's engineering, the importation of the "strategy and tactics of artifice" into biology, with the consequence that "biologists began to see themselves as designers and inventors of new things."[76] In biomimetics and biomimicry, however, life is always already engineered—an exemplar, crafted by evolution, which engineers and other designers copy. This is more than the engineering of biological life; this is engineering of social life *by* "life itself," the biologization of engineering.[77] Perhaps, then, we might think of biological modernism inverted, of designers and inventors of new things who see themselves as biologists, creating new life-like forms in technical substrates. Seen alongside conventional biotechnologies, bio-inspiration suggests a multiplicity of modernities; it evokes not biological modernism, as a monolith, but biological modernisms. This chapter in the book of nature, remade, thus ends with human technology remade through life—making (re)naturalized.

PART THREE

Envisioning

10 * *Strains of Andromeda: The Cosmic Potential Hazards of Genetic Engineering*

LUIS A. CAMPOS

Speculations about colonization, containment, and contamination were widespread in the engineering of biology in the 1960s and 1970s, as scientists, journalists, and others struggled to describe the possibilities and potential risks of newly available techniques. Even commonplace discussions of space colonization resonated with concerns that Earth itself might be colonized—or contaminated—by life from outer space. The most popular touchstone for such concerns emerged out of a new genre of science fiction, of ersatz reports from the near-future, exemplified by Michael Crichton's new style of biomedical thriller. Though entirely fictional, his *The Andromeda Strain* (1969) gave shape to fears of what is now known as "backward contamination" (contamination from outer space to Earth). And even as such fictional narratives of an "Andromeda-like scenario" impacted exobiological discussions—and perhaps even aspects of the public understanding of an ascendant space program—they also came to be explicitly and repeatedly invoked in the mid-1970s by biologists with more terrestrial concerns about the engineering of novel life forms in the laboratory through the use of recombinant DNA technologies.

As American exobiologists and space enthusiasts called for the exploration of potential habitats for life elsewhere in the heavens, and perhaps even some form of colonization of the heavens, laboratory molecular biologists began to narrate the potential dangers of encounters with new forms of life. From the superlunary to the sublunary, a shared discourse of breaching "barriers"—evolutionary barriers, species barriers, safety barriers, and containment barriers—structured many molecular biological debates of the 1970s. From the birth of exobiology in the 1960s to the dawn of genetic engineering in the 1970s, fictional fears of alien life from space mirrored concerns about engineered organisms in the laboratory. Even biologists gathered at the famed 1975 Asilomar conference addressing the potential biohazards of recombinant DNA research explicitly came to frame their own concerns around such breaches with the language of science fiction—and with reference to Michael Crichton's *Andromeda Strain* in particular. While many involved derided such invocations of intergalactic fiction as sensationalism far removed from terrestrial concerns and fomenting a new kind of "molecular politics," these sorts of invocations suggest that we would do well to seek to understand the unexpected and sometimes unruly cultural narratives that condition and may even deeply structure the development of new possibilities for the engineering of life at the molecular level.

This intertwining of planetary protection and laboratory biosafety arrangements suggests that the conceptual, scientific, and regulatory frameworks governing scientific commerce with the heavens above resonated contemporaneously with questions concerning the protocols (and governance) for dealing with potential hazards in the engineering of life here on Earth—a curious double helix of scientific laboratory practice and biosafety policy that passes straight through the realm of science fiction. Cosmic narratives about the Andromeda strain not only intruded into earthly decisions about laboratory biocontainment strategies, but were in fact part of the very construction of new and contested futures for biological engineering, showing the entanglements of earthly lessons about biological colonization and invasion and the planetary protection of heavenly bodies, scaling from the molecular to the galactic. There is a tale to be told here, in other words, of how fictions come to matter.[1]

A for Andromeda

And so we should begin with a consideration of Andromeda itself. Named after the Greek myth of the Ethiopian princess tied to a storm-tossed rock and tortured by Poseidon—and who was later set free by Perseus holding aloft the head of Medusa—the Andromeda Galaxy has long represented the entanglements of earthly and heavenly affairs.[2] As early as the year of Sputnik's launch, it served as the site of paleontologist Ivan Yefremov's communist utopia *Andromeda*, which literary critic Fredric Jameson has called "one of the most single-minded and extreme attempts to produce a full representation of a future, classless, harmonious, world-wide utopian society."[3] The solution to capitalist alienation was a communist colony on an alien world. As the next nearest galactic neighbor worth consideration when thinking about the extent of the life in the universe and our place within it, the Andromeda Galaxy has been to Earth in the galactic context what Mars has been to Earth in more local comparisons within the solar system. Indeed, much as H. G. Wells' *The War of the Worlds* (1897) introduced the notion of planetary contamination from Mars, and as Arthur C. Clarke's "Before Eden" (1960) drew on G. A. Tikhov's mid-century ideas of astrobotany to imagine life on Venus, within a short time Andromeda became a go-to resource for illuminating exobiology.[4]

At a prominent symposium on origins of life research held in New York City in June 1959, the astrophysicist Philip Morrison began his talk by introducing the Andromeda Galaxy as a way to think about life in a cosmic context. The prospect of a vertical biology—from Hubertus Strughold's *The Red and the Green* (1953) to H. J. Muller and Carl Sagan's envisioning of the "red thread"[5] taking life from Earth to space—has long implied the converse: the arrival on Earth of life from elsewhere with its attendant effects. As early as 1946, science fiction authors noted that "medical men have suggested several times in recent years that life in virus—and infectious—form may come to this planet from other worlds."[6] In a 1960s BBC science fiction series entitled "A for Andromeda," the astronomer Fred Hoyle had brought the prospects of Andromeda down to Earth, outlining the possibility of "a radio signal that might contain a detailed biochemical recipe that would enable us to reconstruct an alien being here on Earth."[7] This implicit reference to a novel strain of life engineered here on Earth was inextricably intertwined in Hoyle's story with the question of alien life elsewhere—a reinterpretation of older theories of panspermia with novel possibilities envisioned at the frontiers of biological engineering. Astrobiological prospects were inescapably synthetic biological at the same time.[8]

More often, however, these overlapping prospects involved questions of inadvertency and how best to engineer around unwanted invasion or infection. Mid-century discussions about what is now understood as "planetary protection"—ensuring that celestial bodies would not be contaminated with Earth life before scientists had a chance to study any indigenous life that might be found there—were deeply engaged with such questions of "biological missteps." In the same year as the symposium in which Morrison connected Andromeda to earthly concerns, WESTEX (the Western Group on Planetary Biology) had concluded that while "the political, moral, legal, and economic consequences of premature contamination of celestial objects can hardly be estimated on the basis of present knowledge," nevertheless "biological missteps would do irreversible harm." This was rather unlike "misjudgments in the physical sciences, [which] though costly and exasperating, might still be remediable."[9] "Contamination by extra-terrestrial exploration" became the watchword of the day. In 1959, a second meeting of CETEX (Contamination by Extraterrestrial Exploration), held under the auspices of the International Council of Scientific Union's Committee on Space Research, concluded that "surprises are certain and unlikely possibilities must be borne in mind when dealing with the problem of contamination which is better defined as the problem of reducing the risk whereby one experiment may spoil the situation or other subsequent enquiries."[10] Joshua Lederberg—who had coined the very term "exobiology"—made such concerns of extraterrestrial forward contamination a priority: "If a planet can be infected by terrestrial organisms," he noted, then "planetary microbiology must therefore be accepted, both on a scientific and an explorational context, as a major issue in the space program, not an afterthought or diversion to be accommodated so long as the annoying questions it raises do not distract from the seeming urgencies of the moment. History gives us some pale analogies; the rabbit in Australia, smallpox in America, *Treponema* in Europe."[11]

Indeed, formal studies of the patterns and effects of biological colonization had long been a focus in biology. One prominent meeting on "The Genetics of Colonizing Species" was held in 1965 at the Asilomar conference center on the Monterey Peninsula of California. Here, the architects of the evolutionary synthesis and other prominent biologists—among them Ernst Mayr, Theodosius Dobzhansky, Ledyard Stebbins, C. H. Waddington, and Richard Lewontin—gathered to speculate about *Myxoma* virus (which had been introduced to control and contain Australia's invasive rabbit population), the "'alien' habitats of invading species," the dominance of "hypervirulent" strains, and whether such "infections" would take.[12]

Engineering biological security here on Earth and ensuring planetary protection elsewhere would soon come to be seen as interrelated issues. Even as Lederberg had led efforts to spare celestial bodies from earthly contamination, the reverse prospect—fears of "germs from space"[13]—had emerged from the haunts of Wellsian science fiction to become a central topic of concern for many, including NASA, which developed lunar quarantine precautions. Indeed, the international and biosafety ramifications of such "germs from space" scenarios were clearly identified in prospect by exobiologists, who saw nationalism in space as "obsolete":

> There is a chance—slight but hardly negligible in view of the possible consequences—that vast epidemics could be set off on earth by organisms inadvertently imported from the moon—or later from some other planet. Shouldn't all the world's governments and scientists have some voice in deciding whether that risk ought to be taken—not merely the rulers and scientists of either the United States or the Soviet Union?[14]

An international consensus was clearly necessary when dealing with a cosmic threat to avoid exobiological doom,[15] and international space diplomacy would parallel efforts to develop mechanisms for dealing with issues of biological contamination and how to avoid inadvertently engineering an epidemic here on Earth. These worlds of the astrobiological and the synthetic biological would draw ever closer with the publication of Michael Crichton's *The Andromeda Strain* (1969) and the famed 1975 Asilomar meeting on the potential biohazards of work with recombinant DNA.

The Andromeda Strain

The Andromeda Strain (1969) was Michael Crichton's first novel,[16] the story of a military satellite descending to Earth in northern Arizona with a bug from space. Crichton credited the idea for the novel—which was written while he was still in medical school—to a tutorial he had taken at Harvard. It was then that he had read the paleontologist G. G. Simpson's work *The Major Features of Evolution* (1953); Simpson had "inserted an uncharacteristically lighthearted note that organisms in the upper atmosphere had never been used in a science fiction story. I set out to do that," Crichton noted.[17] Where Simpson had been dismissive of the idea, Crichton saw a promising prospect and turned it into a bestselling novel.[18] Selling over two million copies when it was first released, *The Andromeda Strain* drew the American public's attention from the imminent prospect of atomic apocalypse to a new realm of potential biological catastrophe.

As Joanna Radin has shown, Crichton's new narrative style depended on verisimilitude, and consisted of "authenticating detail ripped from leading academic scholarship," providing reports from the near-future of technology that often "blurred the line between fiction and reality," even as they also blurred the line between exobiology and the engineering of biology.[19] (Joshua Lederberg even protested to the publisher that his own identity in the story was not sufficiently fictionalized.[20]) Although preceded by H. G. Wells, Arthur C. Clarke, and many others, Crichton's *Andromeda Strain* reinvigorated a genre: according to one scholar of science fiction, "almost no science fiction confronted questions of evolution and genetics in any depth until the excitement about recombinant DNA reignited interest in the mid-1970s."[21] The book also did not escape NASA's notice. As Michael Meltzer, historian of NASA's own effort at planetary protection, has noted, "Although events in the book were fiction, its premise has become a metric for worst-case scenarios resulting from back contamination."[22]

In short order, "Andromeda strain" scenarios even became common currency among biologists invited to formally consider planetary protection issues. In June 1974, the molecular biologist Norton Zinder had been tasked by Norman Horowitz of the National Academy of Science's Space Science Board to consider the issues of back contamination in a risk assessment for samples coming from a Mars sample return mission. Zinder replied that his "main concern" was "what to do with the samples, as the number of variables in attempting to grow unknown organisms is enormous. Personally I don't believe in universal parasites or Andromeda strains."[23]

But the same interests in containment of contamination (such as quarantine) that had affected considerations of space invaders in exobiology were also soon at play in the development of biocontainment strategies for newly emerging recombinant DNA techniques of "genetic engineering." Stanley Cohen—who with Herbert Boyer discovered the new gene-splicing techniques for creating recombinant DNA—would soon make the analogy with planetary protection explicit. Just as "physical containment barriers have long been used . . . to minimize the possibility of contamination of this planet by extra-terrestrial microbes," Cohen noted, "the concept of biological barriers, which was formulated in some detail at the Asilomar meeting, and which involves fastidious bacterial hosts unable to survive in natural environment and equally fastidious vehicles able to grow only in specified hosts, will contribute significantly to the safety of gene manipulation experiments."[24]

A for Asilomar

From Mary Shelley's *Frankenstein* to H. G. Well's *The War of the Worlds* to Orson Welles' radio broadcast adaptation, fictionalized Hollywood depictions of potential laboratory mishaps had become commonplace by the early 1970s, leading one newspaper to report:

> We can see it all now. The laboratory is dark and deserted but inside a large bottle there is restless stirring. Suddenly the stopper pops out and slim blobs of deadly new germs slither to the floor, out the door and down the street to the apartment of a beautiful young lab technician. She gives the alarm. After a harrowing week or so during which humanity is threatened with extinction, a handsome young technologist neutralizes the all-powerful bugs by the simple device of spraying them with mountain spring water. (He then marries the lab technician almost as an afterthought.)[25]

By 1971, Crichton's *Andromeda Strain* had been made by Robert Wise (of "The Sound of Music" fame) into a blockbuster film, the first to show a mass American audience the wearing of hazmat suits to deal with biological contamination (see fig. 10.1).[26]

And throughout the early 1970s, talk of an "Andromeda strain" had clearly begun to infect discussions of the potential biohazards of recombinant DNA. "Hopefully fact can be separated from fiction, rumor and myth," molecular biologist Paul Berg wrote to virologist Sir Michael Stoker. "In

Figure 10.1. Investigators in hazmat suits arrive in the fictional small town of Piedmont, New Mexico, in Robert Wise's 1971 film adaptation of *The Andromeda Strain.*

essence we want to distinguish hard fact from fuzzy impression, etc. Most likely the conclusion will be that we know very little for sure."[27]

Even the relatively sober *Chemical and Engineering News* noted concern about the possible risks of an emerging biotechnology based on recombinant DNA, reporting on serious concerns about the prospect of a laboratory-made Andromeda strain: "With images of an 'Andromeda Strain' lurking in the background, a group of research scientists, in concert with the National Research Councils, has urged a moratorium on genetic manipulation of microorganisms. And general agreement by workers in the area that it is a good idea lends credence to the note of alarm."[28]

An initial moratorium on certain classes of potentially hazardous experiments had led to a call for an international conference to be held in February 1975 "at which the nature and magnitude of the risks could be assessed," the molecular biologist Paul Berg noted. He would later recall, as one the conference's primary organizers: "Scientists were only just learning how to manipulate DNA from various sources into combinations that were not known to exist naturally. Although they were confident that the new technology offered considerable opportunities, the potential health and environmental risks were unclear." Ultimately held at the very same Asilomar conference center where evolutionary biologists had debated invasive species and alien introductions ten years earlier, the conference aimed to set "standards allowing geneticists to push research to its limits without endangering public health." The goal, as Berg recalled, "was to consider whether to lift the voluntary moratorium and, if so, what conditions to impose to ensure that the research could proceed safely."[29]

Even before "Asilomar," however, in the very earliest days of genetic engineering, journalists struggled to describe the possibilities and the risks of newly available engineering techniques, and often turned to Crichton's tale to capture the issues at stake. "Scientists involved in biological research are becoming more worried that dangerous organisms made in laboratories could escape—à la the movie 'Andromeda Strain,'" the *Boston Sunday Globe* reported in November 1973.[30] But even professional science journalists from sober scientific serials used the trope: "ANDROMEDA STRAIN?" ran the all-caps headline of a 1974 editorial in *Bioscience*:

> It's happened. Specifically, molecular geneticists have discovered relatively simple methods for the introduction of specific genes from vertebrates and other organisms into the genome of bacteria such as the common *E. coli* of the human gut. Further, alien genes from other species of bacteria and even viral DNA can now be grafted into a bacterium where it was not previously present . . . appalling dangers to the human races are inherent in this discovery.[31]

Written just a half year before Asilomar, the editorial called for "formal action at the earliest opportunity to support the resolution of the Committee on Recombinant DNA Molecules. . . . Everyone concerned should do whatever is possible to further the holding of the proposed international conference on this danger."[32]

The Andromeda Strain was also a ready reference point for the *Medical World News,*[33] and even for mass-media sources like *Time,* which ran a story on "The Andromeda Fear" that same summer (five years to the week of the moon landing).[34] This pattern continued right up to the days just before the 1975 Asilomar meeting, with journalists repeatedly recognizing the ways in which "microbiological agents have figured prominently in science fiction."[35] Long before Asilomar, science fiction—and *The Andromeda Strain* in particular—set the frame for considerations of the future of genetic engineering.

This framing did not substantively change with the Asilomar meeting.[36] "Scientists to Resume Risky Work on Genes: Danger of 'Andromeda Strain' Posed," blared a front-page headline of the *Boston Globe,* just as the meeting concluded.[37] Another early report on the meeting asked: "Could these unknown forms of life—'novel biotypes,' the scientists called them—threaten the world as an 'Andromeda Strain'? (The scientists here blanched at even the casual mention of the term.)"[38] And more than a year later, after a unanimous city council vote in Massachusetts to hold a public hearing on the risks of recombinant DNA on June 23, 1976, one onlooker noted: "It seems science fiction has arrived unannounced to Cambridge—the vision of the future has been dumped on our city."[39] The science fictional continued to intersect with the exobiological in the years after Asilomar, with one newspaper reporting in 1977: "There is the prospect that some scientist, now tinkering with the genes of germs will inadvertently—or even purposely—create strange new organisms that are more dangerous and more damaging to human health than anything the moon might have spawned."[40] "There is a dark side to the genetic moon too, at least in the minds of some scientists," warned the *Los Angeles Times*.[41]

While journalists came to frame Asilomar alternately in terms of *Frankenstein*, *Jaws*, and even the celebrity of Farrah Fawcett-Majors, *The Andromeda Strain* remained a primary point of reference.[42] Rather than some sort of misguided post hoc misinterpretation of events, "Andromeda Strain" scenarios for the future of engineered life served as both a prompt for the Asilomar discussion and a ready means for summarizing its significance afterward—both agonist and aftermath.

Sensational Speculations

That's not how it felt to most scientists involved with Asilomar, however, who became increasingly concerned about what they viewed as sensationalist coverage of the prospects for recombinant DNA technologies, rather than coverage properly attentive to actual engineerable futures for biology. At the time of Asilomar, however, even calling work with recombinant DNA "genetic engineering"—a term which had eugenical overtones—would have sounded sensationalist to most American molecular biologists.[43] How to emphasize novel prospects for the engineering of life at the molecular level and without seeming sensationalistic when even the phrase "genetic engineering" sounded alarms? (And how to do so when the mere presence of over a dozen journalists at the 1975 meeting—an unusual state of affairs in the world of scientific conferences in the mid-seventies—was as much a cause for notice?) If Asilomar were constantly framed by Andromeda, then the journalists were obviously to blame for this misprision.[44]

Indeed, just a few months after Asilomar, Paul Berg himself was quite disappointed with the journalistic coverage:

> I thought, in fact, many of the reporters had missed the point. I thought that they were playing up the theme . . . that we were trying to head off the prospects of some misuse of genetics and that it could create all kinds of monsters and Andromeda strains, and that we were warning the world that this line of research should not be done. And there were very few who seemed to perceive that what we were talking about was that this was an important line of research; we were in favor of doing this research.
>
> [All these] banner headlines . . . and these—they just really fried me,

> because when I saw them, I realized that yes, what we had done was being misconstrued, and if scientists were reading that, then they would clearly, legitimately be upset about it.[45]

Berg's perspective was echoed by others working on the follow-up process of developing more formal guidelines inside the National Institutes of Health. One meeting participant similarly bemoaned the influence of fancies of space invaders on the adoption of powerful new techniques:

> It's fundamentally irrational. It's because of the Andromeda strain idea and it has some intimate connection with heredity and geneticists have always been misunderstood. There's something irrational about people's reaction to the idea that inheritance might be altered as opposed to being simply analyzed. It took about a hundred years for people to get used to the idea of analysis. The idea of genetic manipulations [is] the issue and a political problem.[46]

These Andromedan-inspired fears of the public became the disease that the scientists themselves had to grapple with:

> People do care, and I think that's one reason why we are sitting here today. I didn't get a chance to say this, but I think we are sitting on a very important political problem here. Unless we can really come to grips with the DNA thing—somehow get it out of the headlines and into reality—this disease of "what you might be doing to us," is going to spread and science itself may be very threatened. You are going to be doing your cholera research in glove boxes with housewives looking over your shoulder to make sure you don't.[47]

And circulation of these Andromedan fears would have a hard time competing against scientifically understood risks:

> If they say, "Look at the risk you are putting me under of getting your epidemic Andromeda disease," somehow we have to communicate to them, "Listen, lady. You are under a risk of developing cervical carcinoma or your grandchild is and that risk is finite and real and horrible and if we do these experiments we can guarantee that, within generations, at least, we will reduce that risk. How about putting up with this risk?" It's risk-risk analysis and nobody is looking at the real risk.

The concept of "real risk" involved in the engineering of life was what was at stake—but how to calculate the "real risk" when no one knew what the risks even were? Even as Jim Watson had denounced the gathered assembly at Asilomar for pretending to be able to assess the risks of any potential hazard ("We can't even measure the fucking risks!"), he ultimately regretted (and resented) even having brought up any such safety concerns at all when he saw what had happened in the aftermath of Asilomar.[48] Scientists could be their own worst enemies, he suggested, with some prominent scientists actively spreading what he saw as unwarranted concern with sensationalist Andromedan tropes.

Chief among those Watson had in mind was the biochemist Liebe Cavalieri. Cavalieri had published a piece in the *New York Times* Sunday magazine in August 1976 entitled "New Strains of Life—or Death," casting Andromedan strains on the new laboratory techniques. As if on cue, Jeremy Rifkin and colleagues picked up on Cavalieri's claims for *Mother Jones*, further fanning the flames:

> Cavalieri and his colleagues are deeply concerned over the possibility that new Andromeda-type virus, for which there is no known immunization, might accidentally be developed in a laboratory somewhere and spread a deadly epidemic across the planet, killing hundreds of millions of people. They also fear that a new, highly resistant plant might be developed that could wipe out all other vegetation and animal life in its path.[49]

Watson was not the only one bothered at Cavalieri's intervention (going so far as to say, "Well, that article at best was crap"): the molecular biologist Norton Zinder, who had consulted with Norman Horowitz about exobiological concerns, was likewise incensed:

> Touted on the magazine's cover as a potential holocaust, it continued inside with many pages of demagogic diatribe. All of recombinant DNA research and all the scientists involved were trashed. Cancer would be set rampant in the population by self-serving, elitist, Nobel-seeking scientists. / A month later, the article was introduced into the Congressional Record . . .

After a bit of sleuthing, Zinder discovered that "none of the science writers of the *Times* had seen the article, and that they were all as dismayed as the rest of us."[50] But nevertheless, he noted, the media environment had transformed, and was being overtaken by these worst of fictions:

> Despite the continued relatively balanced texts, the tone and the mood emanating from the media now changed drastically . . . something scientifically important and, despite all, perhaps beautiful had happened was lost in the transmission. . . . There is a class of newspaper in this country, epitomized

> by the *National Enquirer*, which thrives on sensationalism. Recombinant DNA chimeras were made to order for them. The monsters formed and the horrors caused were only limited by the imagination of the writers.

That a prominent scientist would promulgate Andromedan claims to an increasingly concerned public was also a source of deep concern for scientists at NIH actively trying to draft the recombinant DNA guidelines:

> The people who talk about the Andromeda strain have media access and they have something to say to the media that you can't say. . . . I mean, it really is molecular politics. It's got nothing to do with science; nothing to do with rationality; I am reasonably safe on those grounds. It's molecular politics, not molecular biology and I think we have to consider both, because a lot of science is at stake.[51]

For his part, Cavalieri responded to his critics with a simple assessment: "Some of us have been called sensationalist for bringing up those possibilities; but in fact one need not imagine or embroider; the subject of recombinant DNA *is* sensational."[52]

The "Andromeda Strain" had been fully enfolded into public representations of genetic engineering, and this blurring of the line between fact and fiction—the stylistic move precisely at the heart of Crichton's success—is precisely what infuriated molecular biologists like Bernard Davis. The suggestion that a "scientist working in this field may produce yet another Andromeda strain" presumed that "the first strain existed in fact rather than in fancy," he noted.[53] "The chimeras of mythology are not about to descend from the realm of art into that of technology":

> The Golem, like the Andromeda strain and the chimera, remains a product of man's literary imagination and not his technology. As Philip Handler recently stated, "Those who have inflamed the public imagination [over recombinant DNA] have raised fears that rest on no factual basis but their own science fiction."[54]

In an article in the *American Scientist* entitled "The Recombinant DNA Scenarios: Andromeda Strain, Chimera, and Golem,"[55] at a public lecture at Harvard University on "Darwin, Pasteur, and the Andromeda Strain,"[56] and in the pages of *Time* magazine, under the headline "Doomsday," Davis further took such scientific sensationalism to task: "Those who claim we are letting loose an Andromeda strain are either hysterics or are trying to wreck a whole new field of research."[57] "The Andromeda Strain remains entertaining science fiction," Davis insisted, but "even the most favorable experience [with various recombinants] will not eliminate the specter of a future Andromeda strain unless we interpret it in terms of epidemiological principles," and offered to drink recombinant DNA to make the point.[58]

By 1977, Andromedan themes were front and center at a National Academies Forum on "Research with Recombinant DNA," a two-and-a-half day vituperous meeting that included "accusations, insults, and even a feeble protest demonstration."[59] While the *Paramecium* expert Tracy Sonneborn denounced "pessimists" who "set forth a great catalogue of imagined catastrophes, modeled on the fiction of the Andromeda strain," the biochemist Daniel E. Koshland Jr. suggested: "In the current dispute, some individuals refer to the 'Andromeda strain' as though it had been created in the laboratory and was not merely a work of fiction. Failure to distinguish fact from fiction can lead to fundamental disagreement even among those who believe that both safety and progress are desirable goals."[60] Asilomar co-organizer Paul Berg found himself befuddled by the continued focus on fantastical plague-bearing plasmids:

> Andromeda strain? I remember reading that book and seeing the movie, but I haven't thought of any situation where one would create something that would have that similarity. We're talking about methodology which effectively moves genes around but doesn't create a new gene. I would offhand guess that a lot of the regulatory mechanisms that hold in check most of the biological world quite effectively, would still act even in these recombinant structures so that an Andromeda strain, taking your question literally, an Andromeda strain is unlikely—something that can do something new, like chew up plastic—and therefore, not be containable and grow on the plastic or whatever the Andromeda strain grew on.[61]

Andromeda Infects Asilomar

Journalists repeated claims that science fiction possibilities for the engineering of life were becoming reality even as scientists tried to deny substance to the Andromedan themes used to present these prospects. But behind closed doors, as recombinant DNA policy was being made, the strain was already infecting the scientific community: many scientists themselves continued to think through the possibility of unanticipated biohazards by making explicit reference to *The Andromeda Strain*. Some saw clear parallels with the checkered history surrounding the introduction of larger-scale organisms, as had been discussed at the 1965 Asilomar meeting. As one correspondent of Sinsheimer's wrote in 1976:

> if we are to monkey with genetics and possibly produce something dangerous, I would say it would be difficult with anything smaller than a horse. In other words, what we need is visibility and size. Along this same line, have you happened to have read Michael Crichton's *The Andromeda Strain*? It is science fiction, dealing with an unknown and highly dangerous strain of microorganism.[62]

Or as Stanley Falkow noted in a National Academy of Sciences Plasmid Subcommittee meeting:

> It has been a common practice for over ten years to perform intergeneric matings between classical *E. coli* Hfr or F prime donors and *Salmonella* and *Shigella*. Most of these experiments were performed to elucidate the virulence of these enteric pathogens or to examine their overall genetic organization. . . . The likelihood of a serious biological hazard akin to the fictional Andromeda strain emerging from either Type 1 experiments or from the laboratory manipulation of naturally occurring plasmids seems remote. There is, however, a certain degree of potential hazard.[63]

At a later 1977 workshop, E. S. Anderson from the Enteric Reference Laboratory of London offered a new option "substantially to reduce the alarm provoked by the postulated science-fiction type of holocausts, based on no evidence of possibility, to which the lay public has reacted so vigorously."[64] Even Norton Zinder invoked *The Andromeda Strain* in a testy message to Joshua Lederberg entitled "The Truth" (taking him to task for writing an article on microbiology without mentioning transduction):

> The unmentioned truce is over! . . . You should know that with all of the bacterial sequences, the major conclusion is that it was transduction that was the means for horizontal gene transfer as they are all full of pieces of phages. As you know cholera toxin is controlled by genes on a filamentous phage. If you give the talk, as I said, you can talk about Andromeda strains and emerging this or that as you choose but if you don't mention transduction, I will have no choice but to correct you.[65]

The tension was clear: the risk of biological escape and potential biohazard of newly engineered forms of life was a scientifically grounded possibility worth considering but it was still somehow useful to distinguish such speculations from mere sensationalism or problematic science fiction. A level of care was called for. At the same time, there were also clear and obvious parallels for considerations of biological contamination that came directly from the exobiological world of planetary protection—indeed, much of Asilomar's framing and turn to engineered biological barriers has exobiological antecedents.[66] How then to talk in a carefully speculative scientific manner about everything raised in *The Andromeda Strain* without talking about *The Andromeda Strain*?

Fears of engineered ecological relations run amok have long existed. How to put together longstanding fears of invasive species and feral ecosystems with the potential promise and merit of genetically engineered organisms (and perhaps someday artificially constructed ecosystems) that might fundamentally transform human life? One option was to find another analogy. Maxine Singer, a second of the Asilomar co-organizers, and in a discussion with the other members of the self-constituted plasmid group, made explicit the connection between such historical concerns about biological introductions across national borders and planetary protection issues with the kinds of concerns being discussed surrounding genetic engineering of cosmids and suggested turning to antibiotic resistance as another useful mode for thought:

> The worry over possibilities such as these is not new; it has been expressed through legislation to prevent the transportation of certain plant and animal species between countries and between certain states in the US, and it has been expressed in the elaborate decontamination procedures to which leaving and re-entering space vehicles have been subjected. However, there has been little more than anguished hand-wringing over the antibiotic-induced spread of resistance plasmids. Perhaps the actions recommended in these pages to minimize the potential hazards of novel recombinant microorganisms will serve to stimulate similar actions to control the existent serious problem of antibiotic induced plasmid spread.[67]

Widespread antibiotic resistance was not yet common in 1975, and yet riffing on speculative scenarios was somehow supposed to be beyond the pale for Singer. In a remarkable rhetorical twist, she characterized irresponsible riffing itself, not some particular potentially hazardous experiment utilizing recombinant DNA, as the "uncontainable" problem that might spread virally. Singer found in the consideration of history a deeper lesson from all the sensationalism about the importance of academic freedom: "The consequences of attempts to restrain the search for knowledge have been even more fearsome than the science fiction scenarios constructed by genetic fear-mongers."[68] Singer was exhausted by the effort: "My phones didn't quit. I would patiently explain molecular biology research and explain that the scary, doomsday scenarios were pure fiction and why. And still the headlines screamed. It was frustrating and impossible. It was an absolutely astonishing nightmare."[69] Singer's position highlights a further inherent tension—speculative fictions of possible futures are necessary and important in science and for public health, as some potential biohazards involved with the engineering of life at the molecular level may be real and in need of careful consideration and addressing. But other sensationalistic speculations are unwarranted for provoking unsubstantiated and irresponsible fears. But which were which? (Or rather, whose?)

Andromeda Ascendant

Time and again, it was repeated, the idea of an "Andromeda strain" was just "science fiction." And yet at the same time it was also widely reported that "genetic engineering . . . is not science fiction any more. It's here."[70] These tensions over the fictional status of the Andromeda strain and the real-world status of speculative futures for work with recombinant DNA reached into the halls of Congress by November 1977. At one Senate hearing on proposed regulation of recombinant DNA research, Oliver Smithies, a professor from the University of Wisconsin and a past president of Genetics Society of America, began his remarks by noting:

> I think we are here because of a very popular modern science fiction novel by Michael Crichton—*The Andromeda Strain.* The novel, and the successful movie made from it, was entertaining and credible. Crichton described an imaginary bacterial strain which caused an uncontrollable epidemic of serious consequences. It took little or no imagination for the popular press and the general public to extrapolate the behavior of this imaginary bacterial strain to hypothetically dangerous bacterial products of recombinant DNA research.

Another witness bemoaned the state of affairs, and concluded the general public "have great difficulty knowing what is fact and what is fiction, consequently it is difficult to come up with a reasonable consensus."[71]

Senator Harrison "Jack" Schmitt (R-NM), seemed to agree. After pointedly correcting a witness' testimony that had referred to an outbreak of plague in New Mexico, Schmitt blamed "a lack of general understanding among the electorate" for "the visions of the Andromeda strain and other things that may in this case be unrealistic." But he also held out his own interpretation of the novel for reaching a conclusion about the likelihood of *E. coli* K-12 bacteria surviving in natural environments beyond the laboratory: "I think for the sake of the record, it is important to remember the conclusion of the novel, the 'Andromeda Strain,' where the particular pathogen was destroyed by being out of its environmental niche. . . . We have been talking about the 'Andromeda Strain' without finishing the story. Even though it was fiction, I think it does make the point that should be made in this case." It took a former astronaut and senator from New Mexico to insert a literary interpretation of fiction about a pathogen from space landing in the American Southwest into a discussion about the reality of an Andromeda strain scenario and its likely outcome in the real world. *The Andromeda Strain* had infected Congress.

"The problem is not solely because of the technicalities of the issue," MIT biophysicist Alexander Rich noted in the Congressional hearing, "but rather because there is an apparent continuum in the discussions between at the one extreme the Andromeda strain, the fiction of the Andromeda strain, and at the other extreme, the recombinant DNA."[72] And it was precisely this continuum between fact and fiction that made the issue so bedeviling. President of the National Academies of Science Philip Handler had denounced such fantastical characterizations for their irresponsibility:

> The statements of a handful of scientists, few of whom were currently close to this field of research, were taken as evidence that "the scientific community is itself divided"—although no pollster ascertained just how divided we were. And the public was frightened by tales of imaginary hazards reminiscent of the *Andromeda Strain*—a book written to be entertaining, not believed.[73]

But from the very beginning, the exploration of risk scenarios in some kinds of work involving recombinant DNA—the envisioning of possible worlds that might come to exist—was an act of both science and science fiction. As Jay Clayton has noted, when scientists engage in such efforts they

> engage in the kind of extrapolation that is the hallmark of SF. Their underlying syntax is the question "What if?" They ask us to "frame and test experiences as if they were aspects of science fiction" while enjoying the trust accorded to nonfiction. They constitute a rhetorical genre of science writing, the nonfiction cousin of science fiction, while borrowing their authority from the sciences.[74]

The Andromeda Strain had infected even the US Department of Health, Education, and Welfare, with Secretary Joseph Califano noting: "The most serious furtherest [*sic*] out possibility is a kind of Andromeda strain possibility that people conjure up. We see no evidence of that. But there is always the danger of something like that, the perceived danger of something like that. That is one of the reasons why we feel there has to be such strict guidelines in the conduct of recombinant DNA research."[75] And so, despite its indisputable existence as science fiction, the exobiologically inspired *The Andromeda Strain* (and the perceived danger it represented) was widely acknowledged—even to the highest levels of government—as a core reason for promulgating earthbound guidelines for new work with recombinant DNA in the first place.

This dual-use status of science fiction as somehow inseparable from science was clearly evident to contemporaries:

> As easily as we can dream up catastrophes flowing from the malevolent future uses of recombinant DNA technology, we can imagine scenarios in which changes in our natural environment will require, for the very survival of the human species, a capacity to control genetic codes that is just now becoming possible. Extinction, or other disaster, *could* be the result of failing to develop that control. Science fiction can be written in many ways.[76]

For activists, on the other hand, even what were at one moment in time speculative prospects of concern—that "recombinant DNA techniques would come to be used in industrial production as well as in research and that scientists would design organisms to be introduced into the environment"—were often proven all too real, as predictions "once deemed science fiction by other scientists" had come to pass.[77]

Even as laboratory design became a matter of public debate and public record—by 1979, "Andromeda Strain Design Syndrome" had been defined as: "the inclination to over-complicate the design problem"[78] of biomedical research facilities in the interest of proper control—strains of Andromeda had also already begun to emerge in internal memos of Stanford's Office of Technology Licensing as early as 1976. One internal memo, "The Technology and the Threat," described its aim to use licensing to "inhibit scientists from conducting research that might result in an Andromeda Strain being unleashed upon the world . . . the fact that the university can be perceived to be profiting . . . may cause the university to be tarred with the same brush as the researcher . . . [that] develops the Andromeda Strain."[79] *The Andromeda Strain* had even infected top-rank research universities.

And ten years later—1987—little had changed on the regulatory front in Congress, with Senator Max Baucus (D-MT) noting: "Nobody wants an Andromeda strain, the spread of some new organism that could be harmful to people, animals, to crops, or to the environment," and yet several proposed bills never became law. As one university vice president for research, himself a microbiologist, testified: "The emergence of an 'andromeda strain' has not occurred and the fears of the most devoted science fiction advocates regarding a 'monster bug' have failed to materialize."[80]

But the idea of viruses from space landing on Earth and wreaking havoc had not disappeared. It was neither just a metaphor nor evanescent fiction—*The Andromeda Strain* even appears to have served as a tool for further scientific investigation. In 1974 Hiromitsu Yokoo and Tairo Oshima published a remarkable article about ΦX174,[81] asking "Is Bacteriophage ΦX174 DNA a Message from an Extraterrestrial Intelligence?" (This article would in turn be picked up by a later generation of synthetic biologists at MIT around 2004 interested in the idea of reconstructing life from scratch from a basic coded message.) Implicitly reinvigorating Hoyle's original story of life from Andromeda being encoded for reconstruction on Earth, Yokoo and Oshima's work showed the persistence of a deeply embedded imaginary.[82] Indeed, the idea of astrobiological life from Andromeda fueled overlapping and sometimes competing imaginaries for the synthetic engineering of life on Earth: the fevered dreams of critics of recombinant DNA research, elected officials, institutional bureaucrats, as well as scientific researchers. Infecting everyone it reached, the *Andromeda Strain* brought together the worlds of the astrobiological with the firmly terrestrial. Science fiction possibilities and scientific futures (not to mention legal and reputational liabilities) were never easily disentangled in the engineering of life at the molecular—or was it intergalactic?—level.[83]

Antivirals

Well into the 1980s, popular science authors writing on a variety of topics would continue to refer to "Andromeda strains" in the context of discussions about planetary protection, quarantine, and related issues. But its deep cultural valences for considerations about potential biohazards were fading from relevance in discussions about new forms of emerging industrial biotechnology.[84] The tremendous capitalization and perceived success of the biotechnology industry raises shades of Hoyle's original story in which he had warned of a future in which "greed threatens to blind humanity to the risk of a subtle form of alien invasion."[85] (Or perhaps even its absence from the biotechnological scene might be viewed as proof of the virus' integration into earthly affairs.)

Under the headline "Goodbye to Guidelines," *Nature* reported in 1982 that "the hazards were always hypothetical, involving processes not then (or since) demonstrated in the real world."[86] The Scottish novelist Naomi Mitchison, the more talented and literary sister of J. B. S. Haldane, included a deflating reference to astrovirological scenarios in her 1983 novel, *Not by Bread Alone.*[87] And by 1988, even as the industry newsletter *Bio/technology* noted that "the biotechnology industry's image is still suffering from the 'Andromeda Strain' syndrome,"[88] David Baltimore was suggesting that "speculations about the possibility of inadvertent development of a destructive organism like the fictitious Andromeda Strain" were now irrelevant: "I am personally satisfied that most of such talk is simply science fiction and that the research can be made as safe as any other research."[89] By 1994, Bernard Dixon would sing a new tune in *Bio/technology:* "I also really wonder where the evidence is for 'the harm done by the public impression that biotechnologists are about to release the "Andromeda strain."'"[90]

In a retrospective published in 1995, Berg and Singer concluded that "after twenty years of research and risk assessment, most recombinant DNA experiments are, today, unregulated. . . . The fear of 'Andromeda strains' has disappeared."[91] While throughout this period there were always others who sought to seriously treat the prospect of an "Andromeda strain" from within the realms of labor safety and of microbiology,[92] for others the prospect had been delayed to a distant future. (One forward-looking account published in *Nature* in 2000 and referring to events in the year 2441, claimed that "biomedical researchers remind us, how-ever, that not all microbes have given up their war on humans—many deadly species remain unreconstructed. The so-called Andromeda strain, for example, is still under the sway of an unstable dictator who vacillates between homicidal frenzy and paranoid isolation."[93])

How did it come to pass that, for all its force—as a ready resource drawn upon before, during, and after Asilomar—these "strains of Andromeda" in time ceased to serve a central reference when envisioning futures for the engineering of life?[94] "The technology that seemed like science fiction in 1975 is now commonplace," intoned *Science* in 2000.[95] A span of time interpreted as a demonstration of general safety was another factor. As Arthur Kornberg noted: "I want to say it as loudly and clearly and as frequently as I can, that today, in 1997, twenty-five years after these techniques became available and used in millions of experiments by the clumsiest and most inexperienced people, there hasn't been a single incident of any harm done or anything questionable regarding anybody's welfare. . . . It has clearly been less hazardous than slicing bagels."[96]

Perhaps it is that questions of inadvertent invasion or infection—which had so motivated exobiological and early recombinant DNA concerns alike—were in the fullness of time increasingly becoming substituted for questions of engineered and designed intent in ever more carefully modified organisms, as shotguns have become scissors (and as engineers turned into editors). Or perhaps it is that the metaphors have been gradually changing—by 2004, an occasional reference can be found instead to "a biological Chernobyl."[97]

And yet another reason that might be in keeping with this tale is the emergence of *Jurassic Park* (1990), an undisputed cultural touchstone for the next generation to grow up enmeshed in a world of contemporary biotechnology (and not coincidentally a novel also written by Michael Crichton, and turned into a major motion picture film). Jason Kelly, CEO of the Boston-based synthetic biology company Ginkgo Bioworks, has publicly recounted how the film inspired him to want to become a genetic engineer, and his company has even produced corporate T-shirts riffing on the logo of *Jurassic Park*. While the intended moral of *Jurassic Park* might be encapsulated in Jeff Goldblum's famous line "Life, um, finds a way," Ginkgo has pursued the idea of designing organisms for industrial and consumer purposes with a different end in view: "There will be dragons," reads the back of one company T-shirt design, pulling the engineering of biology simultaneously both backwards and forwards in time, rather than up and down in exobiological space. A different sort of emerging molecular politics is at play here, structuring ever-new possibilities for the further engineering of life at the molecular level and activating other imaginaries of *de*-extinction rather than of epidemics leading to extinction—imaginaries of invention, resurrection, entertainment, and capital, rather than of invasion or infection.[98] Though both of these unruly stories ply the line between fact and fiction, between irresponsible sensationalism and responsible ventured speculation on biofutures, *Jurassic Park* has clearly not constructed the same sort of fear for the prospects of the molecular engineering of life as *The Andromeda Strain* once did. But this is a story about *space*, which has nearly run out, and that is a story about and for other times.[99]

Afterword: On the Descent of Heavenly Strains

The finishing touches on this essay were completed in early 2020 while avoiding a highly contagious and potentially deadly infectious agent that had arrived at a city in the southwestern United States. Or in other words, this essay was completed just as the novel coronavirus SARS-CoV-2 was identified as the infectious agent responsible for a worldwide pandemic. "What could be more bone-chilling than a seemingly out-of-control virus leaping from region to region around the globe, without a known vaccine to prevent it or slow it down, causing death and economic mayhem along the way?" reported the *New York Times*: "The coronavirus narrative has the texture and feel of 'The Andromeda Strain,' the 1969 science-fiction thriller, come to life."[100] Perhaps the Andromeda strain narrative had been only in suspension, as the fictional pandemic was once again resurrected to inform our all-too-real one.

In a month when American deaths due to COVID-19 passed the 100 thousand mark, *New Yorker* film critic Anthony Lane offered for contemporary audiences a fresh perspective on Robert Wise's 1971 cinematic adaptation of *The Andromeda Strain*, which contained what Lane called "the most alarming thing I've come across, in this trade-off between the real and the imagined." In an eerie presage of our present pandemical predicament, he described a moment in the film when the experts' concerns are relayed to the White House:

> "By then, the disease could spread into a worldwide epidemic."
> "It's because of rash statements like that the President doesn't trust scientists."

"That's a little *too* close to the bone, I reckon," Lane remarked, but he felt that the film presciently captured "how harshly politics and medicine can scrape against each other, whenever peril impends"—just as science and science fiction can.[101] While hellish narratives about alien-induced pandemics repeatedly intruded into conversations on Capitol Hill in the 1970s, with many deriding such references to intergalactic fiction as mere sensationalism, these sorts of invocations—framing governance responses within the language not only of science but of science fiction—suggest that we would do well to seek to understand the unexpected and sometimes unruly cultural narratives in which the engineering of life is always situated. Because, sometimes, those unruly cultural narratives come to reflect political and medical realities emerging with warp speed. Though the referents may change, or old ones take on strikingly new resonance, framing our real-world tragedies in terms lent to us from the speculative near-future never quite goes out of style. Familiar echoes from far-off constellations will resonate, new recombinations will be cultured, and novel strains of Andromeda will doubtless emerge.

11 * *Engineering Human Nature in the Genome Age: A Long View*

NATHANIEL COMFORT

Modern biology is fundamentally an experimental and not a descriptive science. . . . Its results . . . always assume one of two forms: it is either possible to control a life phenomenon to such an extent that we can produce it at desire . . . or we succeed in finding the numerical relation between the conditions of the experiment and the biological result. . . . Biology as far as it is based on these two principles cannot retrogress, but must advance. —Jacques Loeb, 1912[1]

Wherever man has begun to know scientifically, he has found himself also, better than before, able to predict; he has gained the power to control.
—Lewellys Barker, 1914[2]

If we don't play God, who will? —James Watson, 1996[3]

The Control of Human Nature

We understand DNA better as engineers than as biologists. We can describe its structure down to the atomic level. We can surgically manipulate an organism's DNA, sometimes with predictable results. By tweaking the DNA of an embryo, we can alter its developmental program. We can recreate (tiny, simple) biological genomes and nudge them into "booting up," as the biologist Craig Venter calls it.[4] Others have built simple computers and nanobots out of DNA.[5] It's tempting to think that, as Barker implies above, if we can manipulate it we understand it. Yet for all the technical power of twenty-first-century experimental biology, our understanding of the biology of genomes is rudimentary. The more we learn about the genome, the more complex the association between genotype and phenotype becomes. Any two human genomes are about 99 percent identical, yet in that remaining one percent the amount of known genetic variation is growing rapidly. From laboratory biology to popular culture, we tend to take it for granted that the essence of what makes us human—and what makes us individuals—must somehow lie in our DNA.[6] Most of our genome isn't genes. It isn't "junk," either. What is it? How does it relate to phenotype? What is its relationship to environment and experience? We don't know. Despite explosive growth in knowledge of the genome, on human nature we have little in the way of deeper explanations.[7]

I want to examine and contextualize two twenty-first-century technologies of the genome: so-called gene editing, especially with the technology known as CRISPR; and genome-wide association studies (GWAS), in particular the statistical method of calculating "polygenic risk scores" (PRS) for complex traits. I use "complex trait" as an actor's category, to encompass anything from physiology and disease, to personality and intellect, to social behavior. Researchers in this area often call their field "social and behavioral genetics"; it has also been called "sociogenomics." In the popular imagination, these technologies allow us to grasp, and to engineer, the essence of "what it means to be human" and what makes each of us individuals.[8] In various times and places, this has meant health, constitution, temperament, consciousness, intelligence, personality, and behavior; no doubt, one could add to that list. For convenience, I will refer to these ideas as "human nature," not to reify a philosophical concept but to name a cultural construct.

In some ways, these technologies are not as revolutionary as their champions would have us believe. They show continuities across centuries: the broad appeal of "reading" one's deepest nature on the surface of the body; the urge to separate that nature from the effects and experiences of life; the desire to locate in that nature simple explanations for hugely complex suites of human social behavior; the drive to shape that nature; and the will to social control that shaping human nature implies.[9] Debates over polygenic scores and gene editing with CRISPR often seem to recapitulate earlier debates over heredity, eugenics, genetic engineering with recombinant DNA, and cloning.

But these continuities take place in new contexts, which modulate their meanings and give them distinct resonances. I'm interested here in two such discontinuities: the shift in thinking from genes to genomes—what some scholars have called "postgenomics"[10]—and the contrast between the collectivist ethos of social reform in the Progressive era and the market-based individualism of the neoliberal era of the late twentieth–early twenty-first centuries. Mindful of the extensive debate among Americanists over whether there was a Progressive era in American history, I find a fairly traditional view to map well onto the history of science: a period spanning roughly from the 1890s to the 1920s, in which the nation was reform minded, mildly collectivist, with a belief in a strong, benevolent government as a necessary check on capitalist exploitation, and scientistic.[11]

I take neoliberalism in a broad sense as an approach to political economy having its modern origins in the 1970s, which denotes the privatization of public services, weakening of government, shrinking of the welfare or social state, deregulation of industry, disempowering of labor, and erosion of the social contract, as well as the globalization of markets, with corporate power becoming increasingly supra-national, stateless. Its roots plunge deep into the twentieth century. The term was coined in 1938. Nine years later, Friedrich Hayek founded the Mont Pelerin Society, to combat collectivism, central government, and the decline of absolute moral standards—what we would today call "moral relativism."[12] After midcentury, the economist and Pelerin Milton Friedman emerged as a leading voice in what, by the 1970s, became known as the "Freshwater school" of macroeconomics. Interpreting Hayek, they pressed for deregulation of business, privatization of public institutions, services, and lands, and a Darwinian competition for wealth that, in theory, would "trickle down" and stimulate the entire economy. The Freshwater school's advice to the regime of the Chilean dictator Augusto Pinochet in the 1970s and to those of Ronald ("Government is not the solution to the problem; government is the problem") Reagan and Margaret Thatcher in the 1980s tipped the scales, bringing about a period of increasing dominance of neoliberal policies, in the US, Britain, and around the world. Neoliberalism has evolved in modern society to the point where it would be largely unrecognizable to a Hayek.[13]

Working with Sequence

Over that same span—from the 1970s through the 2010s—the technologies of the gene also transformed dramatically. Twentieth-century hereditary science was all about genes; the twenty-first is all about DNA sequence. The completion of the Human Genome Project, in the first years of the millennium, unleashed a torrent of sequence. Dramatic increases in the speed of DNA sequencing, and corresponding lower cost of sequencing, as well as the development of methods of analyzing and comparing sequence have put genomic information in the hands of medical professionals, the courts, advertisers, and the general public. *The* genome has been replaced by *your* genome. A new industry of "biobanks" has emerged, some public, like the NIH data or UK BioBank, and some private, like the databases of direct-to-consumer genome sequencing companies, such as 23andMe and Ancestry.com. The private companies charge customers for the service of providing them with a genome profile, and then sell the data in a rapidly growing marketplace; the sequence is their primary product. DNA differences are now measured not in terms of "the gene for" a trait but in variations of individual DNA bases: single-nucleotide polymorphisms, or SNPs ("Snips"). At the same time, the estimated number of genes in the human genome has dropped by about eighty percent since 2000, from 100,000 to 20,000 or so.[14] Some authors foresee the decline or even demise of the gene concept as it becomes buried in blizzards of sequence. Twenty-first-century heredity is granular, probabilistic, less concerned with finding genetic traits and more concerned with finding the genetic component of *all* traits.[15]

Even though any two human beings are about 99 percent similar genetically, out of the three-billion-nucleotide human genome, millions of DNA letters will be different between any two people. Now, rather than whole DNA "words" one can work with individual letters, A, C, T, G. Early methods of gene editing, such as "zinc fingers" and "TALENS," proved difficult to use and to generalize. Far and away, the most successful method of gene editing to date is known, somewhat misleadingly, as "CRISPR." "Clustered, regularly interspaced, short palindromic repeats" are one component of a complex natural microbial immune system against invading viruses. In 2012, Jennifer Doudna at UC Berkeley and Emmanuelle Charpentier at Umeå University engineered the several components of the natural CRISPR system into a laboratory tool for precisely cutting DNA at any site that can be specified with a twenty-base "guide" sequence, determined by the researcher. In 2014, Feng Zhang's group at MIT's Broad Institute further engineered the system for use on human cells. "CRISPR" is now the consensus shorthand for this multiply engineered, highly artificial construct, this molecular Swiss army knife. Both more precise and more accurate than previous methods of genetic engineering, CRISPR technology has become ubiquitous in biomedical laboratories. Its most revolutionary qualities may be its relative ease of use and low cost, which put it within reach of almost any laboratory and even do-it-yourself gene-hackers working out of the garage. In 2020, Doudna and her collaborator Emmanuelle Charpentier were awarded the Nobel Prize in Physiology or Medicine.[16]

The sequence inundation has brought with it new analytical methods. Recent years have seen an explosive growth of biobanks, both public and private—databases containing DNA sequence from thousands, ultimately millions of genomes. The gold standard for analyzing DNA big data is the Genome-Wide Association Study, or GWAS—essentially a correlation matrix between a trait of interest and any SNPs that co-vary with it. The polygenic risk score (PRS) is a statistical analysis performed on GWAS data that weights and sums each SNP's contribution to the genetic trait, to yield a measure of the heritability of the trait. It says nothing directly about the mechanisms that produce the trait. But it produces very robust data that can be used predictively—without knowledge of the underlying biochemistry. Polygenic scores have already proven valuable in the diagnosis of certain cancers and the technique is often cited as a cornerstone of "precision medicine." Social scientists, psychologists, educators, economists, and political theorists are also exploring PRS as a way of understanding the genetics of complex social behaviors.[17] A critical scholarly literature has begun to situate this work within the sociology of science, but being so new, it has so far received little attention from historians; hence, it has yet to be placed in long-term perspective.[18]

If Jacques Loeb was right that the control of nature is the driving force of experimental biology, then the goal of human genetic engineering is ultimately to control human nature. And controlling human nature has distinctive, often chilling valences in the neoliberal context.[19]

Making Human Nature Legible

Twenty centuries ago and more, Hippocratic physicians read one's nature, one's constitution, as a mixture of the four humors expressed idiosyncratically in the face and body. In the eighteenth century, Enlightenment thinkers sought to make this legibility scientific, reliable, repeatable. The Swiss cleric Johann Caspar Lavater (1741–1801) developed face-reading as a science.[20] He gave it a Greek-rooted name—"physiognomy"—and insisted, more than a little defensively, that it was "as capable of becoming a science as any one of the sciences, mathematics excepted." Using silhouettes, profiles typically made as busts cut out of black paper and pasted on a light background—a cheap technology that made portraiture accessible to people of little means, Lavater found clues to the inner character in the line of skull, brow, nose, mouth, chin, and neck. Easily accessible to lay audiences, physiognomy became wildly popular. Scholars have suggested that through much of the nineteenth century, any literate European or American was likely to have some general knowledge of Lavater and physiognomy."[21]

Lavater insisted that he would use his technology only for noble means. "No man has anything to fear from my inspection," he wrote, "as it is my endeavour to find good in man, nor are there any men in whom good is not to be found."[22] Like a sort of moral biofeedback, physiognomy would help one become aware of strengths, weaknesses, proclivities, and risks. The technology permitted the betterment of oneself and others, as well as the evaluation of others with whom one might socialize or conduct business. Prediction opened the way to human improvement, a biosocial engineering—albeit (we insist, over Lavater's protestations) a highly subjective, qualitative engineering. Lavater, an optimist, imagined that such judgments would be rendered beneficently, generously. Any insidious motives, he thought, surely would be few and easily swamped by mankind's overall charitable nature.

Others approached physiognomy in a less Christian spirit. In the 1850s, the French asylum director Benedict Morel noticed correlations between mental and physical characteristics in his captive population of people deemed insane. He suggested that a flawed constitution *caused* visible defects and illness. A goiter or a twisted face was not merely a sign but a symptom of constitutional or temperamental defect.[23] Moreover, defects would accumulate over the generations, leading, in his view, to a widespread degeneration of the "physical, intellectual, and moral" traits of the human species. Morel's book was widely read and influential. It spread, for example, to Italy, where the prison director Cesare Lombroso conjured physiognomic types that he argued correlated with specific criminal tendencies. Similarly, in the 1870s, the American Robert Dugdale, a regional prison inspector for upstate New York, noticed family ties among the inmates. In 1877, he published a salacious, seven-generation account of alcoholism, promiscuity, crime, and general shiftlessness in a single lineage he pseudonymed as "the Jukes." Dugdale's account was widely repeated as a "genetic morality tale," in the words of the historian Paul Lombardo.[24]

About the same time, the British director of prisons, Edmund DuCane, was looking for new and efficient means of fighting crime. "We ought to try and track it out to its source," he said in 1879, "and see if we cannot check it there instead of waiting till it has developed and then striking at it." DuCane presumed that criminality was hereditary: Family history, he said, would "contain the true clue to their criminal career."[25] He consulted Francis Galton, the statistical pioneer, traditional father of eugenics, and cousin of Charles Darwin, about the matter. Galton promptly invented a high-tech means of reading human nature from faces: composite photography. Collecting front-on photographic portraits of people who shared a behavior—such as having committed the same crime—he made faint exposures and superimposed them on a single plate, yielding a fuzzy "average" image of, say, the larcenist or murderer. The intended application of this averaged portrait, of course, would be to compare real individuals against the various criminal types identified by Galton's method. (Galton also developed fingerprinting as a criminological tool.)[26]

Today, physiognomy is an emblem of pseudoscience, but in its day it was hard science.[27] In their obituary of Dugdale in 1884, the *New York Times* wrote that his achievement was to show that "the whole question of crime and pauperism rests strictly upon a physiological basis." The popularity of physiognomy surely rested in part on its appeal to the cultural authority of science—an authority that increased dramatically through the twentieth century and into the twenty-first.[28]

Nature, Nurture, and Social Control

"Nature" and "nurture" are ancient coinages, but their explicit opposition is owed to Galton's developing thoughts about human hereditary improvement. "There is nothing in what I am about to say that shall underrate the sterling value of nurture," he wrote in 1873, preparing to slash the price of silver. "Nevertheless, I look upon race"—nature—"as far more important than nurture." In 1892, August Weismann proposed a rigid "barrier" between the germ line, which carried heredity, and the somatic line, or the ephemeral cells of the body. The nature-nurture dichotomy was an Occam's razor that severed Lamarckian inheritance of acquired characteristics from the science of heredity.[29]

For more than a century, this hard separation of nature and nurture was enormously productive for making knowledge about heredity. It undergirded the typological "biometry" developed by the Galton school, and like a wildfire it cleared away a thorny underbrush of speculation about heredity and permitted the seeds of Mendelism, dropped from Mendel's pea pods in the 1860s, to germinate in the years after 1900.[30] Mendel's theory flourished in part because of its simplicity. The English naturalist William Bateson characterized the Mendelian theory in terms of "unit characters." One gene, one trait. Mendelism was the first theory of heredity that *predicted* the outcome of a well set up breeding experiment. In 1903, Mendel's determiners were located physically, arrayed along the chromosomes like dried peas on a string. After Alfred Sturtevant drew the first gene map, in 1913, classical genetics generated experiments, in Robert Kohler's apt, punny metaphor for Thomas Hunt Morgan's fruit fly lab, like a "breeder reactor." Classical genetics took off.[31]

The Mendelian conception of the gene, simple and binary, facilitated a strict, deterministic view of human nature, which contributed to the burgeoning popularity of eugenics.[32] In 1910, the zoologist Charles Davenport opened the Eugenics Record Office (ERO) at Cold Spring Harbor, NY. It quickly became the epicenter of a burgeoning nationwide eugenics movement. American eugenics had both a nature and a nurture; part of the former was the simplicity of the binary: either you had "the gene for" a trait or you did not. In the Progressive era, eugenicist ideologues made pedigrees that seemed to show simple Mendelian inheritance for subnormal intelligence, criminality, promiscuity, alcoholism, laziness, mental disorders, disease predisposition—all were framed as single-gene, on or off traits. The ERO's Arthur Estabrook revisited and expanded Dugdale's study, supplying a Mendelian framework. Similar studies on other families with bad genes followed.[33]

Progressive-era eugenicists really had but one option for shaping the gene pool: encourage or discourage reproduction, by immigration and marriage laws, and laws providing for the institutionalization and/or sexual sterilization of those deemed "unfit." We rightly remember eugenic sterilization laws with horror and shame, but it's also important to remember that in their heyday, they were democratically enacted and, like physiognomy, grounded in contemporary science. State laws restricting certain marriages and providing for eugenic sterilization enjoyed broad political support, as did the federal Johnson Immigration Act of 1924. Eugenics was not an anomaly of the Progressive era, it was an emblem of it.[34]

"Classical" eugenics, then, was a kind of social engineering by means of socially manipulating heredity—nurture via nature via nurture. Eugenics reinforced class structure, biologized the status quo, and subordinated the interests of the individual to those of the state.[35] True, by the late 1920s, consensus among professional geneticists was that heredity was more complex than they had thought and it was too early to pursue active eugenic programs. But among physicians, public health workers, birth control reformers, and the general public, the social appeal of a simple scientific explanation for disease, birth defects, and mental deficiency grew through the 1920s and 1930s. Pseudoscience may not be recognizable until after the fact.[36]

Neoliberal Human Nature

Genetic engineering is a product of the neoliberal era as much as of the molecular era. In the 1930s, Franklin Delano Roosevelt's New Deal inaugurated four decades of predominantly Keynesian monetary policy, with broad public support for mild social programs (public schools, libraries, Social Security) and not others (health care) in a capitalist market. In the 1950s and 1960s, the rise of molecular biology redoubled and retripled the belief that the material of the genes, DNA, was the "secret of life." So tightly tied was this rush of genetic knowledge to the relative stability and civic-mindedness of the time that when the administrative state was challenged in the late 1960s, one scientific pioneer thought that molecular biology was ending, too.[37] It was just beginning.

In the 1970s the balance of political-economic power tipped toward neoliberalism, and, with some fluctuations, there it has since remained. Not coincidentally, though for complicated reasons, this tilt coincided with the rise of genetic engineering. In 1974, Stanford University and the

University of California filed for a patent on the gene-splicing technology developed by Stanley Cohen, of Stanford, and Herb Boyer, of UC San Francisco; later that year, Hayek received the Nobel Prize in Economics. By the time of the Asilomar conference on recombinant DNA in 1975, Boyer had already been approached by a venture capitalist interested in funding a for-profit company based on the technology; this became Genentech, the first company based on recombinant DNA technology.[38] The Bayh-Dole Act of 1980 permitted the results of government-funded science—until then, seen as public property—to be privatized and monetized and was an enormous stimulus to both the biotechnology and computer industries. The *Diamond v. Chakrabarty* decision of the same year, which established precedent for the patenting of novel recombinant organisms, strengthened this gesture. These landmark events in the history of biotechnology were also neoliberal efforts to rekindle a smoldering economy and indeed the health sciences and information technology led a booming economy in the 1980s and 1990s.[39]

Almost from the invention of recombinant DNA technology, researchers began to dream of using it to perform "genetic surgery" on humans. The idea was to treat genetic disease by inserting a therapeutic gene into a modified virus and then "infecting" the patient with it; the virus would do the tricky part of inserting the gene into the chromosome. Through the 1990s, gene therapy was so hyped it could sound like a comic-book secret super power. "Gene guns" would cure cancer with an injection. Corner drugstores would peddle "universal cells" that would enable us to "kiss high cholesterol, high blood pressure, even AIDS, goodbye."[40] But the early history of gene therapy was marred by gene cowboys performing unethical human trials and tragic patient deaths as a result of ethical lapses, vainglory, rushed protocols, and unforeseen consequences. In 2002, the Recombinant DNA Advisory Committee issued a moratorium on gene therapy trials. Gene therapy has recovered to an extent; it now has a real but tightly circumscribed role in therapeutics.[41]

Molecular biology promised new ways of making human nature legible for surveillance, and the logic of neoliberalism easily encompassed this biologizing trend. For example, in 1991, Frederick Goodwin, head of the Alcohol, Drug Abuse, and Mental Health Administration, launched the Violence Initiative to identify and pre-treat "violence-prone individuals"—particularly in the inner cities (read: black people) "before they have become criminalized."[42] Turning social problems into problems of genetic prediction conveniently bracketed the neoliberal policies that generated white flight into the suburbs and created the violent urban ghetto in the first place. Focusing solely on the genetic "roots" of social behavior tabled social solutions to social problems—education, a social safety net, safeguards against overtly racist hiring and housing practices, and so forth—favoring instead high-tech solutions from the for-profit private sector that located the problems in the individuals, not society.

With the Human Genome Project—often touted as a triumph of public science—the idea of scientific knowledge as a public good took another heavy blow.[43] Within months of the announcement that the HGP was complete, NIH Director Elias Zerhouni laid out a "road map" for the future, "a set of bold initiatives aimed at accelerating medical research." The Roadmap identified "compelling opportunities," including "new pathways to discovery" and "reengineering the clinical research enterprise." While the NIH sustained its mission of improving the nation's health, the benefits of research would henceforth be realized with the methods of the free market. Bristling with corporate jargon like "synergize," "stakeholders," and "public-private partnerships," Zerhouni's roadmap positioned biomedicine as a major growth field of the twenty-first century and declared the NIH open for business.[44]

Genomes quickly became manipulable, global commodities. When CRISPR burst on the scene in 2015–2016, both scientists and the public sought to use first-generation genetic engineering as a template. Sensationalist media quickly recycled the usual promises and perils: the "end of disease," "designer babies," a "new eugenics."[45] But it soon became clear that CRISPR had emerged in a new world, global and competitive. Asilomar veterans Paul Berg and David Baltimore joined Doudna and others in urging a "prudent path forward" on issues such as germline engineering—taboo since the recombinant DNA era—and organized the first International Summit on Human Gene Editing, in Washington, DC, in late 2015, but they were scooped by Chinese researchers announcing that they had edited nonviable embryos. Doudna and Baltimore's attempt, soon afterward, to convene a "new Asilomar" at a Napa Valley winery smelled more of elitism rather than scientific democracy. The US and other countries declared a moratorium on germline CRISPR. By the Second International Gene Editing Summit, in Hong Kong three years later, it was broken: Another Chinese researcher, He Jianqui, stunned the scientific community and the public by announcing (via press conference and YouTube) that he had not only used CRISPR on viable human embryos, but that a pair of twin girls had recently been born bearing an alteration intended to make them immune to HIV infection.[46] Clearly, in today's global economy, American scientists can no longer be assured of controlling a scientific narrative, let alone dictating policy.[47]

CRISPR has opened up a new Wild West of genetic engineering, in many ways more freewheeling than the first. Asilomar-era fears about an engineered germ accidentally escaping[48] now seem quaint in contrast to the efforts of the MIT geneticist Kevin Esvelt and others to *deliberately* release bugs that are not only engineered, but souped up with "gene drive," which spreads the engineered genes through wild populations at faster-than-Darwinism warp speed.[49] And we have Josiah Zayner, a self-styled "biohacker," part entrepreneur and part libertarian activist, selling do-it-yourself CRISPR kits to private citizens (Zayner cheered He Jianqui's reckless achievement).[50] The other major taboo established in the recombinant DNA era, against any gene therapy aimed at enhancement, has also fallen, partly a result of blurring boundaries between therapy and enhancement, and partly as a result of market forces. This has had the eyebrow-raising consequence that some grandstanding bioethicists and others have come to openly advocate eugenics, so long as it is based in neoliberal individualism rather than Progressive-era collectivism. In the genome age, many principles of bioethics seem to be upheld only to the extent the market will bear them.[51]

For all the romance and derring-do of the CRISPR cowboys, the more transformative biosocial engineering may come from genetic prediction and surveillance. The legibility of the genomic self reanimates and amplifies concerns about surveillance à la Lombroso, Dugdale, DuCane, and Goodwin. Although it failed, the same impulse was at work in 2012, after the tragic school shooting at Sandy Hook Elementary School. Wayne Carver, the state medical examiner for the state of Connecticut, floated the idea of analyzing the genome of murderer Adam Lanza for clues to the biology of mass shooting. In an uncanny echo of Goodwin and DuCane, Carver said, "I think it's great to consider if there's something here that would help people understand this behavior."[52]

Continuity and change. Early GWAS were criticized for making simplistic "gene for" claims based on correlation of complex behaviors with single "candidate" genes. The result was a new round of genetic determinism that mapped the concerns of Progressive-era eugenicists note for note: a "slut gene"; a "laziness gene"; a gene for "being a jerk"; a gene for violent crime; a gene that raises your IQ by six points.[53] Polygenic risk scores bring more complexity to the analysis. The principle of a polygenic score is that complex behaviors like being a jerk do not have a single "gene for." Rather, researchers now understand that complex behaviors result from webs of interactions among and between many environmental factors and tiny DNA differences. Some researchers estimate that a complex trait like intelligence could involve a thousand or more genes—roughly one in five.[54]

The difference is that modern technology (potentially; it's early yet) permits this "reading of the self" to be done prenatally, via prenatal genetic screening, or even at conception, if one uses in vitro fertilization. Naturally, these predictive powers are being commodified. In 2017, there formed a company bluntly named Genomic Prediction, which aimed to create a sort of prenatal genetic "report card" of polygenic scores. The cards would include scores for Type 1 and Type 2 diabetes, coronary artery disease, dwarfism, several cancers, height, and intellectual disability (the contemporary equivalent of "feeblemindedness"). Marry this business model to sociogenomics and you might offer predictions for years of schooling, income, sexuality, and other aspects of temperament.[55] The collectivist eugenics of the Progressive era sought to weed out the unfit in the lowest social strata; in today's individualistic, market-driven world, the control of human heredity targets those at the top, with the money and resources and education to choose the best versions of themselves. That is, if it works.

Engineering Our Futures

Genetic engineering and prediction, like cloning, eugenics, physiognomy, and other sciences of human identity, are predicated on the idea that if we can read out someone's hidden true nature, we can engineer it, master it, bend it to our purposes. Although researchers working with PRS often nod to the fact that environment is important and warn against genetic determinism, the method is designed to tease out genetic effects from other "confounding" effects. The nature of polygenic scores is still to separate nature from nurture. This fact does not make PRS geneticists evil eugenicists. From long and ongoing conversations with some of the leading workers in this field, I find that politically they lean progressive; they do not want to perpetuate racist or sexist stereotypes; they want their science to be used for good, humane ends. Understanding the genetic component of social behaviors, they say, can help us understand environments so that we can tailor them to each individual's needs.[56] Yet they too are products of their environments. In the context of the neoliberalizing trajectory of society over the past fifty years—privatizing, globalizing, market-driven, individualist, corporatist, technocratic—the engineering of human biology incentivizes focusing on individual defect rather than collective strength, favoring "precision" therapy or punishment over social amelioration, and seeking high-tech solutions to social problems instead of social ones.

I began with the familiar observation that our technological prowess in genetics and biotechnology greatly exceeds our understanding of the processes we are manipulating. That's not obviously problematic—we encounter a similar situation every time we drive a car or pull out our cell phones. It becomes troublesome because of the specific values we attach to DNA. Those values, like all values, are malleable, context dependent, historically contingent.

The current framing of DNA as the engineerable seat of human nature, then, leads to an irony. GWAS and polygenic scores represent science's best effort to date at taking into account biological complexity and the interplay of genes and environmental variables. Scientists and journalists are more careful than ever to avoid deterministic wording and disavow deterministic interpretations. And yet, for both intellectual and economic reasons, this science, as it has developed in our culture, is ushering in a new, structural genetic determinism that is in some ways stronger and more pervasive than any before. Precision medicine focuses on the genetic dimension of all disease. Millions of consumers buy their genetic identity from private companies, which pitch genome profiles not just as a measure of your nature, but as the *nature* of your nature. DNA, ubiquitous in popular culture and increasing rapidly, derives its value from the shared assumption that it is the "secret of life." The idea that DNA cracks open ineffable human nature, laying it bare for analysis, repair, and design, coexists happily with the idea of DNA as a commodity, as intellectual property, because the conceit of self-understanding sells. We have never understood the environment so well. We have never been more hereditarian.

12 * *Engineering Uplift: Black Eugenics as Black Liberation*

AYAH NURIDDIN

Engineering metaphors proliferated in the intellectual landscape of Progressive-era reform efforts. By the early twentieth century, engineering emerged as a branch of applied science and its conceptual applications extended beyond science, technology, and industry and into the social and political. Progressive-era reformers imagined that bodies and populations could also be engineered to create a society based upon scientific principles, standardization, control, and management. Progressives mobilized the cultural authority of science to apply an engineering approach to a whole host of problems that plagued American society. This engineering approach shaped eugenics, euthenics, education, the social sciences, and public health.[1] For Progressive-era eugenicists, engineering metaphors of making, intervening, controlling, maintaining, and envisioning create opportunities to imagine eugenics as a form of human engineering to be wielded in service of social and biological improvement.[2] (See Nathaniel Comfort's contribution to this volume.)

American eugenicists were varied in their commitments, politically and scientifically, but they shared an interest in the social and biological engineering of populations to improve their collective future. Eugenics could be mobilized for a whole host of purposes, whether it was the "science of human improvement through better breeding" or the "self-direction of human evolution."[3] Much of the historiography of American eugenics emphasizes its function as an intellectual project and focuses on the largely white architects of the broader American eugenics movement. Rooted in a Progressive-era ethos of improvement through science, American eugenics as a project was about using science to address social problems. Many eugenicists argued that intervening in the heredity of the unfit would save society from degeneration and race suicide.[4] These assertions were part of a longer history of concerns about the dysgenic effects of racial mixing and "mongrelization." Eugenicists were especially concerned with blackness as a dysgenic trait, which led to segregation and anti-miscegenation laws.[5] In *The Passing of the Great Race,* eugenicist Madison Grant stated that

"Negroes have demonstrated throughout recorded time that they are stationary species and that they do not possess the potentiality of progress or initiative from within."[6] The historiography of eugenics has largely focused on the ways that eugenicists like Grant weaponized eugenic measures of institutionalization and compulsory sterilization against the poor, people of color, and people with disabilities.[7] Growing scholarship on African American engagement with eugenics shows how eugenic measures disproportionately targeted marginalized populations. However, historian Gregory Dorr and others have illustrated the ways that marginalized groups reinterpreted eugenics as useful for their own liberation.[8]

For some African American physicians, scientists, and reformers in the first half of the twentieth century, eugenics offered a scientific solution to problems facing the race. I call their approach "black eugenics" and define it as a hereditarian approach to racial uplift that focused on reproductive control, public health, and social reform. Racial uplift as a broader idea developed in the late nineteenth century after Reconstruction and was tied to the emergence of a black middle class. It emphasized the importance of morality and respectability for racial improvement, and promoted education, religiosity, chastity, and temperance as essential to the struggle for racial justice.[9] African Americans applied black eugenics to various racial problems and used it to advocate for racial equality and challenge scientific racism. As part of a broader Progressive-era impulse to mobilize scientific strategies for addressing social and cultural issues, black eugenics emphasized the importance of improving the collective stock of African Americans socially and biologically to ensure the future of the race. Engineering uplift through black eugenics encompassed both social and biological uplift in order to remake and reimagine how race could function in the face of unrelenting racism.

Black eugenics and the broader American eugenics movement shared some pivotal figures, concepts, and strategies. However, black eugenics emphasized the ways that African Americans intervene upon innate and acquired racial characteristics for specific purposes. It exemplified the ways in which African Americans understood the relationship between the biological and the social, and how eugenic and euthenic interventions could be mobilized for racial improvement. Similar to John Harvey Kellogg's vision of eugenics, black eugenics offered a capacious view of racial improvement that encompassed heredity and environment. Kellogg was a physician, cereal king, director of the Battle Creek Sanitarium in Battle Creek, Michigan, and cofounder of the Race Betterment Foundation. Kellogg asserted that eugenics as "the science of race hygiene" and euthenics as "the science of personal hygiene" were both connected and necessary to the project of human improvement.[10] At its core, black eugenics was about engineering nature *and* nurture through interventions in reproductive control, public health, and social reform. Engineering metaphors therefore become useful for engaging how black eugenics developed as a multifaceted strategy for addressing the many problems that faced African Americans. Metaphors of intervention and control are useful for examining the ways that African Americans interpreted public health and reproduction in service of racial improvement. Metaphors of maintaining and envisioning helped to frame the broader goals of black eugenics as a project of liberation. Engineering metaphors provide a powerful analytic for studying how African Americans understood the ways that eugenics could operate on individual bodies and on the race as a whole.

African Americans sought to engineer the composition of the race by intervening on its heredity to imagine new possibilities and futures. Black eugenics operated on the individual and population scale, and intervened upon perceived biosocial traits of the entire race. African Americans became invested in determining which characteristics of the race could be altered, improved, or eliminated. In a eugenic context, plasticity is crucial for thinking about how heritable traits can be altered within a population.[11] Plasticity was a key feature of black eugenics because African Americans emphasized that perceived negative traits were not natural qualities of the race and therefore could be addressed through eugenic interventions. Black eugenics then was altering the traits that white eugenicists considered fixed and innate qualities of the race such as promiscuity, laziness, and susceptibility to disease. By mobilizing black eugenics to direct their own evolution, African Americans sought to engineer the composition of the race as a way to recognize and undo the damage done by slavery and thus advocate for racial improvement and equality.

To trace the project of engineering uplift, this chapter argues that black eugenics was primarily about intervening, controlling, and improving the race socially and biologically. It will examine the ways in which African Americans mobilized eugenics to make arguments about their racial composition and eugenic potential. First, it will engage how African Americans developed their own visions of racial science and sought to define the composition and characteristics of the race. Defining the race was essential to determine where and how to deploy eugenic interventions. Second, the chapter will analyze how African Americans engineered their uplift through eugenic interventions in behavior and morality, public health work, and reproductive control. Together, these cases illustrate the broad and varied ways that African Americans mobilized black eugenics in service of racial uplift. Engineering uplift through black eugenics thus functioned as a set of strategies to develop the race to its fullest eugenic potential.

Defining the Race

Racial science in the United States functioned as a part of a longer desire to understand human origins and to order the natural world. Polygenesis emerged in the nineteenth century as a prevalent theory for explaining both human origin and racial difference. Polygenist theory argued that different races had different moments of creation and origin, as opposed to monogenism, which argued that humans had a single creation moment and origin (Adam and Eve from Genesis). By the mid-nineteenth century, polygenesis had a prominent position in American racial science and was used to justify slavery.[12] African Americans became particularly invested in challenging polygenist theory as a justification for slavery as part of their engagement with racial science. For African Americans, racial science had different stakes than it did for white people: it became imperative to understand and define the origins of race in order to prove their humanity. Slavery and other forms of racial violence relied on beliefs that African Americans were not fully human, and therefore were deserving of oppression. As Mia Bay has shown in *The White Image in the Black Mind*, African Americans theorized about race to challenge white assumptions of their inferiority in order to assert their humanity.[13] In an 1854 address entitled "The Claims of the Negro Ethnologically Considered," for example, Frederick Douglass argued for the "humanity of the Negro." Douglass encouraged his audience to denounce "the scientific moonshine that would connect men with monkeys" and recognize that African Americans were in fact human beings equal with their white counterparts, not subhuman or intermediates between humans and animals.[14] Douglass' address was largely a critique of polygenesis: "The argument to-day, is to the unity of, as against that theory, which affirms the diversity of human origin."[15] He offered a scathing critique of why polygenesis was both scientifically and morally wrong. Challenging polygenesis and other forms of dehumanizing racial science was vital for dismantling justifications for slavery and extralegal violence.

Beyond responding to and refuting white racial science, African Americans crafted their own theories of race to describe themselves and to define blackness in relation to other racial groups. They theorized about the composition and characteristics of the race, and how racial characteristics were explained by both nature and nurture. This is exemplified by the work of Hightower Theodore Kealing in his essay entitled "The Characteristics of the Negro People." Kealing, professor and later president of Paul Quinn College in Dallas, theorized about the different types of traits in African American people. Kealing's essay appeared in a collection published by Booker T. Washington called *The Negro Problem* in 1903, which examined questions of race relations broadly known as the "Negro Problem," and which also proposed solutions that aligned with Washington's vision of racial uplift. Its authors included some of the most significant African American figures of the twentieth century, including W. E. B. Du Bois and Paul Laurence Dunbar. Each author was tasked with examining a particular facet of the Negro Problem, such as the role of industrial education or legal discrimination. Kealing's essay analyzed the traits of African Americans and their relationship to the broader issues of black life.

Kealing's essay argued that African Americans had what he defined as inborn and inbred characteristics. Inborn characteristics were "ineradicable; they belong to the blood; they constitute individuality; they are independent, or nearly so, of time and habit." Inbred, or removable, traits were defined as "acquired, and are the result of experience. They may be overcome by a reversal of the process which created them."[16] Both inborn and inbred characteristics were heritable, and he essentially argued for the heritability of acquired characteristics as an explanation for the existence of removable traits. Kealing's use of the terms inborn and inbred differed somewhat from their use by breeders in agriculture and animal husbandry. He likely referred to the history and legacy of slave breeding, where enslaved people were forced into sexual relations and pregnancies to bear offspring to benefit their enslavers. After slavery's end, some African Americans argued that the legacy of this abhorrent practice continued to affect the race as a whole.[17] Kealing's engagement with the heritability of innate and acquired traits, and their relationship to questions of nature and nurture, reflected broader discourses of heredity in the twentieth century. However, for Kealing and other African Americans interested in theorizing about race, understanding which traits were hereditary and which were acquired was necessary for determining how to engineer social and biological racial uplift.

Kealing framed the inborn characteristics as positive and the inbred as negative. Inborn characteristics, such as religiosity, imagination, and affectionateness, were innate qualities of the race. Inbred characteristics, such as shiftlessness, dishonesty, and promiscuity, were acquired as a result of slavery and racial discrimination. Kealing argued that although slavery had been an important civilizing force in African American life, it had also been deeply destructive: "Slavery has been called the Negro's great schoolmaster, because it took him a savage and released him civilized; took him a heathen and released him a Christian; took him an idler and released him a laborer."[18] Kealing used the term "plasticity" to think about how African Americans responded to the historical and cultural contexts that developed these negative inbred characteristics. What ultimately saved African Americans from extinction under the brutality of slavery was their "plasticity to moulding forces and [their] resiliency against crushing ones."[19] Though Kealing was not trained in the biological sciences, his arguments reflected the pervasiveness of Progressive-era science and eugenic thinking. His vision of inborn characteristics as innate and fixed while inbred characteristics were acquired and plastic illustrated the ways in which he understood the nature of race and the possibilities of racial improvement. His theorization of racial characteristics, without specific reference to major white hereditarian thinkers, exemplified how African Americans understood race as a concept and how it could be intervened upon in service of racial equality.

One approach to defining the race in service of uplift was through the study and location of racial characteristics in the physical body, exemplified by the work of Howard University physician and physical anthropologist William Montague Cobb, who spent much of his career studying anatomy to analyze and locate racial characteristics in the physical body. Cobb received his MD from Howard University in 1929 and then went on to a PhD in physical anthropology at Western Reserve University (now Case Western Reserve University) under renowned physical anthropologist T. Wingate Todd in 1932.[20] In 1926, Cobb was part of a National Research Council initiative called the Committee on the Negro that endeavored to "promote anthropological and psychological studies on the American Negro." The committee included prominent figures like Franz Boas, Charles Davenport, Earnest Hooton, Aleš Hrdlička, and Cobb's advisor Todd. Cobb's position on this committee in proximity to famous proponents of scientific racism was certainly unusual, but largely illustrative of the ways he navigated questions of race and racism throughout his career. He cited and was cited by anthropologists and scientists who made arguments about the anatomy of race, even when those authors would have believed in his racial inferiority. Cobb occupied a prominent and unique position in the field of physical anthropology, which influenced how he navigated questions of race in scholarship and practice.

From his time on the Committee, Cobb published some of its findings as an article in the *Journal of Negro Education*. He argued that the purpose of studying race was not the "elaboration of politically significant interpretations" but was instead concerned with "understanding the biological phenomena of heredity, growth and development, and environmental influence."[21] The article relied heavily upon anthropometric data from Todd, Davenport, Hrdlička, and others to analyze the physical racial characteristics of black bodies. Cobb included measurements of craniofacial features, limbs, the pelvis, and viscera to argue which characteristics were inherent to the race and which were caused by social conditions. He also posited that there were no negative consequences to racial mixing and "hybridization" despite broader concerns about the dysgenic effects of race-crossing. After cataloging a whole host of anthropometric data on a numerous physical features, Cobb concluded that "racial characters are largely variations of form which have no distinct functional survival value in modern civilization."[22] Cobb's work as an anatomist and physical anthropologist primarily focused on defining and classifying racial characteristics to develop a more rigorous scientific study of race. In a later article "The Negro as a Biological Element in the American Population," Cobb made historical and biological arguments about the composition and characteristics of the race. This piece was less focused on anthropometric data, but still discussed physical differences. Cobb noted that the vital organs were the same across racial groups, except for the spleen of the American Negro. He also added that

> the vital or lung capacity of the Negro has been frequently reported to be less than that of the white, and his feet have been stated to be typically low-arched or flat, characteristics difficultly reconcilable with his known endurance and fleetness.[23]

Concerns about the racial differences in lung capacity have a much longer history, and were often mobilized in arguments about racial susceptibility to tuberculosis in the late nineteenth and early twentieth centuries.[24] In addition to discussion of these physical differences, Cobb also emphasized other aspects of African Americans, including their strong constitution, fertility, mental ability, and "special aptitudes" like an "acute perception of pitch and rhythm."[25] Cobb reiterated his belief that racial mixing did not have dysgenic effects, and argued that the American Negro was an "intermediate between the parent African and white stocks." He ended the article with a call for greater African American inclusion in the rights of American citizenship. For Cobb, studying racial characteristics was an important way to advocate for racial equality on scientific grounds.

Locating racial characteristics in the physical body was not limited to anatomy and physical anthropology, but also included questions of disease, pathology, and environmental factors. Pathologist Julian Herman Lewis published his magnum opus *The Biology of the Negro* in 1943, in which he described the physical and medical characteristics of African Americans. In 1915, Lewis had become the first African American to be awarded a PhD in physiology from the University of Chicago, and he completed his MD at Rush Medical College in 1917, before joining the faculty of the physiology department at the University of Chicago.[26] Lewis was interested in what he called comparative racial pathology, or the study of racial differences through the manifestation of disease. He argued that comparative racial pathology should become its own field under the title "anthropathology."[27] *The Biology of the Negro* was a detailed guide to numerous racial characteristics including anthropometric measurements; racial differences in viscera, physiology, and biochemistry; and disease categories. Like Kealing, Lewis differentiated between innate and acquired racial characteristics in the expression of disease. However, Lewis essentially argued that racial characteristics could not be attributed to the nature of black bodies without first ruling out the possibility of environment as the cause of disease.[28] He was not interested in using the text to make specific arguments. Rather he hoped that it would provide a guide to a robust literature on racial characteristics. He was also explicit about excluding material that he considered "biased and so obviously propaganda."[29] His interest in racial characteristics illustrated that even though he was clearly invested in notions of biological racial difference, he also believed that anthropathology could be tainted by racism.

For Kealing, Cobb, Lewis, and other African Americans in the first half of the twentieth century, it was essential to fully understand how race operated in medicine, science, and culture. They understood that grappling with the race concept would be vital to making arguments about the humanity of African Americans. Even though they came from various disciplines and articulated different interpretations of race, the larger aim of their work was to challenge the scientific racism that contributed to rampant extralegal violence. What then became crucial to the project of racial uplift was not only understanding race as an idea, but also determining how eugenics could be mobilized to intervene upon plastic racial characteristics. As African Americans studied the race and its many characteristics, they also mobilized the tools of black eugenics to engineer its social and biological composition.

Engineering the Race

One of the fundamental goals of eugenics, particularly negative eugenics, was to alter, improve, or eliminate undesirable traits from the race. Black eugenics emphasized that through interventions that addressed disease, reproduction, and behavior, African Americans could engineer the race socially and biologically. This impulse was part of a broader interest in using science to address the problems facing African Americans and was exemplified by the work of Howard University botanist Thomas Wyatt Turner. The son of formerly enslaved people, Turner was a graduate of Howard University and Cornell University, and conducted biological research at Cold Spring Harbor under Charles Davenport in 1904. While on the faculty at Howard University, Turner taught biology and eugenics using *Applied Eugenics* by Paul Popenoe and Roswell Johnson as well as Charles Davenport's 1911 essay "Eugenics and Euthenics."[30] Turner believed that science, especially the biological sciences, had an important role to play in addressing social problems. He argued that the biological laboratory in particular was the site for improving humanity by improving human health and hygiene and fostering "race development." Among the other "ideals" of the biological laboratory, he acknowledged, were "the vigorous movements which we see on all sides for social betterment and social uplift." Turner's vision of the biological laboratory also included "trying to establish the true relationship of races and nations to each other."[31] Turner's articulation of the goals and ideals of the biological laboratory were the foundation for his eugenic thinking: if the laboratory could be mobilized for social improvement and racial development, then the race could be both socially *and* biologically uplifted. His mobilization of the biological laboratory to address the problems facing African Americans illustrates the broader thinking that undergirded the project of black eugenics. Building from this vision, African Americans used black eugenics to address questions of behavior, public health, and reproductive control.

Behavior and morality were central to twentieth-century racial uplift ideology, and as such were essential to engineering uplift through eugenics. Members of the African American middle class embraced a vision of racial uplift in which they were vanguards of the race destined to uplift their poorer counterparts. Racial uplift ideology emphasized the importance of projecting and performing respectability as part of advocating for racial equality. Racial assumptions about African Americans included ideas about behavior and morality, which were treated as racial characteristics. Kealing's essay focused almost exclusively on moral characteristics and argued that many positive characteristics were innate to the race. He also argued that negative characteristics were inbred through slavery and racism. Kealing asserted that education and hard work would "cure the evils entailed upon him by an unhappy past," and argued specifically that "industrial education with constant application, is the slogan of his rise from racial pauperism to productive manliness."[32] Even though education and hard work are not usually considered eugenic interventions, they could be deployed to undo acquired racial characteristics and thus have a eugenic effect on the race. By emphasizing the importance of chastity, temperance, education, and hygiene, the reformation of behavior became an important facet of engineering uplift through black eugenics.

Engineering racial uplift through black eugenics also relied on public health and hygiene efforts, which had long been a significant part of American public health work. These efforts had influenced other fields that also emphasized the importance of the environment in health and heredity, such as hygiene and euthenics. At its core, public health was about the improvement of health and heredity on a population level through environmental controls.[33] Therefore, public health was a eugenic project that was fundamentally about engineering health and heredity on a large scale. Black public health work emphasized improving the race by addressing environmental conditions and providing access to health care. It functioned as a way to improve the constitutions of African Americans and thus their collective heredity and eugenic fitness. Black public health was animated by concerns about high mortality from infectious disease, and whether it could lead to the degeneration or extinction of the race. For their part, African American physicians focused much of their practice on addressing high mortality and identifying its causes. John A. Kenney was the general secretary of the National Medical Association (NMA) and founding editor of its journal. The NMA was formed in 1895 as a professional organization for African American physicians, dentists, and pharmacists, who were excluded from joining the American Medical Association (AMA).[34] In his 1912 book *The Negro in Medicine*, Kenney argued for the urgency of black public health work to improve the health conditions of African Americans:

> In many places, without quibbling over such academic questions as whether the Negro is dying as rapidly as some other people, or whether there is some racial inherency productive of its high mortality, or whether it is due to environment, the race is realizing that its death-rate is high; that certain diseases are taking more than their fair toll of human life from its ranks, and that many of these diseases are preventable. With this realization, many Negroes have set to work to improve their living conditions and reduce the mortality.[35]

Kenney and many of his contemporaries believed that better public health would save African Americans from racial extinction. His concerns about extinction, or "race suicide," were part of broader conversations about the urgency of using eugenics to address public health problems.[36] Addressing potential extinction was vital regardless of whether high mortality was caused by heredity, the environment, or both. For African American physicians and other health workers, public health activism and interventions provided a way to engineer the composition of the race by addressing its collective health.

African American public health work was animated by the need to address high disease mortality, and its relationship to racial susceptibility. Questions of racial susceptibility were rooted in a longer history of racial science, and often deployed to explain the health conditions of

African Americans. African American physicians were particularly concerned about the prevalence of syphilis and tuberculosis, and the racial assumptions associated with those diseases. While white physicians and eugenicists considered susceptibility to be an innate quality of the Negro race, African American physicians countered that high disease morbidity and mortality were primarily the result of poor living conditions and a lack of access to health care. They were especially keen on dismantling racial assumptions about susceptibility to syphilis. African Americans were largely considered more susceptible because of perceived differences in the anatomy of black reproductive organs, which bolstered beliefs in the innate hypersexuality and sexual depravity of African Americans.[37] Black public health work was also rooted in challenging ideas of racial susceptibility that were part of a broader culture of scientific racism. "Social diseases" like syphilis were the primary cause of "racial decay," Kenney told the teachers of the Tuskegee Institute Summer School, in an address later published as "Eugenics and the Schoolteacher." "Eugenics would prevent any condition which means race deterioration," he said.[38] Even though Kenney understood syphilis as a problem that could be solved with eugenics, he did not argue that susceptibility was an innate racial quality. Rather, he understood syphilis as a moral and health problem that affected the less fit members of the race. Earlier, in 1910, Kenney had published a response to an article that reinforced the racial assumptions around syphilis, which challenged that African Americans had high rates of syphilis due to sexual depravity:

> Has there ever been an article more filled with wholesale condemnation of the morals of "all classes" of the race? And what is most unfair is his attempt to condemn even our refined classes along with the slum product among whom he undoubtedly has had large experience.[39]

By situating syphilis as a problem for the "slum product" of the race, Kenney asserted that syphilis was a problem of eugenic fitness and not a problem of race. He differentiated between the "refined classes" of the race as those who considered themselves uplifted, and the "slum product" as those who needed to be uplifted. He showed that, as with other races, African Americans as a race also had unfit members who were detrimental to the collective stock of the race. The prevalence of syphilis functioned as an indicator of the eugenic unfitness in the "slum product" of the race and was therefore not an inherent quality of all of its "refined classes."

As with syphilis, racial assumptions about tuberculosis were based on beliefs in the fundamental differences between African American and white bodies. African American and Native American people were considered highly susceptible to tuberculosis while white Americans and Jewish people were considered highly resistant.[40] Many African American physicians were eager to show that the prevalence of tuberculosis was not an innate quality of the race, or a marker of racial degeneration. They instead argued that environmental conditions and poor health care were the primary causes of tuberculosis morbidity and mortality. A 1921 editorial in the *Journal of the National Medical Association* on a study of tuberculosis in eleven states concluded that poor environmental conditions and health care caused high tuberculosis morbidity and mortality rather than racial susceptibility:

> We believe that it is more generally conceded than formerly, that . . . there is no special inherent tendency on the part of the Negro people to tuberculosis. We believe firmly that this higher death rate on our part is due to ignorance, superstition, unwholesome and unhygienic living conditions. We believe that careful study and investigations will support us in these conclusions.[41]

For Kenney, prevalence of tuberculosis did not equate to an innate racial susceptibility but instead was a symptom of much broader problems facing African Americans in the early twentieth century. The author also credited National Negro Health Week for the significant decline in the death rate, further illustrating that it was not rooted in a racial susceptibility to tuberculosis. African Americans were able to use public health efforts to argue that they were not eugenically unfit as a race.

Concerns about tuberculosis, syphilis, and other health problems that faced African Americans in the early twentieth century led to the development of various public health efforts, and one of the largest of these efforts was National Negro Health Week (NNHW). NNHW originated with a group of Hampton University alumni who organized themselves into the Negro Organization Society in 1910, and then collaborated with Tuskegee Institute president Booker T. Washington to create their health campaign. Washington also drew upon the scholarship of sociologist Monroe Work. Work joined Tuskegee's Department of Records and Research in 1908, where he did extensive studies of African American vital statistics that he compiled into his major publication, the *Negro Year Book*. Work documented high death rates but was able to use the data to demonstrate that almost half of the recorded deaths could have been prevented with public health interventions, undercutting arguments about racial susceptibility.[42] He exhibited parts of the *Negro Year Book* at the Second International Eugenics Congress in 1921, and even corresponded with Charles Davenport about his other scholarship in the same year.[43] NNHW launched in 1915 and continued through the 1940s in cities across the country. It was cosponsored by the United States Public Health Service (USPHS) as well as state and local health agencies. NNHW offered a whole range of programs and projects, including vaccinations, physical exams, tuberculosis and venereal disease screenings, and neighborhood cleanups. NNHW also held health exhibitions that included better baby contests, which were a staple of the eugenics movement.[44] Alongside many other black public health efforts in the first half of the twentieth century, NNHW was a strategy to improve African American health and thus uplift the race, while also refuting assumptions of the innate inferiority of African Americans.

Reproductive control was a significant component of black eugenics because reproduction was essential to the promotion of good traits, such as strength and good health, and elimination of undesirable ones, such as weakness and feeblemindedness. For African Americans, targeting their reproduction toward racial uplift was crucial for engineering the composition of the race and ensuring its future. Birth control inflected with eugenics became embedded in the ways that African Americans imagined the race and the ways in which it could be improved. In addition to uplifting the race through education and morality, Mitchell shows how questions of heredity, sexuality, and reproductive control shaped the ways that African Americans envisioned uplifting the race.[45] These ideas made their way into African American health practice, social science, and activism, as well as various forms of literature, print media, and art.[46] The fundamental difference in the ways that African Americans understood the possibilities of birth control rested upon what was at stake. African Americans saw birth control as useful for addressing public health and social inequality, but there was additional significance for African American women who saw birth control as a way to claim reproductive autonomy and as a vindication of black womanhood. As part of this vision of birth control, African Americans embraced the birth control movement and its advocacy to articulate strategies of black eugenics.

African Americans published some of their perspectives on birth control in the *Birth Control Review*, which was the primary periodical for Margaret Sanger's American Birth Control League (ABCL). It was an important tool for the dissemination of information about birth control, as well as the activist work of the larger movement. In June 1932, the *Birth Control Review* dedicated its second special issue to African Americans and birth control. Entitled "A Negro Number," the issue featured prominent African American leaders including W. E. B. Du Bois, Charles S. Johnson of Fisk University, and Midian O. Bousfield of the Negro Health Division of the Rosenwald Fund (the philanthropic arm of the Sears-Roebuck Company).

The articles emphasized the importance of "quality" reproduction so that African Americans could improve their collective stock. Ideas of quality reproduction over sheer quantity were central to how African Americans invested in eugenics imagined the possibilities of birth control. They also centered the ways in which birth control offered solutions to public health problems and socioeconomic disparities. W. E. B. Du Bois contributed an article entitled "Black Folks and Birth Control" in which he argued that poor and eugenically unfit African Americans were having too many children and the uplifted classes were having too few. These socially and biologically uplifted classes, what Du Bois envisioned as the "Talented Tenth" of African Americans, were not having enough children to improve the collective stock of the race.[47] He asserted that the least fit members of the race "breed carelessly and disastrously," and were thus damaging the collective stock. He concluded the article by stating, "They must learn that among human races and groups, as among vegetables, quality and not mere quantity really counts."[48] George Schuyler of the *Pittsburgh Courier* raised similar points in his article, "Quantity or Quality." African Americans needed to embrace "scientific birth control" for racial improvement:

> The question for Negroes is this. Shall they go in for quantity or quality in children? Shall they bring children into the world to enrich the undertakers, the physicians and furnish work for social workers and jailers, or shall they produce children who are going to be an asset to the group and to American society. Most Negroes, especially the women, would go in for quality production if they only knew how.[49]

Engineering the composition of the race through quality reproduction was another central feature of black eugenics. Du Bois and Schuyler contributed just two of several articles to "A Negro Number." These articles not only emphasized the significance of birth control for targeting African American reproduction for racial improvement, they also included discussions of the ways in which birth control was necessary to address public health concerns and socioeconomic depravation. Together, the articles in the "Negro Number" exemplify the ways in which African Americans understood the relationship between reproductive control and racial uplift in the twentieth century.

The *Journal of the National Medical Association* became a mouthpiece for one of the most notorious tools of the broader eugenics movement. Compulsory sterilization as a form of reproductive control was argued to be essential for biosocial racial uplift. In an address delivered before the Cape May County Medical Society in 1922, Philadelphia physician

Alfred Gordon asserted that eugenics was necessary for racial improvement specifically when it concerned matters of mental health. In a "general outline of eugenic endeavor," he focused on the prevention of mental illness as part of a eugenic project for racial uplift:

> The problem of race betterment embraces the two fundamental elements of eugenics, namely the knowledge of the laws of heredity and sterilization of the mentally unfit. The beneficent results of education in that direction are too obvious to dwell upon.[50]

For Gordon, eugenics offered a solution to the prevalence of mental illness and its impact on the population. He argued that a eugenic vision of prevention could not only address a variety of mental disorders, but also have a broader effect through the biological improvement of the race. For Gordon, mental disorders were real entities, medically, but not fixed characteristics of the race. He saw sterilization as a way to eliminate traits that afflicted members of the race but were not innate to the race as a whole. Sterilization of the unfit then functioned as a tool for engineering the race.

Conversations about compulsory sterilization also made their way into the black press, where African Americans of different social strata understood and mobilized eugenics for racial uplift. Thelma Berlack Boozer wrote for two major African American newspapers, the *Pittsburgh Courier* and the *New York Amsterdam News*, during the 1920s and 1930s under the pen name "T.E.B." In 1934, she authored a piece entitled "Birth Control Gains Sanction" in a column called "The Feminist Viewpoint." She argued in favor of both birth control and compulsory sterilization for addressing social problems, and even explained surgical sterilization procedures. She argued that "we need compulsory sterilization of the mentally and physically unfit. Too, we should welcome birth control as a national measure."[51] She concluded the article with a call to action: "More well-born babies, fewer ill-born babies, and sterilization of those unfit to become parents will aid society in solving some of its major problems. And it is about time that our lawmakers realize this and act accordingly."

Educator, activist, and writer Rebecca Stiles Taylor made similar arguments about the importance of sterilization for African Americans. Taylor was a regular contributor to the *Chicago Defender* and led the Savannah chapter of the National Association of Colored Women's Clubs (NACWC). In a 1937 piece entitled "As A Woman Thinks," Taylor explained the nature of heredity and argued for the use of compulsory sterilization to alleviate the social and health problems that faced African Americans. Taylor argued that African Americans needed to use the knowledge of heredity to prevent "racial decay" and uplift the race:

> In our mad race for education (book learning) and our pathetic but insistent struggle for economic stability, we have somewhat overlooked the importance of improving our group in its natural, moral, mental, and physical qualities—its inborn endowment of body and mind, for to do this we must plan first for the increasing of our WELL-BORN and second for the decreasing of our ILL-BORN.[52]

Taylor's vision for engineering uplift emphasized developing the best qualities of the race and limiting the propagation of the unfit. She argued that African Americans should work toward improving their inborn characteristics, similar to Kealing's arguments about the race and the ways in which he understood the plasticity of its characteristics. Taylor also described who counted as well-born or ill-born and argued that eugenic marriage was the best tool for cultivating the well-born among African Americans. In her view, compulsory sterilization of African Americans was necessary, so long as the black community had not fully embraced eugenic thinking. Once they did, the need for compulsion would evaporate.

> With such an education and such an elevation of their [African Americans'] tastes there will be little need for sterilization for they will seek their mental, moral, social, and physical equal or superior and their children will of a consequence, all things being equal, not swell the ranks of defectives.[53]

Taylor was hopeful that African Americans would internalize the value of eugenics for uplifting the race. If African Americans made eugenic choices about their marriage and reproduction, they would not need external and coercive measures like compulsory sterilization to engineer the race. Instead, African Americans of different social strata would socially and biologically uplift themselves by embracing the possibilities of eugenics in their lives.

Together, Taylor and Boozer exemplified the gendered politics of racial uplift and black eugenics. Reproductive control not only served as a tangible application of black eugenics, but also had additional importance for African American women. Birth control became a way to advocate for reproductive autonomy, which was especially significant when African American women were subjected to racialized sexual violence during and after slavery. Violence against African American women was rooted in racial assumptions about their sexualities.[54] In order to dismantle the justifications for this violence, African American women embraced and mobilized respectable womanhood. They asserted this vision of womanhood through the performance of middle-class values and through reproductive control, which in the early twentieth century was a radical articulation of black politics.[55] In this way, reproductive control as part of black eugenics not only provided visions and strategies for uplifting the race, but also served as a vindication of black womanhood. Engineering uplift through black eugenics thus had broad and tremendous possibilities for improving African American life.

Conclusion

Engineering metaphors illustrate the ways that African Americans envisioned the social and biological possibilities of racial improvement. These metaphors provide the intellectual framework for the application of black eugenics to the social and biological problems that faced the race. For African American figures like Kealing, Cobb, Du Bois, Boozer, Taylor, and others, black eugenics was an essential part of a larger project of black liberation. Black eugenics was fundamentally about intervening upon and controlling innate and acquired racial characteristics to repair the damage done by structural racism and thus envision new futures for the race. This would effectively dismantle all of the social and biological justifications for racism. It ultimately functioned as a way to argue for the humanity of African Americans, which was central to advocating for racial equality and challenging scientific racism. Engineering uplift was thus eugenic and euthenic, and relied on the relationship between the two as the basis for racial improvement. As a concept, black eugenics creates space to rethink the ways in which eugenics was understood and mobilized as a social and biological strategy for marginalized peoples to advocate for equality. Engineering uplift through black eugenics therefore shows how African Americans envisioned the ways they could use science to imagine a world without racism.

13 * *Terraforming Planets, Geoengineering Earth*

JAMES RODGER FLEMING

Can humanity survive on Earth into the indefinite future without taking control of the climate system and the biosphere? In the distant future, will we learn to engineer the sun and other planets? Given the physiological limits of the human body, what protective equipment will we need? Or will robotic landers or human cyborgs able to withstand harsh environments do the work? Such is the stuff of science fiction and science fantasy. So-called geoengineers argue that, to stay within planetary limits, it is imperative to learn how to manipulate life and the Earth's environment at grandiose scales. Do we know how to do that? Can we do that with short-sighted, often temporary and gear-headed technological fixes? Is geoengineering even engineering? Or is it merely "geoscientific speculation"? The professional practice of engineers involves designing, building, operating, and maintaining reliable technological infrastructures; it requires precision more than speculation. Engineers may dream of intervening at grandiose scales, but they lack both the technical capability and the wisdom to terraform other planets and geoengineer the Earth. There are too many people with diverse backgrounds and interests, and too many ecosystems in the way. This essay links the long and checkered history of planetary manipulation fantasies with actual attempts to terraform planets and geoengineer Earth. Historical perspective shines unflattering light on these fantasies, yet they persist and multiply, Hydra-like.

Geoengineering

The race to save the planet is very real, and the urgency to do so is growing rapidly, fueled by a combination of climate angst and the wildly unrealistic proposals of the climate engineers.[1] The bundled term "geoengineering" comprises two very different aspects: carbon dioxide removal characterized by increasing biological sinks or using chemical processes to bind it with minerals, and solar radiation modification proposals involving the intentional manipulation of the Earth's shortwave radiative budget. Shoot sulfates or reflective nanoparticles into the upper atmosphere; turn the blue sky milky white. Make the clouds thicker and brighter, or eliminate some altogether. Fertilize the oceans with iron to stimulate massive algae blooms; turn the blue seas soupy green. Suck carbon dioxide out of the air with hundreds of thousands of giant, artificial trees and store it underground or in the oceans for millennia. While these proposals sound edgy and exciting, they are not engineering; they represent patchwork attempts to use engineering or technology to solve problems, even societal problems, in an extraordinarily large and complex system. They are clearly unreasonable approaches to climate deficiencies that do not fall within standard definitions of climate mitigation and adaptation. Their proponents have been called "barking mad."[2] Motivated by the quest for ultimate control of nature, advocates of geoengineering are, with very few exceptions, WEIRD: Western, Educated, Industrialized, Rich, and Democratic individuals with superman complexes.[3] These are two bold claims, and I make them intentionally to be provocative.

The editors of the venerable *Oxford English Dictionary* recently proposed to define geoengineering (*n.*) as "the modification of the global environment or the climate in order to counter or ameliorate climate change." To assign a specific goal to the activity, however, does not make sense, since, first of all, the discipline does not yet exist. Second, an engineering practice defined by its scale (geo) need not be constrained by the good that might result from it, such as the amelioration of climate change. The *Urban Dictionary* definition drops the statement of purpose and simply defines geoengineering as "the intentional large-scale manipulation of the global environment; planetary tinkering; a subset of terraforming or planetary engineering." Of course, any manipulation techniques deployed on such a grandiose scale, like any engineering practice, could be used for both good and ill—or they may result in huge unintended consequences—a planetary Oops! What might go wrong with such speculative macro-engineering schemes? Are they precise enough even to be called "engineering?" The US National Academies did not think so, and in 2015 renamed them "interventions."[4]

The study of carbon dioxide and climate has historical roots in the work of Swedish scientists Svante Arrhenius (1859–1927) and Nils Ekholm (1848–1923). In 1896 Arrhenius published a landmark paper that examined the effect of different levels of atmospheric carbon dioxide concentration on the temperature of the planet.[5] His energy budget model estimated the change of temperature that would follow if the quantity of carbon dioxide in the atmosphere was two-thirds, double, or even triple its present value. He reported that a doubling of carbon dioxide would raise global temperatures by about 3 to 3.5°C while a reduction of carbon dioxide by one-third would lower temperatures by roughly the same amount. His was in essence a geological model, and the onset of ice ages and interglacials was his primary concern.

In 1901 Ekholm suggested that human activity might someday play a major role in controlling Earth's temperature. He pointed out that over the course of a millennium the accumulation in the atmosphere of carbonic acid from the burning of pit coal would "undoubtedly cause a very obvious rise of the mean temperature of the Earth." He also thought this effect could be accelerated by the "digging of deep fountains pouring out carbonic acid" or perhaps decreased "by protecting the weathering layers of silicates from the influence of the air and by ruling the growth of plants." By such means, Ekholm pointed to the grand possibility that it might someday be possible "efficaciously to regulate the future climate of the Earth and consequently prevent the arrival of a new Ice Age." In this scenario, climate warming by enhanced coal burning would be pitted against the natural changes in the Earth's orbital elements or the secular cooling of the sun. Ekholm concluded with a grand rhetorical flourish, "It is too early to judge of how far Man might be capable of thus regulating the future climate. But already the view of such a possibility seems to me so grand that I cannot help thinking that it will afford Mankind hitherto unforeseen means of evolution."[6]

A half-century later, at a time when many scientists were beginning to express concern about the enhanced greenhouse effect, the Caltech geochemist and futurist Harrison Brown imagined feeding a hungry world by *increasing* the carbon dioxide concentration of the atmosphere to stimulate plant growth:

> We have seen that plants grow more rapidly in an atmosphere that is rich in carbon dioxide. . . . If, in some manner, the carbon-dioxide content of the atmosphere could be increased threefold, world food production might be doubled. One can visualize, on a world scale, huge carbon-dioxide generators pouring the gas into the atmosphere. . . . In order to double the amount in the atmosphere, at least 500 billion tons of coal would have to be burned—an amount six times greater than that which has been consumed during all of human history. In the absence of coal . . . the carbon dioxide could be produced by heating limestone.[7]

The fantasy persists. Columbia University scientist Wallace Broecker and science writer Robert Kunzig end their book, *Fixing Climate* (2008), with a vision of future climate stabilized by machines that both capture and release carbon dioxide.

> Our children and grandchildren, having stabilized the CO_2 level at 500 or 600 ppm, may decide, consulting their history books, that it was more agreeable at 280 ppm. No doubt our more distant descendants will choose if they can avert the next ice age; perhaps, seeing an abrupt climate change on the horizon, they will prevent it by adjusting the carbon dioxide level in the greenhouse. By then they will no longer be burning fossil fuels, so they would have to deploy some kind of carbon dioxide generator to operate in tandem with the carbon dioxide scrubbers.[8]

Until recently, carbon dioxide removal methods were considered aspects of geoengineering. Now, there is an attempt to classify them as a special type of mitigation. Such proposals constitute technological imaginaries and represent a new category of discussion that deflects attention away from the failures of more reasonable things we need to do now such as adaptation, fuel efficiency, and fuel switching, along with adopting widely shared social, cultural, and environmental goals.

Militarized geoengineering experiments actually occurred. In the late 1950s and early 1960s, the United States and the Soviet Union conducted nuclear weapons tests in the upper atmosphere in pursuit of military advantage.[9] On May 1, 1958, at the National Academy of Sciences, University of Iowa physicist James A. Van Allen announced that Geiger counters aboard the JPL Explorer 1 and Explorer 3 satellites had picked up high readings at certain points in their orbits, indicating that powerful radiation belts (later known as the Van Allen belts) surround Earth. This was the first major scientific discovery of the space age. Ironically, and on that very same day, Van Allen joined Operation Argus—the US military's top-secret project to detonate atomic bombs in space, with the goal of generating an artificial radiation belt and disrupting the ionosphere. This was planetary-scale engineering—or "geoengineering."

"Space is radioactive," noted Van Allen's colleague Ernie Ray. The military wanted to make space *even more* radioactive by nuclear detonations that, in time of war, could disrupt enemy radio communications from half a world away and damage or destroy enemy satellites and intercontinental ballistic missiles. In late August and early September 1958 a specially equipped naval convoy launched and detonated three 1.7-kiloton atomic bombs in near space above the South Atlantic Ocean to "seed" the ionosphere with high-energy nuclear particles and radioactive debris. Van Allen's Explorer 4 satellite, launched a month earlier, carried lead-shielded Geiger counters designed to withstand the blasts and document the tests. The Soviet Union went on to detonate four small space-bombs in 1961, and then three larger ones in 1962, during the height of the Cuban Missile Crisis. One of the tests, conducted over Kazakhstan and Kyrgyzstan, started a fire that burned down a power plant and destroyed electrical and telephone lines.

The largest and highest US test was the 1.4-megaton Starfish Prime H-bomb detonation in July 1962 at an altitude of 400 km over Johnston Island, which disrupted the natural Van Allen belts, destroyed several communication satellites, and damaged electrical equipment in Hawaii, almost 1,500 km away. This led British radio astronomer Bernard Lovell, along with the International Astronomical Union, to protest, "No government has the right to change the environment in any significant way without prior international study and agreement." Van Allen, who had eagerly participated in the tests, was thus one of the world's first geoengineers. He later regretted his involvement since the blast destabilized Earth's magnetosphere for at least a decade.

It should be clear by now that the concept of geoengineering the climate is not new. Tables 13.1 and 13.2 list proposed and attempted weather and climate control projects over the past two centuries.

There are many more entries possible in these tables, which include scientific fantasies, but not science fiction. For example, in 1971, when a cinematic cloud bearing the Andromeda Strain heads out over the Pacific Ocean, scientists seed it with silver iodide to make it rain out into the ocean, ending the crisis.[10] Would that COVID-19 was susceptible to such techniques.

Terraformation

Engineers have long dreamed of manipulating the sun and the stars and terraforming distant planets.[11] In his book *Terraforming: Engineering Planetary Environments* (1995) Martyn J. Fogg reviewed the history and some of the technical aspects of "orchestrated planetary change." He defined "planetary engineering" as the application of technology for the purpose of influencing the global properties of a planet and "terraforming" as the process of enhancing the capacity of a planetary environment to support life. The ultimate in terraforming would be to create an uncontained planetary biosphere emulating all the functions of the biosphere of the Earth—one that would be fully habitable for human beings. "Astroengineering," or modifying the properties of the sun or a star, by intervening in its nuclear reactions is admittedly hyper-speculative now, but who can say in the future?

Fogg described how ecological engineering techniques might be used someday to implant life on other planets and how geoengineering might be used to ameliorate (or perhaps exacerbate) the currently "corrosive process" of global change on the Earth. He presented order-of-magnitude

Table 13.1. Proposed Weather and Climate Control

1841	J. P. Espy proposes lighting giant fires to make rain
1877	N. Shaler suggests re-routing the Pacific's warm Kuroshio Current through the Bering Strait to melt the Arctic ice
1901	N. Ekholm proposes burning coal seams to prevent the return of an ice age
1912	C. L. Riker and W. M. Calder suggest a 200-mile jetty in the Atlantic Ocean to divert the Gulf Stream and improve the climate
1929	H. Oberth proposes giant mirrors in space to warm the Earth
1945	J. Huxley suggests nuclear weapons could dissolve polar ice cap
1945	V. Zworykin proposes perfect prediction/ control with digital computer
1950s	Soviets "declare war" on permafrost and seek an ice-free Arctic Ocean
1954	H. Brown envisions CO_2 generators and scrubbers to regulate climate
1955	I. Langmuir proposes Pacific Basin cloud seeding
1957	P. M. Borisov suggests damming the Bering Strait would melt the Arctic sea ice
1992	US National Academy suggests shooting sulfates into the stratosphere
2006	P. Crutzen's "Modest Proposal" to inject sulfur into the stratosphere to combat global warming
Since 2006	A plethora of proposals from geoengineers including fertilizing the oceans, capturing and sequestering all carbon dioxide emissions, genetically modifying crops, painting roofs white, making clouds brighter, suppressing cirrus cloud formation, putting reflective nanoparticles in the stratosphere, launching space mirrors, and surrounding Earth with dust from a pulverized asteroid

Note. Table 13.1 is adapted from Government Accountability Office, "Climate Engineering: Technical Status, Future Directions, and Potential Responses," GAO-11-71 (2011), https://www.gao.gov/products/GAO-11-71, pp. 6–7. Details of all of these geoengineering proposals can be found in Fleming, *Fixing the Sky*.

Table 13.2. Attempted Weather and Climate Control

1891	R. Dyrenforth claims rainmaking by concussion during Texas drought
1890s	Hail shooting widely practiced in Europe
1920s	Experiments with electrified sand for fog clearing and rainmaking
1944	Fog Investigation and Dispersal Operation clears fog at military airfields in Great Britain
1947	Project Cirrus attempts diversion of Atlantic hurricane
1950s	Widespread commercial cloud seeding
1958	Project Argus, three atomic bombs detonated in magnetosphere
1962	Project Stormfury attempts to modify hurricanes with cloud seeding
1962	Starfish Prime, H-Bomb detonated in magnetosphere
1967	Cloud seeding over Vietnam leads to UN Environmental Modification Convention in 1978

Note. Table 13.2 is adapted from Government Accountability Office, "Climate Engineering: Technical Status, Future Directions, and Potential Responses," GAO-11-71 (2011), https://www.gao.gov/products/GAO-11–71, pp. 6–7. Details of all of these geoengineering proposals can be found in Fleming, *Fixing the Sky*.

calculations and the results of some simple computer modeling to assess the plausibility of various planetary-engineering scenarios. He deemed it "rash to proclaim" impossible any scheme that does not "obviously violate the laws of physics."[12] Yet Fogg focused only on possibilities, not on unintended consequences, and left unaddressed questions of whether the schemes are desirable, or even ethical. According to Fogg, fixing an uninhabited planet is easier than fixing the Earth, because people, their politics, and their infrastructures get in the way.

Would it be better to engineer planets for humans or to engineer humans and perhaps cyborgs to withstand harsh environments? In 1929 the physicist and science popularizer J. D. Bernal predicted that humanity's future transcended earthly limits and that humans, with suitable cybernetic enhancements, were destined to colonize space. In his visionary parable, "The World, the Flesh, and the Devil," Bernal imagined the colonization of space in hollow engineered spheres some ten miles in diameter that harness all available sunlight and starlight for their propulsion and maintenance while providing a habitable and completely sustainable artificial environment for their inhabitants. Such a biosphere, Bernal imagined, "takes the place of the whole earth and not of any part of it," with everything recycled and nothing permanently wasted. It was our destiny to harness the stars, that "cannot be allowed to continue in their old way, but will be turned into efficient heat engines."[13]

The flip side of transforming planetary environments is transforming human nature. Bernal envisioned humans being fundamentally altered in the future: "Sooner or later the useless parts of the body must be given more modern functions or dispensed with altogether, and in their place we must incorporate in the effective body the mechanisms of the new functions. Surgery and biochemistry are sciences still too young to predict exactly how this will happen." He wrote about wiring human brains to mechanical devices and human brains to other brains, concluding, "Normal man is an evolutionary dead end; mechanical man, apparently a break in organic evolution, is actually more in the true tradition of a further evolution." In the end, humanity might become "completely etherealized," and attain cosmic consciousness.

Bernal envisioned an ultimate bifurcation of humanity into mechanizers and humanizers. The mechanizers, serving as ultimate eugenicists, will shape the future—colonizing space, transforming the body, and ultimately controlling the universe and unevolved humans. A large and docile population will choose to remain on Earth, seeking happiness and prosperity, "enjoying their bodies, exercising the arts, patronizing the religions," and leaving the machine behind. "From one point of view, the scientist cyborgs would emerge as a new species and leave humanity behind [both figuratively and literally]. . . . The world might, in fact, be transformed into a human zoo, a zoo so intelligently managed that its inhabitants are not aware that they are there merely for the purposes of observation and experiment."[14] In Bernal's scheme, this branch of humanity is ruled by the ten percent or so of the population that are engineers and scientists. For Bernal, the ruling mechanizers, "groping unsteadily beyond," are the heroic pioneers of the post-human condition, "on their way to the stars."

In 2011 the US Defense Advanced Research Projects Agency and NASA Ames Research Center convened an unusual meeting called the "100 Year Starship Symposium" aimed at fostering the goal of achieving interstellar flight and colonization in the next century. The conference brought together science fiction writers, celebrity astronauts, defense intellectuals, and a host of Rube Goldberg–like inventors. Most papers were aimed at designing energy propulsion systems, some that violated the laws of thermodynamics, and spacecraft that could approach or exceed the speed of light; others addressed the challenges of constructing a perfectly sustainable artificial biosphere. A key motivation behind leaving this planet lies in insurance against a catastrophic failure that might render life on Earth unsustainable, for example from environmental collapse, global pandemic, misuse of military technology, or a civilization-ending asteroid impact. Behind the distant prospects for ultimate success of a starship was the predictable mantra of spin-offs from space research and exploration improving daily life through innovations, inventions, and advances.

One of the attendees was Kim Stanley Robinson, whose little-known novel *Icehenge* addresses such issues, concluding that designing a self-contained life-support system for a starship or terraforming Mars were both near impossible undertakings.[15] The five-hundred-year project of the novel's heroine, Emma Weil, is the terraforming of Mars, while starship captain Eric Swann aims to colonize a planet in another system. "What's the big difference?" asked Swann. "About ten or twenty light years," replied Weil (22). That is, both projects were equally speculative.

One of the biggest challenges facing the starship was generating fresh air, fresh water, and food for the crew while recycling wastes with near 100 percent efficiency. The starship is a traveling biosphere, and engineers have to balance the photosynthetic coefficient for algae and the respiratory coefficient for the humans and animals to prevent too much build-up in either carbon dioxide or oxygen: "Light feeds algae. Algae feed plants and fish. Plants feed animals and humans and create oxygen and water. Animals feed humans, and humans and animals create wastes, which sustain microorganisms that mineralize the wastes (to an extent), making it possible to plow them back into the soil" (29). Eighty percent efficiency in this system was good enough for a three-year voyage; 99 percent perhaps for 100 years, but perfect closure in any system, even a planet, is not technically possible. Major problems include mineral deficiencies, the incomplete recycling of wastes, and minute losses of water that would coat the interior of the ship and pool in cracks and crevices. The starship would have to recharge its systems somehow. Even some 600 fictional years from today, Mars remains a hostile environment for humans, and author Robinson envisions humans safely venturing only as far as Pluto, while the fate of the starship, which had left the solar system, remains unknown.

The pace is much quicker, unrealistically so, in Robinson's later more popular writings, *Red Mars*, *Green Mars*, and *Blue Mars*, where it takes less than two generations, beginning in 2026, for the granddaughter of the first Martian colonist to depart from a fully terraformed solar system in an interstellar vessel headed to another star system twenty light-years away. Such is science fiction, but what about its more proximate cousin science fantasy? Fantasy often informs reality (and vice versa). NASA managers know this well, as do Trekkies. The best science fiction authors typically build from the current state of a field to construct futuristic scenarios that reveal and explore the human condition. Scientists as well often venture into flights of fancy. Although not widely documented, the fantasy-reality axis is a prominent aspect of the history of science and technology. The chief distinction is that the fiction writers provide a moral core and compass.

Science fiction and science fantasy meet in such classic works as Olaf Stapledon's *Last and First Men* (1930), a two-billion year "history" of the future in which the human species and its many successors escape the dying Earth and colonize other worlds, until the remnants of humanity are extinguished when the sun becomes a supernova. Near the end the last men, living on Neptune, design an artificial human dust, "capable of being carried forward on the sun's radiation, hardy enough to endure the conditions of a trans-galactic voyage of many millions of years, and yet intricate enough to bear the potentiality of life and of spiritual development." This, for Stapledon, is humanity's final legacy. In a more prosaic vein, Robert Heinlein's *Farmer in the Sky* (1950) concerns the terraformation of Jupiter's moon Ganymede by frontier homesteaders who depart an Earth that is overcrowded and near ecological exhaustion. The colonizing farmers face a super-harsh environment of thin air and biting cold. Not only do the hardscrabble space pioneers have to nurture their crops in such conditions, they have to create their own soil from crushed rock. *The Greening of Mars* (1984), by James Lovelock and Michael Allaby, brings contemporary environmental and social issues into story telling about planetary transformation.[16] Writing before the Montreal Protocol was enacted or the Cold War ended, the authors anticipated using banned chlorofluorocarbon gases to warm the Martian climate, transporting them there with surplus US and Soviet missiles, and paying for the whole operation with funds from the "peace dividend." The colony was populated by "homeless" people who had sold all their Earthly assets in exchange for Martian real estate futures, valuable only in proportion to the progress of terraformation. In other such accounts, it is only carefully selected elites who are allowed to make the trip.

Two ersatz but unsuccessful starship missions have already been launched on Planet Earth, but not by NASA or any other space-faring nation. In the early 1990s, a crew of eight attempted a shakedown cruise of some 16 months in Biosphere 2, a closed-system experimental facility in the Arizona desert, but the life-support systems failed miserably and the mission had to be aborted prematurely.[17] Oxygen levels in the craft, which began at a robust 21 percent, systematically decreased to about 14 percent, causing members of the crew to suffer from high-altitude sickness, sleep apnea, and extreme fatigue. Other life-support systems also went erratic, perhaps due to the cement materials used in the building's foundation. Carbon dioxide levels fluctuated on a daily basis by as much as 600 parts per million (ppm), with much greater seasonal variation. Wintertime levels soared as high as 4,500 ppm, or close to a lethal concentration.[18] Although the facility was huge, enclosing 3.5 acres, it was not huge enough, and fluctuating plant photosynthesis alternating with system respiration threatened to overwhelm the carbon dioxide scrubbers. With the human crew suffering mightily, most of the mammals and birds brought on board died, and with insect pests such as ants and cockroaches flourishing, the mission came to a screeching halt. The airlocks opened, not to the harsh vacuum of outer space, but to the friendly biosphere 1 of Planet Earth.

After a thorough cleaning and tuning of the life-support systems, a second mission was launched a year later with a crew of seven, but it too crashed, this time within six months, due to a severe management dispute, a mutiny that involved monkey-wrenching the craft, and the early departure of two crew members. The closed-system research days of Biosphere 2 were over. For several years the facility was the site of a planned residential development with tours being offered to the public. It is currently managed by the University of Arizona, which uses its the soaring glass vivarium for experiments on dryland grass species. The lessons of the two missions launched in the 1990s indicate that we need to learn how to run a small artificial biosphere successfully before we can ever hope to travel to the stars, terraform a planet, or geoengineer our own. We have a lot to learn in the next 100 years.[19]

NETs

Very recently, in the wake of the failure of international agreements to achieve meaningful emissions reductions, ideas have emerged about "fixing everything," manifest most recently in climate circles as "negative emissions" technologies, or NETs, that promise to control the carbon cycle by drawing down carbon dioxide on a planetary scale at a rate that exceeds global emissions. NETs are intended to be scientifically validated, technically feasible, and societally sensitive pathways for the capture, removal, and reliable long-term storage of significant amounts of atmospheric carbon dioxide. Teams of engineers have proposed massive projects, most of them still on the drawing board, that include bio-capture, mineral capture, and synthetic capture using chemicals. To date, however, very little has been accomplished by these initiatives.

NETs are high-tech additions to existing strategies of mitigation, energy efficiency, renewables, fuel switching, and adaptation. They are expensive and energy and infrastructure intensive, requiring an estimated 30 percent increase in total world energy use to implement. NETs are at vastly different stages of readiness. Some are close to being ready for large-scale deployment, others require basic scientific research, and, like geoengineering, all fall in the rather vague category of future imaginaries. Working at the planetary scale surely will have unknown ecological consequences.

A widely distributed statement on "Geoengineering the Climate System" that I coauthored in 2009 was issued by the American Meteorological Society. It is also applicable to NETs. The document recommended three major initiatives, one technical, one policy oriented, and one on human dimensions. The first calls for "enhanced research on the scientific and technological potential for engineering the climate system, including research on intended and unintended environmental responses." The second recommends "development and analysis of policy options to promote transparency and international cooperation in exploring geoengineering options along with restrictions on reckless efforts to manipulate the climate system." The third, of greatest relevance here, demands "coordinated study of historical, ethical, legal, and social implications of geoengineering (read climate intervention) that integrates international, interdisciplinary, and intergenerational issues and perspectives and includes lessons from past efforts to modify weather and climate."[20]

More than a decade later, the literature on geoengineering and NETs remains exclusively technical, with no incorporation and seemingly little appreciation of work on human perspectives. NETs proposals include massive terrestrial carbon removal and sequestration schemes within forests and agricultural lands—but no one knows how to store this waste on millennial time scales. Moreover, arable land is limited. Food comes first and biodiversity and water supplies would be threatened by massive crop switching. Bioenergy with carbon capture and sequestration is expensive and logistically challenging since all biological waste would have to be utilized, even garden trimmings, to meet the goals; and the program, instituted locally in places, is not scalable to planetary levels. Coastal blue carbon has been proposed to increase the carbon stored in plants or sediments in coastal zones—but this may result in large-scale ecological damage in some of the most expensive and biologically sensitive areas of the planet. A scheme for direct air capture and sequestration, or "artificial trees," is expensive and energy intensive. Carbon mineralization or accelerated weathering constitutes a proposal to expand and accelerate mining of certain types of rocks—but this is energy intensive and could lead to contamination of water supplies and other as yet unexamined problems.[21]

The monitoring and verification needs of NETs would be an immense, perhaps impossible task. What is often overlooked or underemphasized is the need for and challenge of reliable and perpetual carbon dioxide storage, which, according to Ferenc Toth, may be more dangerous than nuclear waste storage.[22] Its governance would require new types of institutions and arrangements the world has never seen. The current teams are almost exclusively based in the United States, and, by their own rosiest estimates, may be able to develop the capacity to meet only about one-fifth of the planetary need. Many of the researchers hope to gain patents on their techniques, but it is abundantly clear that proprietary solutions for planetary problems will not work. What about the rest of the world? With predominantly male WEIRD super heroes involved, do we even know what questions to ask? It will not be simple or inevitable to move humanity from an "extractionist" mode of production to an "injectionist" set of solutions.

Conclusion

Are the collective imaginaries of terraforming planets and geoengineering Earth shaping future possibilities and perhaps inevitabilities for scientists, policymakers, and the public? Can historically informed humanistic scholarship play an important role in such trajectories?[23] Questions of terraformation and geoengineering do not belong only to the scientists and engineers, to white Western male technocrats and their governments, yet many humanists have bought, hook, line, and sinker, into the "end of the Holocene/rupture of time" discourse of the so-called Anthropocene and the planetary emergency it implies. We risk ceding global narration and global governance to the Earth system scientists. Here, with emphasis added, is how Nobel Laureate Paul Crutzen, chief promoter of the Anthropocene meme, speaks about managing the planet:

> A daunting task lies ahead for *scientists and engineers to guide society* towards environmentally sustainable management during the era of the Anthropocene. This will require *appropriate human behaviour* at all scales, and may well involve internationally accepted, large-scale *geo-engineering projects*, for instance to "optimize" climate. At this stage, however, we are still largely treading on *terra incognita*.[24]

Others, notably Hans Joachim Schellnhuber of the Potsdam Institute for Climate Impact Research, argue that he and his team of Earth system observers and modelers are best qualified to watch the human footprints on the ecosphere, model their impacts, and make collective "rational choices" about the governance of the planet.[25] Referring to his specialty as "Earth-system diagnostics," Schellnhuber advocates sketchy and disruptive geoengineering proposals, referring to them as planetary "stabilization" techniques, "to mitigate anthropogenic aberrations of the ecosphere." His favorites included iron fertilization of the oceans, repairing holes in the ozone layer using military-grade orbital lasers, and suppressing future glaciation events by "judicious injection of 'designer greenhouse gases' into the atmosphere." Shades of Harrison Brown. In a recent podcast hosted by Bruno Latour, Schellnhuber, speaking about the need for planetary management and governance, claimed we are developing ever better and sharper instruments to perform "planetary surgery." He also called for "sharp instruments to cut the social tissue." Without missing a beat Latour responded, "In France we call that the Guillotine."[26]

Prolific author, scientist, and futurist James Lovelock often refers to himself as a "geophysiologist," or planetary physician. He diagnoses the Earth as having a fever induced by *disseminated primatemia* (the superabundance of humans). As treatment, he recommends a low-carbon diet combined with nuclear medicine. He likens geoengineering to crude planetary surgery, no more advanced than the practices of the medieval barber surgeons. While the patient would definitely survive, the parasites might not.[27]

Climate intervention is not, in essence, a heroic saga about new scientific discoveries or techniques that can save the planet, as many of its proponents claim, but an enactment of speculative techno-futurism proceeding with overreaching hubris and self-delusion. The noted aeronautical engineer Theodore von Karman once observed that "scientists study the world as it is, engineers create the world that has never been."[28] This quote has an ominous ring, however, when it comes to the terraforming and geoengineering issues raised in this essay, since some interventions should never happen, and some "worlds" should never be. Some have asked if the risk of climate intervention is worse than the risk of global warming. I think that it just might be.

14 * *Resurrecting the Sublime*

ALEXANDRA DAISY GINSBERG

Leafing through folders of pressed plant specimens at Harvard University's herbaria, synthetic biologists Christina Agapakis and Dawn Thompson were hunting flowers that no longer exist. As they cross-referenced a print-out of *The IUCN Red List* of modern plant extinctions against cursive names inked on yellowing labels, they found twenty species from which they cut tiny tissue samples (fig. 14.1).[1]

Three would still contain enough DNA to allow humans to once again experience the smell of their lost flowers. These were diffused in the artwork *Resurrecting the Sublime* (2019), in collaboration with artist Alexandra Daisy Ginsberg and smell researcher and artist Sissel Tolaas.[2]

Those three flowers, endemic to Hawaii, Kentucky, and South Africa respectively, shared a particular trait. Each was extinguished by colonial action: human destruction of its habitat. In 1912, just two years after the indigenous *Maui hau kuahiwi* was first spotted and named *Hibiscadelphus wilderianus* Rock by Austrian-American botanist Joseph F. Rock, the only such tree was found in a dying state.[3] Colonial cattle ranches had decimated its native dry forests on ancient lava fields on the southern slopes of Mount Haleakalā, on the island of Maui, Hawaii (fig. 14.2).

Four thousand miles away and a decade later, the construction of US Dam No. 41 on the Ohio River at Louisville, Kentucky, cemented the disappearance of the delicate Falls-of-the-Ohio Scurfpea, or *Orbexilum stipulatum* (fig. 14.3).

First collected in 1835, the purple flower was last seen in 1881 in its only known locale, the Devonian limestone outcrop of Rock Island, situated at the river's bend. The reason for its loss is unknown; perhaps reducing buffalo populations impacted other species. But when the dam flooded the channel in the 1920s, the island itself was erased (fig. 14.4).[4]

Eight thousand miles away on the southern tip of Africa, eighteenth-century colonial vineyard expansion had already transformed granitic Wynberg Hill in the shadow of Table Mountain (fig. 14.5).

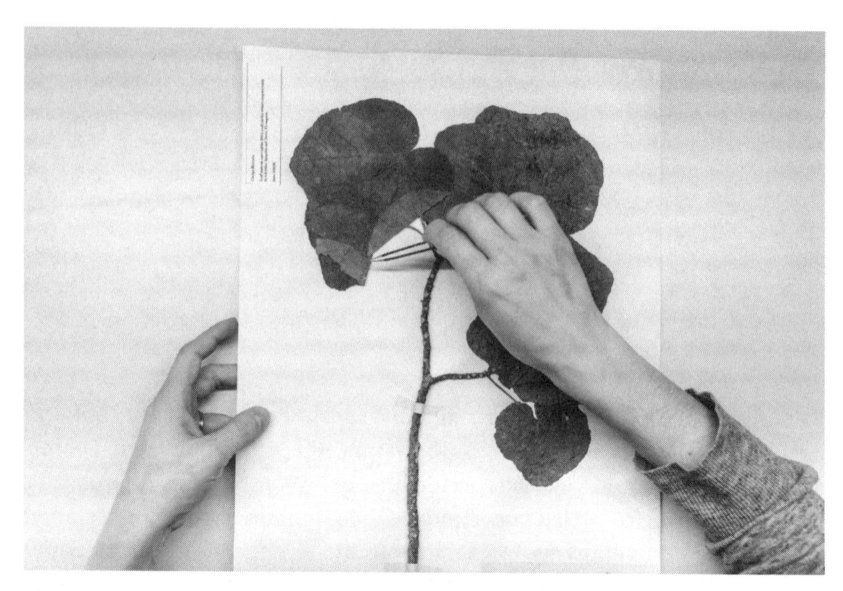

Figure 14.1. Dr. Christina Agapakis taking tissue samples from a specimen of the *Hibiscadelphus wilderianus* Rock at the Harvard University Herbarium. Photograph: Grace Chuang. Courtesy of The Herbarium of the Arnold Arboretum of Harvard University. Photograph © Ginkgo Bioworks, 2018.

Figure 14.2. Google Earth view of the deforested southern slopes of Mount Haleakalā on the island of Maui, Hawaii, once the habitat of the *Hibiscadelphus wilderianus* Rock. Photograph © Google, DigitalGlobe, 2018.

Figure 14.3. Agapakis sampling tissue from a specimen of *Orbexilum stipulatum* from the collection of the Gray Herbarium of Harvard University. Photograph: Grace Chuang. Courtesy of Gray Herbarium of Harvard University. Photo © Ginkgo Bioworks, 2018.

Figure 14.4. Aerial view of Falls of the Ohio and Locks and Dam No. 41 circa 1930s or 1940s, at Louisville, Kentucky. Rock Island was lost as the dam was flooded, and would have been located near the bottom right of the photograph. Image: Public domain. https://commons.wikimedia.org/wiki/File:Aerial_view_of_Falls_of_the_Ohio_and_Locks_and_Dam_No_41_circa_1930s_or_1940s.jpg.

Figure 14.5. Google Earth view toward Table Mountain, with Wynberg Hill in front. This was once the habitat of the extinct "*Leucadendron grandiflorum* (Salisb.) R. Br.," today the suburbs of Cape Town, South Africa. Photograph © Google, Landsat/ Copernicus, DigitalGlobe, 2018.

This was the home of the protea "*Leucadendron grandiflorum* (Salisb.) R. Br.," or Wynberg Conebush. At the turn of the nineteenth century, the botanist Robert Salisbury noted the flower's "strong and disagreeable smell" (fig. 14.6).[5]

However, he had encountered the flower not in Cape Town, but in a collector's garden in London, its extirpation in the wild already likely.[6] Indeed, this flower has a more complex history as the Harvard flower is a cultivated specimen from the 1960s, and thus may be incorrectly labeled: no true sample may exist anywhere (a matter that we are researching) (fig. 14.7).

Only Salisbury's record may prove it ever lived. That three otherwise insignificant organisms in the history of biology happened to be seen, collected, and named by Western botanists before they disappeared is a reminder of the contingency of biological existence, as well as the Western scientific urge to catalog life to confirm that very existence.

Capital helped extinguish these flowers and now capital is required to bring them back. Agapakis and Thompson's scouring of Harvard's archives was the start of a collaboration between synthetic biologists and artists that raises questions about our relationship with nature, and about conservation, colonization, and the complicated role of technology and cap-

Figure 14.6. Robert Salisbury's *Euryspermum grandiflorum* from *The Paradisus Londinensis*, published between 1805 and 1807. This plant shown is now described as "*Leucadendron grandiflorum* (Salisb.) R. Br." Courtesy Biodiversity Heritage Library, provided by Missouri Botanical Garden. Image: Public domain.

Figure 14.7. Dried specimen of what is tentatively being called "*Leucadendron grandiflorum* (Salisb.) R. Br.," collected from a cultivated specimen possibly in 1966. Since the species was last seen around 1805, the true identity of this specimen is being investigated. Photograph © The Herbarium of the Arnold Arboretum of Harvard University, 2018.

ital as it intersects these areas. This short, illustrated essay describes our process and highlights some of the issues raised by an artwork intended as a provocation for reflection on, not as a solution to, our treatment of the natural world.

Agapakis is the creative director of Ginkgo Bioworks, a Boston-based biotechnology company founded in 2009.[7] In its gleaming, robot-assisted "foundries," Ginkgo scientists engineer yeast and bacteria to secrete useful chemicals for humans, from pharmaceuticals to fuel to flavors. Since Ginkgo also produces smell molecules for fragrance companies, divining lost smell molecules from crumbled sections of DNA could potentially be rewarding technically, intellectually, as well as commercially.[8] The project began in 2014 as an internal research project, which Agapakis took up, intrigued to see if it was scientifically possible.

Revealing the flowers' smells from the information encoded in their DNA first required help from paleogenomics experts at the University of California at Santa Cruz, who could extract DNA from the degraded historic samples. Ginkgo's scientists and engineers then analyzed the fragments to predict gene sequences that might encode fragrance-producing enzymes. They compared the DNA with known sequences from other organisms and filled in any gaps using the template genes (fig. 14.8).

This became a large and expensive experiment: some two thousand predicted gene variants were synthesized (the DNA printed) and inserted into yeast, then the yeast were cultured to produce smell molecules and to test what each variant produced. Finally, the team used mass spectrometry to verify the identity of each of the secreted molecules. From the resulting list of smell molecules, in 2018 Tolaas could begin to reconstruct the smell of the three lost flowers in her Berlin laboratory, using identical smell molecules or comparative ones for those that were not commercially available (fig. 14.9).[9]

But while bioengineering can tell us which molecules the plants produced, their quantities—like the flowers—are also lost. The flowers' true smell remains unknowable. This contingency disturbs the solutionist narrative of engineering life: synthetic biologists seek to "build life to understand it," to be able to control it.[10] But here, we cannot know.

Rather than producing a sense of control, using genetic engineering to try to resurrect the smell of extinct flowers—so that humans may again experience something we destroyed—is both romantic and terrifying. This dizzying feeling evokes the sublime, an idea that has preoccupied Western artists and thinkers for centuries. The sublime is an "expression of the unknowable," an aesthetic state reached through exposure to nature and its immensity, encouraging contemplation of humankind's position in it.[11] Artists tried to represent this sensation in nineteenth-century land-

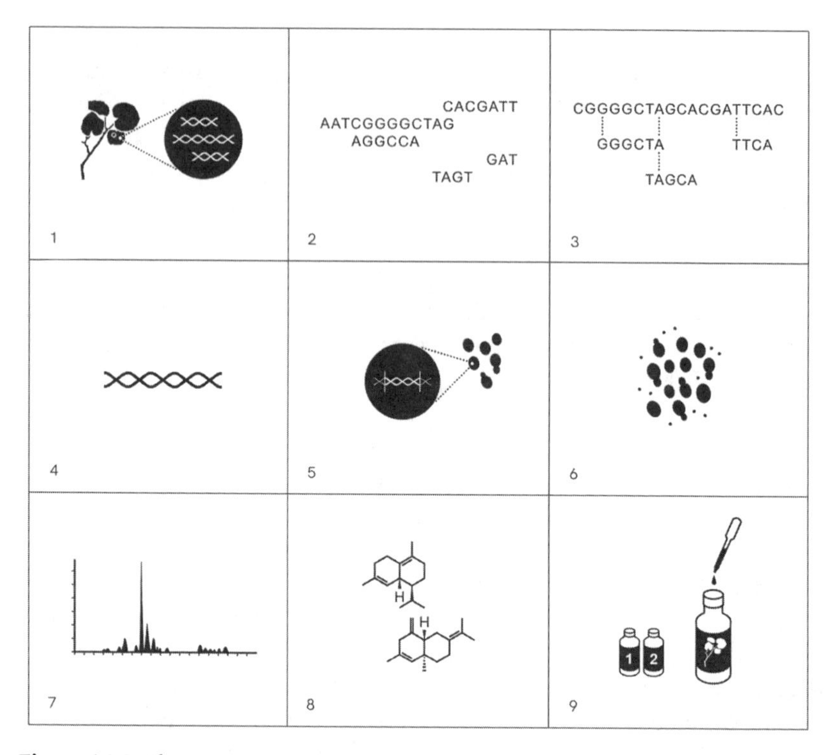

Figure 14.8. The reconstruction process from specimen to smell. 1.Tiny fragments of DNA are extracted from tissue from the dried plant. 2. A DNA sequencing machine reads the fragments, revealing the order of their nucleotide bases: the DNA code. 3. The sequences are compared with a gene from a current organism, to predict genes from the lost flower that encode for fragrance-producing enzymes. 4. The final reconstructed gene sequence, with gaps and errors matched from the template, is printed by a DNA synthesizer. 5. The printed gene is inserted into yeast cells. 6. The yeast is grown, making copies. The inserted gene tells the yeast cells to make the smell molecule. 7. The smell molecule's identity is checked using mass spectrometry, confirming whether the gene works as predicted. 8. The process is repeated for each gene, giving a list of smell molecules the flower may have produced. 9. The flower's smell is reconstructed using identical molecules or comparative ones. We will never know the flower's exact smell: we know which molecules the lost *Hibiscadelphus wilderianus* made, but the amounts of each are also lost. Image © Alexandra Daisy Ginsberg, 2019.

scape paintings; synthetic renderings that captured the violent creativity of nature. Ginkgo's technological feat reverses the "natural order" of time to glimpse a nature that is lost, but like these paintings, even the most advanced biotechnologies can give only an incomplete representation.

Invoking the sublime also connects this work to a changing understanding of the sublime itself: from eighteenth-century efforts to manufacture

Figure 14.9. Sissel Tolaas reconstructing smells in her Berlin laboratory. Photograph © Alexandra Daisy Ginsberg, 2010.

sublime experiences in audiences (such as the spectacular West End theatre sets of Philippe de Loutherbourg[12]), to analysis of the sublime's role in colonial identity-building in the nineteenth century (evident in the Edenic paintings of Frederick Church[13]), to the twentieth century technological sublime of engineered infrastructure,[14] and finally to the sublime's postmodern shift from transcendence to immanence, yet knowingly a constructed experience or illusion.[15]

Acknowledging both the idea of the sublime and this history of its reconstruction, Ginsberg wanted museum visitors to enjoy the total artifice of a "resurrected" smell experienced in a simulated landscape. With her studio team, she designed a series of immersive installations. In the largest version, visitors step inside glazed vitrines, recasting the natural history museum display case as a space for contemplation (fig. 14.10).

Inside each, Tolaas breaks down a lost flower's reconstructed smell into four parts, which are individually diffused from the ceiling. The fragments mix around the visitor, replicating the contingency of biology: there is no one exact smell, since every inhalation is subtly different. A landscape of boulders matched to the geology of the flower's lost habitat completes the diorama of a minimal nature. An ambient soundtrack evokes the lost landscape filled again with buzzing insects and plants in the wind, underpinned by a low frequency rumble that resonates in the gut. As they stand and

Figure 14.10. Installation view of *Resurrecting the Sublime* at the Biennale Internationale Design Saint-Étienne, France, March 2019. The vitrine on the left contains the smell of the *Hibiscadelphus wilderianus* Rock, diffused amidst a landscape of lava boulders, with the reconstructed landscape animated in front. The vitrine on the right contains the smell of the extinct *Orbexilum stipulatum*, the reconstructed landscape completed with limestone boulders. Photograph © Pierre Grasset, 2019.

smell the lost flower in this abstracted environment, the visitor becomes the subject of the nature display. They are no longer just an observer, but part of an observed nature, watched by others looking in (fig. 14.11).

The physical experience induces a connection to otherwise obscure flowers, long ago extinguished in distant places by the actions of earlier colonizers.

This is biotechnology used to engineer a feeling of loss, not to construct a solution. We are not offering de-extinction,[16] but using immersive installations to give a glimpse of a flower blooming in the shadow of a mountain, on a forested volcanic slope, or a wild riverbank; each an interplay of a species and a place that no longer exists (figs. 14.12–14.14).

Is this the reversal of the sublime: the total human domination of nature through the engineering of life? Or does such loss remind us of biology's ambivalence in the face of human efforts to remake nature?

Figure 14.11. The lost landscape is reduced to its geology and the flower's smell: the human connects the two and becomes the specimen on view as they step inside the vitrine. Photograph © Alex Cretey-Systermans, 2019.

Figure 14.12. *Resurrecting the Sublime*: digital reconstruction of the lost landscape of the now extinct *Hibiscadelphus wilderianus* on the southern slopes of Mount Haleakalā on the island of Maui, Hawaii. Image © Alexandra Daisy Ginsberg, 2019.

Figure 14.13. *Resurrecting the Sublime*: digital reconstruction of the extinct *Orbexilum stipulatum* in its lost habitat of Rock Island on the Ohio River, Kentucky, before its extinction in 1881. Image © Alexandra Daisy Ginsberg, 2019.

Figure 14.14. *Resurrecting the Sublime*: digital reconstruction of the lost landscape of the now extinct "*Leucadendron grandiflorum* (Salisb.) R. Br.," Wynberg Hill, Cape Town, imagined some time before 1806. Image © Alexandra Daisy Ginsberg, 2019.

NOTES

Introduction

1. N. Crowe, "The Historiography of Biotechnology," in *Handbook of the Historiography of Biology*, ed. M. Dietrich, M. Borrello, and O. Harman (Cham: Springer, 2018); Finn Bowring, *Science, Seeds, and Cyborgs: Biotechnology and the Appropriation of Life* (London: Verso, 2003); Robert Bud, *The Uses of Life: A History of Biotechnology* (Cambridge: Cambridge University Press, 1993); Melinda Cooper, *Life as Surplus: Biotechnology and Capitalism in the Neoliberal Era* (Seattle: University of Washington Press, 2008); Miguel García-Sancho, *Biology, Computing, and the History of Molecular Sequencing: From Proteins to DNA, 1945–2000* (Palgrave Macmillan, 2012); Sally Smith Hughes, *Genentech: The Beginnings of Biotech* (Chicago: University of Chicago Press, 2011); Peter Keating and Alberto Cambrosio, *Biomedical Platforms: Realigning the Normal and the Pathological in Late-Twentieth-Century Medicine* (Cambridge, MA: MIT Press, 2003); Martin Kenney, *Biotechnology: The University-Industrial Complex* (New Haven: Yale University Press, 1986); Paul Rabinow, *Making PCR: A Story of Biotechnology* (Chicago: University of Chicago Press, 1996); Karen A. Rader, *Making Mice: Standardizing Animals for American Biomedical Research, 1900–1955* (Princeton: Princeton University Press, 2004); Joanna Radin, "Latent Life: Concepts and Practices of Human Tissue Preservation in the International Biological Program," *Social Studies of Science* 43, no. 4 (2013): 484–508; Nick Rasmussen, *Gene Jockeys: Life Science and the Rise of Biotech Enterprise* (Baltimore: Johns Hopkins University Press, 2014); Arnold Thackray, *Private Science: Biotechnology and the Rise of the Molecular Sciences* (Philadelphia: University of Pennsylvania Press, 1998); Doogab Yi, *The Recombinant University: Genetic Engineering and the Emergence of Stanford Biotechnology* (Chicago: University of Chicago Press, 2015).

2. For a recent overview of the mechanist metaphor in the history of biology, see Andrew S. Reynolds, *The Third Lens: Metaphor and the Creation of Modern Cell Biology* (Chicago: University of Chicago Press, 2018). On the limitations of mechanist metaphors, see Maarten Boudry and Massimo Pigliucci, "The Mismeasure of Machine: Synthetic Biology and the Trouble with Engineering Metaphors," *Studies in History and Philosophy of Biological and Biomedical Sciences* 44, no. 4, part B (2013): 660–68.

3. Rheinberger, Hans-Jörg, *Toward a History of Epistemic Things: Synthesizing Proteins in the Test Tube* (Stanford: Stanford University Press, 1997); Hannah Landecker, *Culturing Life: How Cells Became Technologies* (Cambridge, MA: Harvard University Press, 2010).

4. Paul Rabinow, "Artificiality and Enlightenment: From Sociobiology to Biosociality," in *Incorporations*, ed. Jonathan Crary and Stanford Kwinter (New York: Zone Books, 1992).

5. Sheila Jasanoff and Sang-Hyun Kim, eds., *Dreamscapes of Modernity: Sociotechnical Imaginaries and the Fabrication of Power* (Chicago: University of Chicago Press, 2015).

6. H. J. Muller, *Out of the Night: A Biologist's View of the Future* (New York: Vanguard Press, 1935), 69. The alternate title for this eugenical manifesto was *The Remaking of Life.*

7. Philip J. Pauly, *Controlling Life: Jacques Loeb and the Engineering Ideal in Biology* (New York: Oxford University Press, 1987).

8. Sara B. Pritchard, "Joining Environmental History with Science and Technology Studies: Promises, Challenges and Contributions," *New Natures: Joining Environmental History with Science and Technology Studies*, ed. Dolly Jørgensen, Finn Arne Jørgensen, and Sara B. Pritchard (Pittsburgh: University of Pittsburgh Press, 2013).

9. Andrew Lakoff, Paul Rabinow, and Nikolas Rose, *Pharmaceutical Reason: Knowledge and Value in Global Psychiatry* (Cambridge: Cambridge University Press, 2006); Sarah Franklin, "Foucault et les transformations du biopouvoir," in *Histoire des sciences et des savoirs III: Le siècle des technosciences*, ed. Christophe Bonneuil and Dominique Pestre (Paris: Seuil, 2015), 211–31.

10. Antoine Picon, "Engineers and Engineering History: Problems and Perspectives," *History and Technology* 20, no. 4 (2004): 421–36; Jennifer Karns Alexander, *The Mantra of Efficiency: From Waterwheel to Social Control* (Baltimore: Johns Hopkins University Press, 2008).

11. Edward Jones-Imhotep, *The Unreliable Nation: Hostile Nature and Technological Failure in the Cold War*, (Cambridge, MA: MIT Press, 2017).

12. Loeb to Ernst Mach, February 26, 1890; cited in Pauly (1987).

13. M. Norton Wise, *Growing Explanations: Historical Perspectives on Recent Science* (Durham: Duke University Press, 2004).

14. W. Patrick McCray, *The Visioneers: How a Group of Elite Scientists Pursued Space Colonies, Nanotechnologies, and a Limitless Future* (Princeton: Princeton University Press, 2013).

15. Jürgen Renn, *The Evolution of Knowledge: Rethinking Science for the Anthropocene* (Princeton: Princeton University Press, 2020); Christophe Bonneuil and Jean-Baptiste Fressoz, *The Shock of the Anthropocene: The Earth, History, and Us* (London: Verso, 2017).

16. For a similar sensibility in science studies see Anna Lowenhaupt Tsing, *The Mushroom at the End of the World: On the Possibility of Life in Capitalist Ruins* (Princeton: Princeton University Press, 2015).

17. John Tresch, *The Romantic Machine: Utopian Science and Technology after Napoleon* (Chicago: University of Chicago Press, 2012); M. Norton Wise, *Aesthetics, Industry, and Science: Hermann von Helmholtz and the Berlin Physical Society* (Chicago: University of Chicago Press, 2018); Tiago Saraiva and Ana Cardoso De Matos, "Technological Nocturne: The Lisbon Industrial Institute and Romantic Engineering (1849–1888)," *Technology and culture* 58, no. 2 (2017): 422–58.

Chapter 1

1. Vijee Venkatraman, "Turning Point: Kevin Esvelt," *Nature* 536, no. 7614 (2016): 117.

2. Katie Langin, "Genetic Engineering to the Rescue against Invasive Species?," *National Geographic*, July 18, 2014, https://www.nationalgeographic.com/news/2014/7/140717-gene-drives-invasive-species-insects-disease-science-environment/.

3. Such approaches to control in biology can be traced to early twentieth-century biologist Jacques Loeb and his mechanistic conception of life, which viewed "control

of natural phenomena as the essential problem of scientific research" (114). Control required not just an apprehension of biological function in terms of the physico-chemical processes that underpinned them, but also "controlling at will the life phenomena . . . bringing about effects which cannot be expected in Nature" (116). Philip J. Pauly, *Controlling Life: Jacques Loeb and the Engineering Ideal in Biology* (Oxford: Oxford University Press, 1987), 114–28; see also Garland Allen, *Life Sciences in the Twentieth Century* (Hoboken: Wiley & Sons, 1975).

4. The terms "mechanism" or "mechanistic" have historically had variable meanings and uses, but epistemological distinctions among versions of mechanism are seldom used by biologists in practice. Accordingly, this paper is not concerned with referring to a philosophically distinct account of natural phenomena but with a general pattern of explanatory practices in biology broadly construed as mechanistic. See Garland Allen, "Mechanism, Vitalism and Organicism in Late Nineteenth and Twentieth-Century Biology: The Importance of Historical Context," *Studies in History and Philosophy of Science Part C: Studies in History and Philosophy of Biological and Biomedical Sciences* 36, no. 2 (2005): 261–83; Arturo Casadevall and Ferric C. Fang, "Mechanistic Science," *American Society for Microbiology* 77, no. 9 (2009): 3517–19.

For scholarship exploring historical and contemporary conceptual formulations of mechanism, see William Bechtel and Robert C. Richardson, "Emergent Phenomena and Complex Systems," in *Emergence or Reduction? Essays on the Prospects of Nonreductive Physicalism*, ed. Ansgar Beckermann, Hans Flohr, and Jaegwon Kim (Berlin: Walter de Gruyter, 1992), 257–88; Peter Machamer, Lindley Darden, and Carl F. Craver, "Thinking about Mechanisms," *Philosophy of Science* 67, no. 1 (2000): 1–25; Daniel J. Nicholson, "The Concept of Mechanism in Biology," *Studies in History and Philosophy of Science Part C: Studies in History and Philosophy of Biological and Biomedical Sciences* 43, no. 1 (2012): 152–63.

5. Sheldon Krimsky, *Genetic Alchemy* (Cambridge, MA: MIT Press, 1982); Michel Morange, *A History of Molecular Biology* (Cambridge, MA: Harvard University Press, 2000).

6. Soraya De Chadarevian, *Designs for Life: Molecular Biology after World War II* (Cambridge: Cambridge University Press, 2002).

7. Robert Bud, *The Uses of Life: A History of Biotechnology* (Cambridge: Cambridge University Press, 1994); Evelyn Fox Keller, *The Century of the Gene* (Cambridge, MA: Harvard University Press, 2009).

8. Nicolas Rasmussen, *Gene Jockeys: Life Science and the Rise of Biotech Enterprise* (Baltimore: Johns Hopkins University Press, 2014).

9. Thomas F. Lee, *The Human Genome Project: Cracking the Genetic Code of Life* (New York: Plenum Press, 1991).

10. Evelyn Fox Keller, *Refiguring Life: Metaphors of Twentieth-Century Biology* (New York: Columbia University Press, 1995).

11. Kevin M. Esvelt, Andrea L. Smidler, Flaminia Catteruccia, and George M. Church, "Concerning RNA-Guided Gene Drives for the Alteration of Wild Populations," *Elife* 3 (2014): 1–39. A genetic element driven by a gene drive can range in size from short nucleotide sequences to larger genetic cassettes comprising multiple different genes.

12. Bernward Joerges, "Do Politics Have Artefacts?," *Social Studies of Science* 29, no. 3 (1999): 411–31; Langdon Winner, "Do Artifacts Have Politics?," *Daedalus* (1980): 121–36.

13. See Dorothy Nelkin and M. Susan Lindee, *The DNA Mystique: The Gene as Cultural Icon* (New York: W. H. Freeman, 1995); Paul Rabinow, *Essays on the Anthropology of Rea-*

son (Princeton: Princeton University Press, 1996); Alan H. Goodman, Deborah Heath, and M. Susan Lindeee, eds., *Genetic Nature/ Culture: Anthropology and Science beyond the Two-Culture Divide* (Berkely: University of California Press, 2003).

14. Sheila Jasanoff, "Ordering Knowledge, Ordering Society," in *States of Knowledge: The Co-production of Science and the Social Order*, ed. Sheila Jasanoff (Abingdon: Routledge, 2004), 13–45.

15. Tiago Saraiva, *Fascist Pigs: Technoscientific Organisms and the History of Fascism* (Cambridge, MA: MIT Press, 2016); Donna Haraway, "For the Love of a Good Dog: Webs of Action in the World of Dog Genetics," in *Genetic Nature/ Culture: Anthropology and Science beyond the Two-Culture Divide*, ed. Alan H. Goodman, Deborah Heath, and M. Susan Lindee (Berkely: University of California Press, 2003).

16. Bruno Latour, *Politics of Nature* (Cambridge, MA: Harvard University Press, 2004).

17. Nikolas Rose, "The Politics of Life Itself," *Theory, Culture & Society* 18, no. 6 (2001): 1–30.

18. Sarah Franklin and Margaret Lock, eds., *Remaking Life & Death: Toward an Anthropology of the Biosciences* (Santa Fe: School of American Research Press, 2003), 4.

19. See John Desmond Bernal, *The Social Function of Science* (New York: George Routledge & Sons, 1939); Michael Polanyi, "The Republic of Science: Its Political and Economic Theory," *Minerva* 1 (1962): 54–74; Ronald D. Brunner and William Ascher, "Science and Social Responsibility," *Policy Sciences* 25, no. 3 (1992): 295–331; Jack Stilgoe, Richard Owen, and Phil Macnaghten, "Developing a Framework for Responsible Innovation," *Research Policy* 42, no. 9 (2013): 1568–80.

20. Robert K. Merton, *The Sociology of Science: Theoretical and Empirical Investigations* (Chicago: University of Chicago Press, 1973).

21. Daryl E. Chubin, "Misconduct in Research: An Issue of Science Policy and Practice," *Minerva* (1985): 175–202.

22. John M. Braxton, ed., "Perspectives on Research Misconduct," *Journal of Higher Education* 65, no. 3 (1994): 239–400.

23. Carl Mitcham, "Co-responsibility for Research Integrity," *Science and Engineering Ethics* 9, no. 2 (2003): 273–90.

24. Heather E. Douglas, "The Moral Responsibilities of Scientists (Tensions between Autonomy and Responsibility)," *American Philosophical Quarterly* 40, no. 1 (2003): 59–68.

25. Shelia Jasanoff, "Constitutional Moments in Governing Science and Technology," *Science and Engineering Ethics* 17 (4): 621–38; Cecilie Glerup and Maja Horst, "Mapping 'Social Responsibility' in Science," *Journal of Responsible Innovation* 1, no. 1 (2014): 31–50.

26. Ian A. E. Atkinson, "Spread of the Ship Rat (*Rattus r. rattus* L.) III New Zealand," *Journal of the Royal Society of New Zealand* 3, no. 3 (1973): 457–72; Janet M. Wilmshurst, Atholl J. Anderson, Thomas F. G. Higham, and Trevor H. Worthy, "Dating the Late Prehistoric Dispersal of Polynesians to New Zealand Using the Commensal Pacific Rat," *Proceedings of the National Academy of Sciences* 105, no. 22 (2008): 7676–80; Kazimierz A. Wodzicki, *Introduced Mammals of New Zealand: An Ecological and Economic Survey*, Department of Scientific and Industrial Research Bulletin No. 98 (Wellington: Department of Scientific and Industrial Research, 1950).

27. Stefan Helmreich, "How Scientists Think; about 'Natives', for Example. A Problem of Taxonomy among Biologists of Alien Species in Hawaii," *Journal of the Royal Anthropological Institute* 11, no. 1 (2005): 107–28.

28. The designation of kiore as native or invasive in Aotearoa New Zealand has been

disputed in part due to their earlier arrival to the islands, integration into local ecosystems prior to European colonization, and significance to Māori history and culture. However, similar contentions are not made concerning rodent species that accompanied European colonizers as being invasive to Aotearoa New Zealand. Carolyn M. King, *Immigrant Killers: Introduced Predators and the Conservation of Birds in New Zealand* (Oxford: Oxford University Press, 1984); Elsdon Best, *Forest Lore of the Māori* (Wellington: E. C. Keating, Government Printer, 1977).

29. James C. Russell and Keith G. Broome, "Fifty Years of Rodent Eradications in New Zealand: Another Decade of Advances," *New Zealand Journal of Ecology* 40, no. 2 (2016): 197–204; B. W. Thomas and R. H. Taylor, "A History of Ground-Based Rodent Eradication Techniques Developed in New Zealand, 1959–1993," in *Turning the Tide: The Eradication of Invasive Species,* ed. C. R. Veitch and M. N. Clout (Gland, Switzerland: IUCN, 2002), 301–10; David Young, *Our Islands, Our Selves* (Otago: University of Otago Press, 2004); P. J. McClelland, R. Coote, M. Trow, P. Hutchins, H. M. Nevins, J. Adams, J. Newman, and H. Moller, "The Rakiura Tītī Islands Restoration Project: Community Action to Eradicate *Rattus rattus* and *Rattus exulans* for Ecological Restoration and Cultural Wellbeing," *Island Invasives: Eradication and Management* (2011): 451–54.

30. Les Kelly, "Predator Free NZ Feasibility Study" (2008), http://predatorfreenz.com/supporting-documents/.

31. "The History of Predator Free 2050," Department of Conservation, Te Papa Atawbai, https://www.doc.govt.nz/nature/pests-and-threats/predator-free-2050/history/.

32. Paul K. Couchman and Kenneth Fink-Jensen, *Public Attitudes to Genetic Engineering in New Zealand,* (Christchurch: New Zealand Department of Scientific and Industrial Research [DSIR], DSIR Crop Research, 1990); Darryl Macer, Howard Bezar, Nicola Harman, Hiroshi Kamada, and Nobuko Macer, "Attitudes to Biotechnology in Japan and New Zealand in 1997, with International Comparisons," *Eubios Journal of Asian and International Bioethics* 7 (1997): 137–51; Andrew J. Cook, John R. Fairweather, Theresa Satterfield, and Lesley M. Hunt, "New Zealand Public Acceptance of Biotechnology," *Lincoln University: Research Report* 269 (2004).

33. Esvelt et al., "Concerning RNA-Guided Gene Drives," 1–39.

34. Ibid., 15.

35. Elaine C. Murphy, James C. Russell, Keith G. Broome, Grant J. Ryan, and John E. Dowding, "Conserving New Zealand's Native Fauna: A Review of Tools Being Developed for the Predator Free 2050 Programme," *Journal of Ornithology* (2019): 1–10.

36. Esvelt et al., "Concerning RNA-Guided Gene Drives," 3.

37. Austin Burt and Robert Trivers, *Genes in Conflict* (Cambridge, MA: Harvard University Press, 2006).

38. Esvelt et al., "Concerning RNA-Guided Gene Drives," 6–10. For greater detail of CRISPR-Cas9 function as it pertains to gene drive construction, refer to source figures 3 and 4.

39. Ibid., 2.

40. Ibid., 5.

41. Ibid., 13–17.

42. Rodolphe Barrangou, "Cas9 Targeting and the CRISPR Revolution," *Science* 344, no. 6185 (2014): 707–8; Elizabeth Pennisi, "The CRISPR Craze," *Science* 341, no. 6148 (2013): 833–36; Heidi Ledford, "CRISPR: Gene Editing Is Just the Beginning," *Nature News* 531, no. 7593 (2016): 156.

43. Kevin Esvelt, "Molecular Biologist Kevin Esvelt: Gene Drives, CRISPR Critters and Evolutionary Sculpting," interview by Robert Pollie, *The 7th Avenue Project,* November

15, 2015; *NOVA Wonders*, "Can We Make Life?," PBS, May 23, 2018; Michael Specter, "Rewriting the Code of Life," *New Yorker*, December 25, 2016.

44. "Responsive Science," published March 7, 2018, https://www.responsivescience.org/about.

45. Kevin Esvelt, "Scientific Philosophy," *Sculpting Evolution*, published July 21, 2015, http:// www.sculptingevolution.org/philosophy.

46. Kevin M. Esvelt and Neil J. Gemmell, "Conservation Demands Safe Gene Drive," *PLoS Biology* 15, no. 11 (2017): 1–8.

47. Kevin Esvelt, "New Zealand's War on Rats Could Change the World," interview by Ed Yong November 16, 2017, *Atlantic*, https://www.theatlantic.com/science/archive/2017/11/new-zealand-predator-free-2050-rats-gene-drive-ruh-roh/546011/.

48. Esvelt and Gemmell, "Conservation Demands Safe Gene Drive," 2.

49. "Predator Free NZ—Expert Q&A," Science Media Centre, press release published January 17, 2017, http:// www.scoop.co.nz/stories/SC1701/S00024/predator-free-nz-expert-qa.htm.

50. Charleston Noble, John Min, Jason Olejarz, Joanna Buchthal, Alejandro Chavez, Andrea L. Smidler, Erika A. DeBenedictis, George M. Church, Martin A. Nowak, and Kevin M. Esvelt, "Daisy-Chain Gene Drives for the Alteration of Local Populations," *bioRxiv* (2016): 057307. For a more comprehensive explanation of daisy drive design and function, see source figure 1.

51. Noble et al., "Daisy-Chain Gene Drives," 2.

52. Ibid., 11 (emphasis added).

53. Kate Evans, "Life Hackers," *New Zealand Geographic* 148 (2017): 34–57; "Kevin Esvelt—Sculpting Evolution," Saturday Morning Radio New Zealand, aired on September 23, 2017.

54. Kevin Esvelt, "Seeking Community Guidance for a Project Aiming to Humanely Remove Invasive Rodents," presented at University of Otago on September 30, 2017, https://www.media.mit.edu/videos/se-kevin-esvelt-comm-guide-2017-09-30/.

55. Kevin Esvelt, "Could Daisy Drive Help Make New Zealand Predator-Free?," *Responsive Science* (2017): 1–7, https://www.responsivescience.org/pub/daisydrivenz.

56. Under the 1991 Resource Management Act, the Māori gained some legal rights to co-governance of New Zealand's natural and physical resources—including the management of invasive species. The relationship between Māori communities, the New Zealand government, and scientists, however, remains contentious, particularly around environmental, genetic, and biotechnological issues. See New Zealand Legislation (1991), Resource Management Act 1991 No 69.

57. Esvelt, "Seeking Community Guidance."

58. Ibid.

59. Ibid.

60. See Jenny Reardon and Kim TallBear, "'Your DNA Is Our History': Genomics, Anthropology, and the Construction of Whiteness as Property," *Current Anthropology* 53, no. S5 (2012): S233–45; Emma Kowal, Joanna Radin, and Jenny Reardon, "Indigenous Body Parts, Mutating Temporalities, and the Half-Lives of Postcolonial Technoscience," *Social Studies of Science* 43, no. 4 (2013): 465–83; Stefan Helmreich, "How Scientists Think; about 'Natives', for Example. A Problem of Taxonomy among Biologists of Alien Species in Hawaii," *Journal of the Royal Anthropological Institute* 11, no. 1 (2005): 107–28.

61. Roma Mere Roberts, "Walking Backwards into the Future: Māori Views on Genetically Modified Organisms," *Perspectives on Indigenous Knowledge, WINHEC Journal* 2005: 1–10; B. Tipene-Matua, "Having Honest Conversations about the Impact of New Technol-

ogies on Indigenous People's Knowledge and Values," *Mātauranga taketake: Traditional Knowledge Indigenous Indicators of Well-Being: Perspectives, Practices, Solutions* (2006): 155–66; Paul Reynolds, "The Sanctity and Respect for Whakapapa: The Case of Ngati Wairere & AgResearch," in *Pacific Genes and Life Patents: Pacific Indigenous Experiences & Analysis of the Commodification & Ownership of Life*, ed. Aroha Te Pareake and Steven Ratuva (Wellington and Yokohama: Call of the Earth Llamado de la Tierra and The United Nations University Institute of Advanced Studies, 2007), 60–73; Waitangi Tribunal, "The Genetic & Biological Resources of Taonga Species," in *Ko Aotearoa Tēnei: A Report into Claims Concerning New Zealand Law and Policy Affecting Māori Culture and Identity* (Wellington: Te Taumata Tuatahi, 2011), 63–95.

62. Jessica Hutchings and Paul Reynolds, "Māori and the 'McScience' of New Technologies: Biotechnology and Nanotechnology Research and Development," *Rangahau* (2005). Retrieved from http://www.rangahau.co.nz/assets/hutching_renolds/maori_mcscienc.pdf.

63. Latour, *Politics of Nature*, 48.

64. Elizabeth A. Povinelli, *Geontologies: A Requiem to Late Liberalism* (Durham: Duke University Press, 2016); Kim Tallbear, "Beyond the Life/Not Life Binary: A Feminist-Indigenous Reading of Cryopreservation, Interspecies Thinking, and the New Materialisms," in *Cryopolitics: Frozen Life in a Melting World*, ed. Joanna Radin and Emma Kowal (Cambridge, MA: MIT Press, 2017), 179–202.

65. Linda Tuhiwai Smith, *Decolonizing Methodologies: Research and Indigenous Peoples* (London: Zed Books, 2013); Vanessa Watts, "Indigenous Place-Thought & Agency amongst Humans and Non-humans (First Woman and Sky Woman Go on a European World Tour!)," *Decolonization: Indigeneity, Education & Society* 2, no. 1 (2013).

66. Fiona Cram, Leonie Pihama, and Glenis Philip Barbara, "Māori and Genetic Engineering," Tamaki Makaurau: International Research Institute for Māori and Indigenous Education (2000); Roberts, "Walking Backwards into the Future"; Maui Hudson, Aroha Te Pareake Mead, David Chagne, Nick Roskruge, Sandy Morrison, Phillip L. Wilcox, and Andrew C. Allan, "Indigenous Perspectives and Gene Editing in Aotearoa New Zealand," *Frontiers in Bioengineering and Biotechnology* 7, no. 70 (2019): 1–9.

67. Eric Scwhimmer, "Making a World: The Māori of Aotearoa/New Zealand," in *Figured Worlds: Ontological Obstacles in Intercultural Relations*, ed. John R. Clammer, Sylvie Poirier, and Eric Schwimmer (Toronto: University of Toronto Press, 2004), 243–71; Hudson et al., "Indigenous Perspectives."

68. Esvelt and Gemmell, "Conservation Demands Safe Gene Drive," 5.

69. "Biological Heritage National Science Challenge Statement," Biological Heritage National Science Challenge, press release published November 16, 2017, https://www.scoop.co.nz/stories/SC1711/S00041/biological-heritage-national-science-challenge-statement.htm.

70. Melanie Mark Shadbolt, "Caution Urged over Gene Drives for Conservation—Expert Reaction," interviewed by Science Media Centre, published online November 17, 2017, https://www.sciencemediacentre.co.nz/2017/11/17/caution-urged-gene-drives-conservation-expert-reaction/.

71. Ibid.

72. Kevin Esvelt, "Aotearoa: Mistakes and Amends," *Responsive Science* (2017): 1–2, https://www.responsivescience.org/pub/aotearoa-amends.

73. Ibid., 2.

74. Ibid., 2.

75. Anna Boswell, "Stowaway Memory," *Pacific Dynamics: Journal of Interdisciplin-*

ary Research 2 (2018): 89–104; Michel Foucault, *The History of Sexuality*, vol. 1, trans. Robert Hurley (London: Penguin, 1978).

76. Jasanoff, "Ordering Knowledge, Ordering Society."

77. Pauly, *Controlling Life*.

78. Ibid., 114. See commentary at note 1.

Chapter 2

1. S. A. Barnett, *The Story of Rats* (Sydney: Allen and Unwin, 2002); Ken P. Aplin, Terry Chesser, and Jose ten Have, "Evolutionary Biology of the Genus *Rattus*: Profile of an Archetypal Rodent Pest," *Aciar Monograph Series* 96 (2003): 487–98; Ken P. Aplin, H. Suzuki, A. A. Chinen, R. T. Chesser, J. ten Have, et al., "Multiple Geographic Origins of Commensalism and Complex Dispersal History of Black Rats," *PLOS One* 6, no. 11 (2011): e26357.

2. Robert Browning, "The Pied Piper of Hamelin" (1843), in *The Complete Poetic and Dramatic Works of Robert Browning* (Boston: Houghton Mifflin, 1895), 268–71, at 268.

3. Jonathan Burt, *Rat* (London: Reaktion Books, 2006), 130–36; American Fancy Rat and Mouse Association, www.afrma.org, accessed August 10, 2019.

4. Alexandra Ossola, "The Complicated, Inconclusive Truth behind Rat Kings," *Atlas Obscura* 23 (December 2016), https://www.atlasobscura.com/articles/the-complicated-inconclusive-truth-behind-rat-kings.

5. Alison Abbott, "The Renaissance Rat," *Nature* 428 (April 1, 2004): 464–66, at 465; William Cheselden, *The Anatomy of the Human Body*, 3rd ed. (London: W. Bowyer, 1726), 291; J. Russell Lindsey and Henry J. Baker, "Historical Foundations," in *The Laboratory Rat*, ed. Mark Suckow, Steven H. Weisbroth, and Craig L. Franklin (Amsterdam: Elsevier, 2006), 2.

6. Alice Y. T. Feng and Chelsea G. Himsworth, "The Secret Life of the City Rat: A Review of the Ecology of Urban Norway and Black Rats (*Rattus norvegicus* and *Rattus rattus*)," *Urban Ecosystems* 17 (2014): 149–62.

7. Philip L. Armitage, "Unwelcome Companions: Ancient Rats Reviewed," *Antiquity* 68 (1994): 231–40; Aplin et al. (2011).

8. Conrad Gessner, *Historiae animalium*, 4 vol. (Zurich: Christoph Froschauer, 1551–58), vol. 1 (1551), 829. All translations are mine unless otherwise indicated.

9. Ibid., 1104.

10. John Berkenhout, *Outlines of the Natural History of Great Britain and Ireland*, 3 vol. (London: P. Elmsly, 1769), vol. 1: 5.

11. Burt, 29–32.

12. Gessner, 1:830.

13. James Rodwell, *The Rat: its History and Destructive Character* (London: G. Routledge, 1858), 51.

14. Winnie Hu, "Rats Are Taking Over New York City," *New York Times*, online edition, May 22, 2019; see also Kimiko de Freytas-Tamura, "Rats Have Ruled New York for 355 Years. Can a Mystery Bucket Stop Them?" *New York Times*, online edition, September 5, 2019.

15. Mary S. Lovell, *The Sisters: The Saga of the Mitford Family* (New York: W. W. Norton, 2001), 125.

16. Robert Sullivan, *Rats* (New York: Bloomsbury, 2004), 6.

17. Takashi Kuramoto, "Origin of Albino Laboratory Rats," *Bioresource Now!* 8 (November 2012): 1–2.

18. Rodwell, 28–30, 98–111; Gordon Stables, "Rats and Mice as Pets," *Boy's Own*

Paper 2 (1880): 542–43; Gordon Stables, “Doings for the Month,” *Boy’s Own Paper* 20 (1897–98): 143.

19. Lindsey and Baker, 2.

20. Feng and Himsworth, 152–53.

21. Bonnie Tocher Clause, “The Wistar Rat as a Right Choice: Establishing Mammalian Standards and the Ideal of a Standardized Mammal,” *Journal of the History of Biology* 26 (1993): 329–49, at 329.

22. Lindsey and Baker, 3–6; Clause.

23. Lindsey and Baker, 6–12; Clause, 330. For a parallel development in mice, see Karen Rader, *Making Mice* (Princeton: Princeton University Press, 2004).

24. Lindsey and Baker, 6–12, table 1–1, p. 7; Clause.

25. On “technological matter,” Hannah Landecker, *Culturing Life: How Cells Became Technologies* (Cambridge, MA: Harvard University Press, 2010).W. E. Castle, “Piebald Rats and the Theory of Genes,” *Proceedings of the National Academy of Sciences, USA* 5, no. 4 (April 1919): 126–30; Horace W. Feldman, “Linkage of Albino Allelomorphs in Rats and Mice,” *Genetics* 9 (1924): 487–92; J. A. Weir, “Harvard, Agriculture, and the Bussey Institution,” *Genetics* 136 (April 1994): 1227–31; Lindsey and Baker, 23–26.

26. Lindsey and Baker, 34–41.

27. Michael R. Dietrich, Rachel A. Ankeny, and Patrick M. Chen, “Publication Trends in Model Organism Research,” *Genetics* 198 (November 2014):787–94. For an overview of animal welfare legislation, see Anita Guerrini, *Experimenting with Humans and Animals: From Galen to Animal Rights* (Baltimore: Johns Hopkins University Press, 2003), esp. 145–48.

28. Abbott, 465.

29. Richard A. Gibbs et al., “Genome Sequence of the Brown Norway Rat Yields Insights into Mammalian Evolution,” *Nature* 428 (April 1, 2004): 493–521; Yuksel Agca and John K. Critser, “Assisted Reproductive Technologies and Genetic Modifications in Rats,” in *The Laboratory Rat*, ed. Mark Suckow, Steven H. Weisbroth, and Craig L. Franklin (Amsterdam: Elsevier, 2006), 165–89.

30. Abbott; Gibbs et al.

31. Ellen P. Neff, “CRISPR Improves Prospects for Transgenic Rats,” *Lab Animal* 48 (June1, 2019): 167; Abbott.

32. Neff; Abbott; Dietrich et al.

33. Neff.

34. https://www.cyagen.com/us/en/community/promotions/cyagen-summer-savings-promotion.html, accessed August 10, 2019.

35. https://speakingofresearch.com/facts/statistics/, accessed October 27, 2019; https://www.peta.org/issues/animals-used-for-experimentation/animals-laboratories/mice-rats-laboratories/ accessed October 27, 2019; National Research Council, *Guide for the Care and Use of Laboratory Animals*, 8th ed. (Washington, DC: National Academies Press, 2011).

36. http://www.ratlife.org/, accessed August 10, 2019.

37. *Brecon and Radnor Express*, May 21, 1914, 8.

38. Feng and Himsworth, 150; Chelsea G. Himsworth, Kirbee L. Parsons, Claire Jardine, and David M. Patrick, “Rats, Cities, People, and Pathogens: A Systematic Review and Narrative Synthesis of Literature Regarding the Ecology of Rat-Associated Zoonoses in Urban Centers,” *Vector-Borne and Zoonotic Diseases* 13 (2013): 349–59.

39. Feng and Himsworth. The COVID-19 pandemic in 2020 forced rats into different behaviors: see Emma Marris, “Rats Come Out of Hiding as Lockdowns Eliminate Urban

Trash," April 3, 2020, https://www.nationalgeographic.com/animals/2020/03/urban-rats-search-for-food-coronavirus/.

40. On urban ecology, see Menno Schilthuizen, *Darwin Comes to Town: How the Urban Jungle Drives Evolution* (New York: Picador, 2018).

41. Feng and Himsworth, 156; Christine Keiner, "Wartime Rat Control, Rodent Ecology, and the Rise and Fall of Chemical Rodenticides," *Endeavour* 29, no. 3 (September 2005): 119–25, at 119–20, citing Charles Elton, "Research on Rodent Control by the Bureau of Animal Population, September 1939 to July 1947," in *Control of Rats and Mice*, ed. D. Chitty and H. N. Southern, vol. 1 (Oxford: Clarendon Press, 1954), 1–24.

42. Sullivan, 164–83; Keiner, 120.

43. For example, Michael G. Vann, "Of Rats, Rice, and Race: The Great Hanoi Rat Massacre, an Episode in French Colonial History," *French Colonial History* 4 (2003): 191–203.

44. Sullivan, 9, 76–85; Rodwell, 84–93; David E. Lantz, *The Brown Rat in the United States*, US Biological Survey, Bulletin no. 33 (Washington, DC: Government Printing Office, 1909), 41–54. On the ineffectiveness of cats in urban rat control, see Michael H. Parsons, Peter B. Banks, Michael A. Deutsch, and Jason Munshi-South, "Temporal and Space-Use Changes by Rats in Response to Predation by Feral Cats in an Urban Ecosystem," *Frontiers in Ecology and Evolution* 6 (September 27, 2018), article 146.

45. Lantz, 41–54; S.A. Barnett, J.D. Blaxland, F.B. Leech, and Mary M. Spencer, "A Concentrate of Red Squill as a Rat poison, and its Toxicity to Domestic Animals," *Journal of Hygiene,* 47 (1949), 431–33.

46. Keiner, 120; Curt P. Richter, "Experiences of a Reluctant Rat-Catcher: The Common Norway Rat—Friend or Enemy?" *Proceedings of the American Philosophical Society* 112, no. 6 (December 1968): 403–15, at 404.

47. Keiner.

48. Keiner, 120–23; Richter.

49. Richter; Keiner, 121–24; "ANTU," ScienceDirect, accessed 15 August 2019, https://www.sciencedirect.com/topics/pharmacology-toxicology-and-pharmaceutical-science/antu/.

50. Richter, 410; Keiner, 123–24.

51. P. Smith, M. G. Townsend and R. H. Smith, "A Cost of Resistance in the Brown Rat? Reduced Growth Rate in Warfarin-Resistant Lines," *Functional Ecology* 5 (1991): 441–47, at 441; Sullivan 122–23; "Warfarin," ScienceDirect, accessed 15 August 2019, https://www.sciencedirect.com/topics/agricultural-and-biological-sciences/warfarin/. Warfarin is also used as a human therapeutic agent in the prevention of blood clots that could cause strokes.

52. T. C. Boyle, *When the Killing's Done* (New York: Viking, 2011).

53. Gregg Howald, C. Josh Donlan, Kate R. Faulkner, et al., "Eradication of Black Rats *Rattus rattus* from Anacapa Island," *Oryx* 44 (2010): 30–40, at 30.

54. Howald et al.; Kelly M. Newton, Matthew McKown, Coral Wolf, et al., "Response of Native Species 10 Years after Rat Eradication on Anacapa Island, California," *Journal of Fish and Wildlife Management* 7 (2016): 72–85, at 73.

55. Howald et al., 30, 34–36; Newton et al.

56. Global Invasive Species Database (GISD), 2015, Species Profile *Rattus rattus*, accessed August 21, 2019, http://www.iucngisd.org/gisd/species.php?sc=19.

57. James C. Russell and Nick D. Holmes, "Tropical Island Conservation: Rat Eradication for Species Recovery," *Biological Conservation* 185 (2015): 1–7; Tim S. Doherty,

Alistair S. Glen, Dale G. Nimmo, et al., "Invasive Predators and Global Biodiversity Loss," *PNAS* 113, no. 40 (2016), 11261–65.

58. B. W. Thomas and R. H. Taylor, "A History of Ground-Based Rodent Eradication Techniques Developed in New Zealand, 1959–1993," in *Turning the Tide: The Eradication of Invasive Species*, ed. C. R. Veitch and M. N. Clout, IUCN SSC Invasive Species Specialist Group (Gland, Switzerland: IUCN, 2002), 301–10. A good overview of New Zealand predator strategy is James C. Russell, John G. Innes, Philip H. Brown, and Andrea E. Byrom, "Predator-Free New Zealand: Conservation Country," *Bioscience* 65 (2015): 520–35.

59. J. A. Wodzicki, *Introduced Mammals of New Zealand: An Ecological and Economic Survey*, Department of Scientific & Industrial Research Bulletin no. 98 (Wellington, NZ: Department of Scientific & Industrial Research, 1950), Introduction; Thomas and Taylor, 301.

60. Elizabeth A. Bell, Brian D. Bell, and Don V. Merton, "The Legacy of Big South Cape: Rat Irruption to Rat Eradication," *New Zealand Journal of Ecology* 40 (2016): 212–18, at 214.

61. Bell et al.

62. Bell et al.; the quote appears on p. 216.

63. A. T. Proudfoot, S. M. Bradbury, and J. A. Vale, "Sodium Fluoroacetate Poisoning," *Toxicological Reviews* 25 (2006): 213–19; "Too Efficient Poison," *Time* 46, no. 26 (December 24, 1945): 70.

64. Accessed August 12, 2019, https://www.doc.govt.nz/nature/pests-and-threats/methods-of-control/1080/why-we-use-1080/.

65. United States Environmental Protection Agency, "Registration Eligibility Decision, Sodium Flouroacetate," June 1995.

66. David R. Towns, Ian A. E. Atkinson, and Charles H. Daugherty, "Have the Harmful Effects of Introduced Rats on Islands Been Exaggerated?" *Biological Invasions* 8 (2006): 863–91, at 865.

67. M. Le Corre, D. K. Danckwerts, D. Ringler, et al., "Seabird Recovery and Vegetation Dynamics after Norway Rat Eradication at Tromelin Island, Western Indian Ocean," *Biological Conservation* 185 (2015): 85–94.

68. Erika Zavaleta, Richard J. Hobbs, and Harold A. Mooney, "Viewing Invasive Species Removal in a Whole-Ecosystem Context," *Trends in Ecology and Evolution* 16 (2001): 454–59. Abraham Gibson's essay in the present volume provides another example of unintended consequences and of how animals can become vermin.

69. Although on possums rather than rats, the best discussion I have found of the moral quandary of large-scale extermination is Annie Potts, "Kiwis against Possums: A Critical Analysis of Anti-possum Rhetoric in Aotearoa New Zealand," *Society and Animals* 17 (2009): 1–20. For a different view, see Alastair S. Gunn, "Environmental Ethics in a New Zealand Context," *New Zealand Journal of Forestry* (February 2007): 7–12. See also Mark J. Farnsworth, Helen Watson, and Nigel J. Adams, "Understanding Attitudes towards the Control of Nonnative Wild and Feral Mammals: Similarities and Differences in the Opinions of the General Public, Animal Protectionists, and Conservationists in New Zealand (Aotearoa)," *Journal of Applied Animal Welfare Science* 17 (2014): 1–17.

Chapter 3

1. On cloning, see Jane Maienschein, *Whose View of Life? Embryos, Cloning and Stem Cells* (Cambridge, MA: Harvard University Press, 2003); Christina Brandt. "Hybrid

Times: Theses on the Temporalities of Cloning," *History and Philosophy of the Life Sciences* (2013): 75–81; Hannah Landecker, *Culturing Life: How Cells Became Technologies* (Cambridge, MA: Harvard University Press, 2007). Sarah Franklin has already demonstrated the historical significance of placing "Dolly the sheep" in her original agricultural context in *Dolly Mixtures: The Remaking of Genealogy* (Chapel Hill, NC: Duke University Press, 2007). For exemplary histories of engineering incorporating technology, state, and capital see, among many other possible references, Chandra Mukerji, *Impossible Engineering. Technology and Territoriality on the Canal du Midi* (Princeton: Princeton University Press, 2009); Ken Alder, *Engineering the Revolution: Arms and Enlightenment in France, 1763–1815* (Chicago: University of Chicago Press, 2010); Gabrielle Hecht, *The Radiance of France: Nuclear Power and National Identity after World War II* (Cambridge, MA: MIT Press, 2009).

2. Amy E. Slaton, *Race, Rigor, and Selectivity in US Engineering: The History of an Occupational Color Line.* (Cambridge, MA: Harvard University Press, 2010); Projit Bihari Mukharji, "Profiling the Profiloscope: Facialization of Race Technologies and the Rise of Biometric Nationalism in Inter-war British India," *History and Technology* 31, no. 4 (2015): 376–96; Edward Jones-Imhotep, "Malleability and Machines: Glenn Gould and the Technological Self," *Technology and Culture* 57, no. 2 (2016): 287–321.

3. Sarah Franklin and Margaret Lock, eds., *Remaking Life and Death: Towards an Anthropology of the Biosciences* (Santa Fe: School of American Research Press, 2003); Landecker, *Culturing Life*; Maienschein, *Whose View of Life?*; Carrie Friese, *Cloning Wild Life: Zoos, Captivity, and the Future of Endangered Animals* (New York: NYU Press, 2013).

4. "Citrusscape" derives from the more generic notion of "cropscape" defined as "the ever-mutating ecologies, or matrices, comprising assemblages of nonhumans and humans, within which a particular crop in a particular place and time flourishes or fails." Francesca Bray, Barbara Hahn, John Lourdusamy, and Tiago Saraiva, "Cropscapes and History: Reflections on Rootedness and Mobility," *Transfers* 9, no. 1 (2019): 20–41. On the presence of the Californian model in Palestine during the Mandate years see Nahum Karlinsky, *California Dreaming: Ideology, Society, and Technology in the Citrus Industry of Palestine, 1890–1939* (New York: SUNY Press, 2012).

5. Aaron Aaronsohn, *Agricultural and Botanical Explorations in Palestine* (Washington, DC: United States, Department of Agriculture, 1910): 3.

6. On plant hunters at the USDA, see Courtney Fullilove, *The Profit of the Earth: The Global Seeds of American Agriculture* (Chicago: University of Chicago Press, 2017); Philip J. Pauly, "The Beauty and Menace of the Japanese Cherry Trees: Conflicting Visions of American Ecological Independence," *Isis* 87, no. 1 (1996): 51–73.

7. On American capital and Zionism see, for example, Allon Gal, ed., *Envisioning Israel: The Changing Ideals and Images of North American Jews* (Jerusalem: The Hebrew University, Magnes Press, 1996).

8. For biographic details on Aaronsohn and his family, see Shmuel Katz, *The Aaronsohn Saga* (Jerusalem: Gefen, 2007).

9. Derek J. Penslar, *Zionism and Technocracy: The Engineering of Jewish Settlement in Palestine, 1870–1918* (Bloomington: Indiana University Press, 1991); "Zionism, Colonialism and Technocracy: Otto Warburg and the Commission for the Exploration of Palestine," *Journal of Contemporary History* 25 (1990): 143–60; Michael N. Barnett, ed., *Israel in Comparative Perspective: Challenging the Conventional Wisdom* (New York: SUNY Press, 2012); Gabriel Piterberg, "The Zionist Colonization of Palestine in the Context of Comparative Settler Colonialism," *Palestine and the Palestinians in the 21st Century,*

ed. Rochelle Davis and Mimi Kirk (Bloomington: Indiana University Press, 2013); Dana von Suffrin, *Pflanzen für Palästina: Otto Warburg und die Naturwissenschaften im Jischuw* (Tübingen: Mohr Siebeck, 2019).

10. On botanical visions of the German Empire see Andrew Zimmerman, *Alabama in Africa: Booker T. Washington, the German Empire, and the Globalization of the New South* (Princeton: Princeton University Press, 2012)

11. Aaronsohn, *Agricultural and Botanical*, 7

12. Edward W. Said, *Orientalism* (New York: Vintage Books, 1994): 171.

13. Wael B. Hallaq, *Restating Orientalism: A Critique of Modern Knowledge* (New York: Columbia University Press, 2018), 17.

14. For a similar perspective on the importance of orientalism for the transformation of the Palestinian agricultural landscape, see Tamar Novick, "Milk and Honey: Technologies of plenty in the Making of a Holy Land, 1880–1960" (PhD diss., University of Pennsylvania, 2014).

15. Aaronsohn, *Agricultural and Botanical.*

16. Ibid.

17. Ibid., 8–13.

18. Michael A. Osborne, *Nature, the Exotic, and the Science of French Colonialism* (Bloomington, 1994); Warwick Anderson, "Climates of Opinion: Acclimatization in Nineteenth-Century France and England," *Victorian Studies* 35, no. 2 (1992): 135–57; Christophe Bonneuil, "Mettre en ordre et discipliner les tropiques: les sciences du vegetal dans l'empire français, 1870–1940" (PhD diss., Paris VII, 1997).

19. A. Aaronsohn and S. Soskin, "Die Orangengärten von Jaffa," *Der Tropenpflanzer, Zeitschrift für Tropische Landwirtschaft,* 6, no. 7 (1902): 341–61.

20. Ibid.

21. On the difficulties of early Jewish settlements see, for example, Ran Aaronsohn, "The Beginnings of Modern Jewish Agriculture in Palestine: 'Indigenous' versus 'Imported,'" *Agricultural History* 69 (1995): 438–53.

22. On Ottoman reforms and their importance in Palestinian countryside, see Rashid Khalidi, *Palestinian Identity: The Construction of Modern National Consciousness* (New York: Columbia University Press, 2010); Ted Swedenburg, *Memories of Revolt: The 1936–1939 Rebellion and the Palestinian National Past* (Minneapolis: University of Minnesota Press, 1995); Alexander Schölch, *Palestine in Transformation, 1856–1882: Studies in Social, Economic, and Political Development* (Washington, DC: Institute for Palestine Studies, 1993).

23. Jonathan Schneer, *The Balfour Declaration: The Origins of the Arab-Israeli Conflict* (London: Bloomsbury, 2010); Rashid Khalidi, *British Policy towards Syria and Palestine, 1906–1914* (Reading, UK: Ithaca Press, 1980).

24. On details of the acquisition of Petah Tikvah by Jewish settlers and local resistance by Arab peasants, see Khalidi, *Palestinian Identity*, 99–100.

25. Karlinsky, *California Dreaming.*

26. J. Henry Burke, *Recent Changes in the Citrus Industry of Israel* (Washington, DC: USDA, 1951); William Hazen, *The Citrus Industry of Palestine* (Washington, DC: USDA, 1938).

27. Irit Amit-Cohen, *Zionism and Free Enterprise: The Story of Private Entrepreneurs in Citrus Plantations in Palestine in the 1920s and 1930s* (Berlin: De Gruyter, 2012): 25–27.

28. Ibid.

29. Karlinsky, *California Dreaming*; Amit-Cohen, *Zionism and Free Enterprise.*

30. S. Ilan Troen, "American Experts in the Design of Zionist Society: The Reports of

Elwood Mead and Robert Nathan," in *Envisioning Israel: The Changing Ideals and Images of North American Jews*, ed. Allon Gal (Detroit: Wayne State University Press, 1996): 193–218.

31. H. Clark Powell, *The Citrus Industry in Palestine* (Government of Palestine, 1928).

32. L. C. Powell, ed., *H. Clark Powell, 1900–1938: Memoirs of His Life and a Bibliography of His Writings* (Los Angeles: Privately printed, 1939).

33. Tiago Saraiva, "The Scientific Co-op: Cloning Oranges and Democracy in the Progressive Era," in *New Materials: Towards a History of Consistency*, ed. Amy Slaton (Amherst, MA: Lever Press, 2020): 118–44.

34. Saraiva, "Scientific Co-Op."

35. Powell, *Citrus Industry*, 5.

36. Ibid.

37. J. D. Oppenheim, "Memorandum on the Standardization of the Jaffa Orange" (1936), Box 10, Folder 7, Herbert J. Webber papers, UA 059. University of California, Riverside Libraries, Special Collections & Archives, University of California, Riverside.

38. Yuval Ben-Bassat, *Petitioning the Sultan: Protests and Justice in Late Ottoman Palestine* (New York: I. B. Tauris, 2013):165–66.

39. Roza I. M. El-Eini, *Mandated Landscape: British Imperial Rule in Palestine* (New York: Routledge, 2006), 160; Shaul Katz and Joseph Ben-David, "Scientific Research and Agricultural Innovation in Israel," *Minerva* 13, no. 2 (1970): 152–82.

40. Oppenheim, "Memorandum . . ."

41. Powell, *Citrus Industry in Palestine*, 23.

42. Ibid.

43. Robert W. Hodgson, *Analyzing Citrus Orchards by Means of Simple Tree Records* (Berkeley: University of California, College of Agriculture, Agricultural Experiment Station, circular 266, 1923), 2.

44. Karlinsky, *California Dreaming*.

45. S. Tolkowsky, *Achievements and Prospects in Palestine* (London: English Zionist Federation, 1917), 2.

46. Ibid, 2–3.

47. S. Tolkowsky, *The Gateway of Palestine: A History of Jaffa* (New York: Albert & Charles Boni, 1925).

48. This appendix is the origin of a full book dedicated to the history of citrus: *Hesperides*.

49. Tolkowsky, *The Gateway of Palestine*, 178–81.

50. J. D. Oppenheim, "A Preliminary Note on the Origin of the Jaffa Orange," *Genetica* 9 (1927): 516–20, on 519. See also P. Spiegal-Roy, "On the Chimeral Nature of the Shamouti Orange," *Euphytica* 28 (1979): 361–65.

51. Swedenburg, *Memories of Revolt*.

52. Yael Allweil, "Plantation: Modern-Vernacular Housing and Settlement in Ottoman Palestine, 1858–1918," *Architecture beyond Europe* 9–10 (2016).

53. "Jaffa Orange," *Bulletin of Miscellaneous Information (Royal Botanic Gardens, Kew)* 88 (1894): 117–19.

54. For an informed discussion of this historiography and the still dominating orientalist framework, see Sherene Seikaly, *Men of Capital: Scarcity and Economy in Mandate Palestine* (Stanford: Stanford University Press, 2016).

55. Seikaly, *Men of Capital*, 17.

56. Khalil Totah, *Arab Progress in Palestine* (New York: Institute of Arab American Affairs, 1946), 3. For the Anglo-American Committee of Inquiry, see Rashid Khalidi,

The Hundred Years' War on Palestine: A History of Settler Colonialism and Resistance, 1917–2017 (New York: Metropolitan Books, 2020).

57. Ibid, 4–5.

58. For the figures on orchard density, see Hazen, *The Citrus Industry*, 8–9.

59. See Robert W. Hodgson, *Comparison and Contrasts in Citriculture between Palestine and California* (Jaffa: M. Shoham's Press, 1931).

Chapter 4

1. Horse Rittel and Melvin Webber, "Dilemmas in a General Theory of Planning," *Policy Sciences* 4 (1973): 155–69.

2. Philip J. Pauly, *Controlling Life: Jacques Loeb and the Engineering Ideal in Biology* (New York: Oxford University Press, 1987), 4.

3. Melinda A. Zeder, "Pathways to Animal Domestication," in *Biodiversity in Agriculture: Domestication, Evolution, and Sustainability*, ed. P. Gepts et al. (Cambridge: Cambridge University Press, 2012): 241–42.

4. Greger Larson et al., "Worldwide Phylogeography of Wild Boar Reveals Multiple Centers of Pig Domestication," *Science* 307 (March 11, 2005): 1618–21; Greger Larson et al., "Patterns of East Asian Pig Domestication, Migration, and Turnover Revealed by Modern and Ancient DNA," *Proceedings of the National Academy of Sciences* 107 (April 27, 2010): 7686–91; Sam White, "From Globalized Pig Breeds to Capitalist Pigs: A Study in Animal Cultures and Evolutionary History," *Environmental History* 16 (January 2011): 96; Laurent A. Frantz, "Evidence of Long-Term Gene Flow and Selection during Domestication from Analyses of Eurasian Wild and Domestic Pig Genomes," *Nature Genetics* 47 (August 31, 2015): 1141–48.

5. Charles Hudson, *Knights of Spain, Warriors of the Sun: Hernando de Soto and the South's Ancient Chiefdoms* (Athens: University of Georgia Press, 1997), 170, 351.

6. John J. Mayer, "Taxonomy and History of Wild Pigs in the United States," in *Wild Pigs: Biology, Damage, Control Techniques and Management* (Aiken, SC: Savannah River National Laboratory, 2009), 9.

7. Hudson, *Knights of Spain, Warriors of the Sun*, 168, 266, 378, 542.

8. William Strachey, "For the Colony in Virginea Britannia," *Tracts and other Papers Relating Principally to the Origin, Settlement, and Progress of the Colonies in North America from the Discovery of the Country to the Year 1776*, vol. 3, ed. Peter Force (New York: Peter Smith, [1612] 1947), 15; Ralph Hamor, *A True Discourse of the Present State of Virginia*, Virginia State Library Publications, no. 3 (Richmond: Virginia State Library, 1957), 23.

9. "Act VI, March 1629–30," *Hening's Statutes at Large: being a Collection of all the Laws of Virginia from the First Session of the Legislature*, vol. 1, ed. William Waller Hening (Charlottesville: University of Virginia Press, 1969), 152.

10. "Act LXIII," *Hening's Statutes at Large: being a Collection of all the Laws of Virginia from the First Session of the Legislature*, vol. 1, ed. William Waller Hening (Charlottesville: University of Virginia Press, 1969), 176.

11. Virginia DeJohn Anderson, "Animals into the Wilderness: The Development of Livestock Husbandry in the Seventeenth-Century Chesapeake," *William and Mary Quarterly* 59 (April 2002): 398–99; Cary Carson et al., "New World, Real World: Improvising English Culture in Seventeenth-Century Virginia," *Journal of Southern History* 74 (2008): 31–88.

12. Robert Beverley, *The History and Present State of Virginia* (London: R. Parker, 1705), 262–63.

13. Pauly, *Controlling Life*, 4.

14. S. M. Shepard, *The Hog in America, Past and Present* (Indianapolis: Swine Breeders' Journal, 1886), 224–25.

15. White, "From Capitalist Pigs to Globalized Pig Breeds," 94–120; James Westfall Thompson, *A History of Livestock Raising in the United States, 1607–1860* (Washington, DC: US Department of Agriculture, 1942): 10.

16. "The Fence Law," *Augusta Chronicle*, June 25, 1878, 1.

17. Brook Blevins, "Fence/Stock Laws," *New Encyclopedia of Southern Culture*, vol. 11: Agriculture and Industry, ed. Melissa Walker and James C. Cobb (Chapel Hill: University of North Carolina Press, 2008), 159; R. Ben Brown, "Free Men and Free Pigs: Closing the Southern Range and the American Property Tradition," *Radical History Review*, no. 108 (2010): 117–37; J. Crawford King Jr., "The Closing of the Southern Range: An Exploratory Study," *Journal of Southern History* 48 (1982): 53–70; Steven Hahn, *The Roots of Southern Populism: Yeoman Farmers and the Transformation of the Georgia Upcountry, 1850–1890* (New York: Oxford University Press, 1983); Shawn Everett Kantor, *Politics and Property Rights: The Closing of the Open Range in the Postbellum South* (Chicago: University of Chicago Press, 1998).

18. Bass, "How 'bout a Hand for the Hog," *Southern Cultures* 1 (1995): 310; Roger Horowitz, *Putting Meat on the American Table: Taste, Technology, Transformation* (Baltimore: Johns Hopkins University Press, 2006), 43–74; Perry Van Ewing, *Southern Pork Production* (New York: Orange Judd Company, 1918), 8–9, 30.

19. Tobe Hodge, "Razor-Backs," *American Magazine* 7, no. 2 (December 1887): 254.

20. J. M. Murphy, "Florida Razorbacks," *Outing* 19, no. 2 (November 1891): 117.

21. Shepard, *The Hog in America, Past and Present*, 20.

22. John F. Reiger, *American Sportsmen and the Origins of Conservation* (Corvallis: Oregon State University Press, 2001).

23. Scott E. Giltner, *Hunting and Fishing in the New South: Black Labor and White Leisure after the Civil War* (Baltimore: Johns Hopkins University Press, 2008), 118.

24. Leonidas Hubbard, "Hunting Wild Hogs as a Sport," *New York Times* (September 1, 1901), sm9.

25. Frank G. Carpenter, "The Isle of Millionaires," *Los Angeles Times*, December 27, 1896, 23; William Barton McCash and June Hall McCash, *The Jekyll Island Club: Southern Haven for America's Millionaires* (Athens: University of Georgia, 1989), 20; Montgomery M. Folsom, "Game on Jekyl," *Atlanta Constitution*, February 17, 1889, 3; "The Jekyl Island Club," *New York Times*, September 9, 1889, 5.

26. Mayer, "Taxonomy and History of Wild Pigs in the United States," 11; see also: William H. Stiver and E. Kim Delozier, "Great Smoky Mountains National Park Wild Hog Control Program," in *Wild Pigs: Biology, Damage, Control Techniques and Management*, 341–42.

27. T. S. Palmer, "The Danger of Introducing Noxious Animals and Birds," *Forest and Stream* 52 (June 3, 1899): 425.

28. "The Razorback Hog," *Forest and Stream* 74 (March 19, 1910): 2.

29. LeRoy C. Stegeman, "The European Wild Boar in the Cherokee National Forest, Tennessee," *Journal of Mammalogy* 19, no. 3 (August 1938): 282.

30. Charles Elton, *The Ecology of Invasions by Plants and Animals* (Chicago: University of Chicago Press, 1958), 89.

31. Tom McKnight, *Feral Livestock in Anglo-America* (Berkeley: University of California Press, 1964), 43.

32. John Mayer and Lehr Brisbin, "Introduction," in *Wild Pigs: Biology, Damage,*

Control Techniques and Management, ed. John Mayer and Lehr Brisbin (Aiken: Savannah River National Laboratory, 2009), 1.

33. "The Razorback Hog," *Forest and Stream* 74 (March 19, 1910): 2.

34. Blake E. McCann, "Mitochondrial Diversity Supports Multiple Origins for Invasive Pigs," *Journal of Wildlife Management* 78 (2014): 202–13; M. Noelia Barrios-Garcia and Sebastian A. Ballari, "Impact of Wild Boar (*Sus scrofa*) in Its Introduced and Native Range: A Review," *Biological Invasions* 14 (2012): 2283–300; Sarah N. Bevins et al., "Consequences Associated with the Recent Range Expansion of Nonnative Feral Swine," *BioScience* 64 (April 2014): 291–99.

35. James C. Lewis, "Observations of Pen-Reared European Hogs Released for Stocking," *Journal of Wildlife Management* 30 (October 1966): 832–35.

36. Mayer, "Taxonomy and History of Wild Pigs in the United States," 8, 13; Philip S. Gipson, Bill Hlavachick, and Tommie Berger, "Range Expansion by Wild Hogs across the Central United States," *Wildlife Society Bulletin* 26 (Summer 1998): 280.

37. Mayer, "Taxonomy and History of Wild Pigs in the United States," 13.

38. Jeffrey Greene, *The Golden-Bristled Boar: Last Ferocious Beast of the Forest* (Charlottesville: University of Virginia Press, 2012); Ian Frazier, "Hogs Wild," *New Yorker*, December 12, 2005, 71–83; Ted Chamberlain, "Photo in the News: Hogzilla Is No Hogwash," *National Geographic News*, March 22, 2005.

39. Jay Reeves, "Alabama Officials Say 'Monster Pig' Hunt Was Legal," *Daily Times* (Florence, Alabama), June 1, 2007, 2b.

40. Brian Stickland, "Monster Pig Raised on Fruithurst Farm, not a Wild Hog," *Columbus Ledger-Enquirer*, June 1, 2007.

41. John Mayer and Lehr Brisbin, *Wild Pigs of the United States: Their History, Morphology, and Current Status* (Athens: University of Georgia Press, 1991), 2.

42. John Mayer and Lehr Brisbin, "Introduction," in *Wild Pigs: Biology, Damage, Control Techniques and Management*, ed. John Mayer and Lehr Brisbin (Aiken: Savannah River National Laboratory, 2009), 1. See also: Meredith McClure et al., "Modeling and Mapping the Probability of Occurrence of Invasive Wild Pigs across the Contiguous United States," *PLoS ONE* (August 12, 2015): 1–17.

43. Will Brantley, "The Pig Report," *Field & Stream* (September 2015): 78–82.

44. Lehr Brisbin and John J. Mayer, "Problem Pigs in a Poke: A Good Pool of Data," *Science* 294 (November 9, 2011): 1280–81.

45. Ed Crews, "Ossabaw Island Pigs," *Colonial Williamsburg Journal* (2010); Scott Magelssen, "Resuscitating the Extinct: The Backbreeding of Historic Animals at U.S. Living History Museums," *Drama Review* 47 (2003): 98–109.

46. Andrew Russell and Lee Vinsel, "Let's Get Excited about Maintenance!" *New York Times*, July 22, 2017.

Chapter 5

1. Dominic J. Berry, "Making DNA and Its Becoming an Experimental Commodity," *History and Technology* 35 (2019): 375–404.

2. Robert Bud, "Life, DNA and the Model," *British Journal for the History of Science* 46 (June 2013): 311–34.

3. Pablo Schyfter, Emma Frow, and Jane Calvert, "Synthetic Biology: Making Biology into an Engineering Discipline," *Engineering Studies* 5 (2013): 1–5; Massimiliano Simons, "The Diversity of Engineering in Synthetic Biology," *NanoEthics* 14 (2020): 71–91; Deborah Scott, Dominic J. Berry, and Jane Calvert, "Synthetic Biology," in *Routledge*

Handbook of Genomics, Health and Society, ed. Sahra Gibbon, Barbara Prainsack, Stephen Hilgartner, and Janelle Lamoreaux (Abingdon: Routledge, 2018), 300–307; Rami Koskinen, "Synthetic Biology and the Search for Alternative Genetic Systems: Taking How-Possibly Models Seriously," *European Journal for Philosophy of Science* 7 (2017): 493–506.

4. Lily E. Kay, "Life as Technology: Representing, Intervening, and Molecularizing," in *The Philosophy and History of Molecular Biology: New Perspectives* (Dordrecht: Kluwer Academic Publishers, 1996), 87–100; Adele E. Clarke and Joan H. Fujimura, eds., *The Right Tools for the Job: At Work in Twentieth-Century Life Sciences* (Princeton: Princeton University Press, 1992).

5. I here focus exclusively on the relationship between scientists or engineers and their materials, particularly the ways in which they reason about and with those materials. I learned a great deal in this respect from Victoria Lee, "Wild Toxicity, Cultivated Safety: Aflatoxin and Kōji Classification as Knowledge Infrastructure," *History and Technology* 35 (2019): 405–24; Dmitriy Myelnikov, "Tinkering with Genes and Embryos: The Multiple Invention of Transgenic Mice c.1980," *History and Technology* 35 (2019): 425–52; Vivian Ling and Lijing Jiang, "A Different Kind of Synthesis: Artificial Synthesis of Insulin in Socialist China," *History and Technology* 35 (2019): 453–80; Daniel Liu, "The Cell and Protoplasm as Container, Object, and Substance, 1835–1861," *Journal of the History of Biology* 50 (2017): 889–925.

6. Pnina Abir-Am, "From Biochemistry to Molecular Biology: DNA and the Acculturated Journey of the Critic of Science Erwin Chargaff," *History and Philosophy of the Life Sciences* 2 (1980): 3–60; Horace Freeland Judson, "Frederick Sanger, Erwin Chargaff, and the Metamorphosis of Specificity," *Gene* 135 (1993): 19–23; Michel Morange, *A History of Molecular Biology* (Cambridge, MA: Harvard University Press, 1998).

7. Erwin Chargaff, *Essays on Nucleic Acids* (New York: Elsevier, 1963), 174.

8. Erwin Chargaff, "The Coagulation of Blood," *Advances in Enzymology and Related Areas of Molecular Biology* 5 (1945): 31–65; Erwin Chargaff, "The Formation of the Phosphorus Compounds in Egg Yolk," *Journal of Biological Chemistry* 142 (1942): 505–12; Erwin Chargaff, "A Study of Lipoproteins," *Journal of Biological Chemistry* 142 (1942): 491–504.

9. My thanks to Alok Srivastiva for emphasising the importance of his training in this particular school of chemistry.

10. Mary Jo Nye, *From Chemical Philosophy to Theoretical Chemistry: Dynamics of Matter and Dynamics of Disciplines, 1800–1950* (Berkeley: University of California Press, 1993), 68.

11. Erwin Chargaff, "What Really Is DNA? Remarks on the Changing Aspects of a Scientific Concept," *Progress in Nucleic Acid Research and Molecular Biology* 8 (1968): 297–333, 327.

12. On Linus Pauling's "architectural approach to chemistry" see Mary Jo Nye, "Paper Tools and Molecular Architecture in the Chemistry of Linus Pauling," in *Tools and Modes of Representation in the Laboratory Sciences*, ed. Ursula Klein (Dordrecht: Kluwer Academic Publishers, 2001), 95–132.

13. As starting points consider Alan J. Roche's discussion of Kekulé's architectural training: Alan J. Roche, *Image and Reality: Kekulé, Kopp, and the Scientific Imagination* (Chicago: University of Chicago Press, 2010), 65–66. Also the help that Arnold Elioart received from an architecture instructor when creating models for his *The Arrangement of Atoms in Space*, as observed in Eric Francoeur, *The Forgotten Tool: A Socio-Historical Analysis of the Development and Use of Mechanical Molecular Models in Chemistry and Allied Disciplines*, (unpublished PhD diss., McGill University, 1998), 75. Konstantin Kipri-

janov shared with me a copy of his unpublished PhD dissertation that serves as a very comprehensive introduction to the historiography. Konstantin Kiprijanov, *Printing Lines and Letters: How Structural Formulae Became the Standard Notation of Organic Chemistry* (unpublished PhD diss., University of Leeds, 2018).

14. Walter Vincenti, *What Engineers Know and How They Know It* (Baltimore: Johns Hopkins University Press, 1990).

15. Stephen Zamenhof and Erwin Chargaff, "Evidence of the Existence of a Core in Desoxyribonucleic Acids," *Journal of Biological Chemistry* 178 (1949): 531–32.

16. Ibid.

17. Lily E. Kay, "Molecular Biology and Pauling's Immunochemistry: A Neglected Dimension," *History and Philosophy of the Life Sciences* 11 (1989): 211–19.

18. Erwin Chargaff, "Chemical Specificity of Nucleic Acids and Mechanism of their Enzymatic Degradation," *Experientia* 6 (1950): 201–9, 208.

19. George Brawerman and Erwin Chargaff, "On a Deoxyribonuclease from Germinating Barley," *Journal of Biological Chemistry* 210 (1954): 445–54.

20. Erwin Chargaff, "Isolation and Composition of the Deoxypentose Nucleic Acids and of the Corresponding Nucleoproteins," in *The Nucleic Acids: Chemistry and Biology*, vol. 1, ed. Erwin Chargaff and J. N. Davidson (New York: Academic Press, 1955), 340.

21. George Borg, "On 'the Application of Science to Science Itself': Chemistry, Instruments, and the Scientific Labor Process," *Studies in History and Philosophy of Science* 79 (2020): 41–56; Leo B. Slater, "Instruments and Rules: R. B. Woodward and the Tools of Twentieth-Century Organic Chemistry," *Studies in History and Philosophy of Science* 33 (2002): 1–33; Angela Creager, "Wendell Stanley's Dream of a Free-Standing Biochemistry Department at the University of California, Berkeley," *Journal of the History of Biology* 29 (1996): 331–60.

22. Chargaff, "Isolation and Composition of the Deoxypentose Nucleic Acids and of the Corresponding Nucleoproteins," 340.

23. The term "narrative of nature" is used in a draft paper by Robert Meunier, to build up the contrast between narratives that scientists write of their research, as opposed to the narrative of their phenomena of interest. "Research Articles as Narratives: Familiarizing Communities with an Approach," in *Narrative Science: Reasoning, Representing and Knowing since 1800*, ed. Mary S. Morgan, Kim M. Hajek, and Dominic J. Berry (Cambridge: Cambridge University Press, forthcoming 2021).

24. Chargaff, "Isolation and Composition of the Deoxypentose Nucleic Acids and of the Corresponding Nucleoproteins," 345.

25. Christoph Tamm, M. E. Hodes, and Erwin Chargaff, "The Formation of Apurinic Acid from the Desoxyribonucleic Acid of Calf Thymus," *Journal of Biological Chemistry* 195 (1952): 49–63.

26. Christoph Tamm, Herman S. Shapiro, Rakoma Lipshitz, and Erwin Chargaff, "Distribution Density of Nucleotides within a Desoxyribonucleic Acid Chain," *Journal of Biological Chemistry* 203 (1953): 673–88, 685.

27. Herman S. Shapiro and Erwin Chargaff, "Studies on the Nucleotide Arrangement in Deoxyribonucleic Acids, 8: A Comparison of Procedures for the Determination of the Frequency of Pyrimidine Nucleotide Runs," *Biochemica et Biophysic Acta* 91 (1964): 262–70, 263.

28. Herman S. Shapiro, Rivka Rudner, Kin-Ichiro Miura, and Erwin Chargaff, "Inferences from the Distribution of Pyrimidine Isostichs in Deoxyribonucleic Acids," *Nature* (March 13, 1965): 1068–70, 1068.

29. Erwin Chargaff, "What Really Is DNA? Remarks on the Changing Aspects of a Scientific Concept," *Progress in Nucleic Acid Research and Molecular Biology* 8 (1968): 297–333, 319.

30. Chargaff, "What Really Is DNA?" 313.

31. W. Patrick McCray, *The Visioneers: How a Group of Elite Scientists Pursued Space Colonies, Nanotechnologies, and a Limitless Future* (Princeton: Princeton University Press, 2013).

32. K. Eric Drexler, "Molecular Engineering: An Approach to the Development of General Capabilities for Molecular Manipulation," *Proceedings of the National Academy of Sciences* 78 (1981): 5275–78.

33. Berry, "Making DNA and Its Becoming an Experimental Commodity."

34. An earlier example of using synthetic analogues in RNA, pioneered by Mariann Grunberg-Manago, can be read in Jean-Paul Gaudillière, "Molecular Biologists, Biochemists, and Messenger RNA: The Birth of a Scientific Network," *Journal of the History of Biology* 29 (1996): 417–45.

35. Nadrian C. Seeman and Neville R. Kallenbach, "Nucleic Acid Junctions: The Tensors of Life?" in *Nucleic Acids: The Vectors of Life*, ed. Bernard Pullman and Joshua Jortner (Dordrecht: D. Reidel Publishing Company, 1983), 183–200.

36. Nadrian C. Seeman, interview by W. Patrick McCray at New York University, New York City, New York, December 5, 2011 (Philadelphia: Chemical Heritage Foundation, Oral History Transcript # 0693); Nadrian Seeman, "At the Edge of Life: The Autobiography of Nadrian C. Seeman," submitted for the Kavli Prize, accessed September 30, 2019, http://kavliprize.org/prizes-and-laureates/laureates/nadrian-c-seeman; Nadrian C. Seeman, "The Crystallographic Roots of DNA Nanotechnology," American Crystallographic Association website (2014), accessed September 30, 2019, https://www.amercrystalassn.org/h-seeman.

37. Holliday is himself best known to historians of biological research on aging. Lijing Jiang, "Causes of Aging Are Likely to Be Many: Robin Holliday and Changing Molecular Approaches to Cell Aging, 1963–1988," *Journal of the History of Biology* 47 (2014): 547–84; Tiago Moreira and Paolo Palladino, "Squaring the Curve: The Anatomo-politics of Ageing, Life and Death,. *Body & Society* 14 (2008): 21–47.

38. D. B. Stadler and A. M. Towe, "Recombination of Allelic Cysteine Mutants in *Neurospora*," *Genetica* 48 (1963): 1323–44.

39. Robin Holliday, "A Mechanism for Gene Conversion in Funghi," *Genetics Research* 5 (1964): 282–304, 283.

40. Ibid., 284.

41. Oral History Transcript # 0693, 43.

42. Ibid., 41.

43. Ibid., 47. The protein crystallographer was Gregory A. Petsko, the undergrad Kathy McDonough.

44. Ibid., 47–48.

45. Ibid., 31.

46. Nadrian C. Seeman, "Nucleic Acid Junctions: Building Blocks for Genetic Engineering in Three Dimensions," in *Biomolecular Stereodynamics*, vol. 1, ed. Ramaswamy H. Sarma (1981), 269–77.

47. Nadrian C. Seeman, "Nucleic Acid Junctions and Lattices," *Journal of Theoretical Biology* 99 (1982): 237–47.

48. David J. Galas and Albert Schmitz, "DNAase Footprinting: A Simple Method

for the Detection of Protein-DNA Binding Specificity," *Nucleic Acids Research* 5 (1978): 3157–70.

49. Oral History Transcript # 0693, 68–69.

50. My thanks to Mary Morgan for suggesting this further significant parallel. On narrative and inference see Mary S. Morgan, "Models, Stories and the Economic World," *Journal of Economic Methodology* 8 (2001): 361–84.

51. Nadrian C. Seeman and Bruce H. Robinson, "Simulation of Double Stranded Branch Point Migration," in *Biomolecular Stereodynamics*, vol. 1, ed. Ramaswamy H. Sarma (1981), 279–300.

52. Seeman and Robinson, "Simulation of Double Stranded Branch Point Migration," 282–83. A copy of the staircase photograph can be found within the McCray interview transcript from the Science History Institute, and the transcript can be requested online for free; accessed May 18, 2020, https://oh.sciencehistory.org/oral-histories/seeman-nadrian-c.

53. Seeman and Robinson, "Simulation of Double Stranded Branch Point Migration," 283–84.

54. Ibid., 280.

55. Min Lu, Qiu Guo, Nadrian C. Seeman, and Neville R. Kallenbach, "Parallel and Antiparallel Holliday Junctions Differ in Structure and Stability," *Journal of Molecular Biology* 221 (1991): 1419–23.

56. Oral History Transcript # 0693, 34.

57. Mary S. Morgan, "Narrative Ordering and Explanation," *Studies in History and Philosophy of Science* 62 (2017): 86–97.

58. Nye, *From Chemical Philosophy*; Catherine Jackson, "Synthetical Experiments and Alkaloid Analogues: Liebig, Hofmann, and the Origins of Organic Synthesis," *Historical Studies in the Natural Sciences* 44 (2014): 319–63; Evan Hepler-Smith, "'A Way of Thinking Backwards': Computing and Method in Synthetic Organic Chemistry," *Historical Studies in the Natural Sciences* 48 (2018): 300–337; M. Norton Wise, "On the Narrative Form of Simulations," *Studies in History and Philosophy of Science* 62 (2017): 74–85.

Chapter 6

1. "Baseball-Playing Hens Help General Mills Sell Larro Feeds," *Tide*, September 11, 1954, Animal Behavior Enterprise Collection (hereafter hereafter cited as ABE Papers), Cummings Center for the History of Psychology, University of Akron, Box M4270, Folder 4.

2. Keller Breland and Marian Breland, "A Field of Applied Animal Psychology," *American Psychologist* 6 (1951): 202–4.

3. Alexandra Rutherford, *Beyond the Box: B. F. Skinner's Technology of Behavior from Laboratory to Life, 1950s–1970s* (Toronto: University of Toronto Press, 2009), 7.

4. Harvey Mindess, *Makers of Psychology: The Personal Factor* (New York: Human Sciences Press, 1988), 108, quoted in Laurence D. Smith, "Knowledge as Power: The Baconian Roots of Skinner's Social Meliorism," in *B. F. Skinner and Behaviorism in American Culture*, ed. Laurence D. Smith and William R. Woodward (Bethlehem: Lehigh University Press, 1996), 63.

5. B. F. Skinner, *A Matter of Consequences* (New York: Knopf, 1983), 47, quoted in Smith, "Knowledge," 64.

6. Hans-Jörg Rheinberger, *Toward a History of Epistemic Things: Synthesizing Proteins in the Test Tube* (Stanford: Stanford University Press, 1997).

7. Sabina Leonelli, "Growing Weed, Producing Knowledge: An Epistemic History of *Arabidopsis thaliana*," *History and Philosophy of the Life Sciences* 29 (2007):193–224, 197.

8. Canguilhem, *Knowledge of Life* (New York: Fordham University Press, 2008), p. 113. See also Jean Gayon, "The Concept of Individuality in Canguilhem's Philosophy of Biology," *Journal of the History of Biology* 31 (1998): 305–25, and Jean-François Braunstein, "Psychologie et milieu. Éthique et histoire des sciences chez Georges Canguilhem," in *Canguilhem: Histoire des sciences et politique du vivant*, ed. Braunstein (Paris: Presses Universitaires de France, 2007). Canguilhem focused on the history of the reflex, seeking to challenge the mechanistic orthodoxy and the use of mechanical metaphors to understand biological processes more generally, by unearthing the important role of a "certain vitalism." See Georges Canguilhem, *La formation du concept de réflexe aux XVIIe et XVIIIe siècles* (Paris: Presses Univ. France, 1955) and, for a thoughtful and detailed analysis of Canguilhem's thinking on the problem of error, see Samuel Talcott, *Georges Canguilhem and the Problem of Error* (Cham: Springer and Palgrave Macmillan, 2019).

9. Canguilhem, *Knowledge*, 110.

10. Canguilhem, *Knowledge*, 9.

11. Canguilhem, *Knowledge*, 16.

12. Edward Jones-Imhotep, *The Unreliable Nation: Hostile Nature and Technological Failure in the Cold War* (Cambridge: MIT Press, 2017).

13. Jones-Imhotep, *The Unreliable*, 218; K. Breland, "The Philosophy of Organisms," January 2, 1963, ABE Papers, Box M4311, Folder 15.

14. Jones-Imhotep, *The Unreliable*, 17.

15. B. F. Skinner, "Pigeons in a Pelican," *American Psychologist* 15 (1960): 28–37, 28.

16. Skinner, "Pigeons," 31.

17. Fred S. Keller and William N. Schoenfeld, *Principles of Psychology: A Systematic Text in the Science of Behavior* (East Norwalk, CT: Appleton-Century-Crofts, 1950; Reprint, B. F. Skinner Foundation, 1995), 72–74.

18. P. Teitelbaum, "The Use of Operant Methods in the Assessment and Control of Motivational States," in *Operant Behavior: Areas of Research and Application*, ed. W. K. Honig (New York: Appleton, 1966), 566–67.

19. B. F. Skinner, *The Shaping of a Behaviorist: Part Two of an Autobiography* (New York: Knopf, 1979), 274.

20. James H. Capshew, "Engineering Behavior: Project Pigeon World War II, and the Conditioning of B. F. Skinner," in *B. F. Skinner and Behaviorism in American Culture*, 128.

21. B. F. Skinner, "Experimental Psychology," in *Current Trends in Psychology*, ed. Wayne Dennis (Pittsburgh: University of Pittsburgh Press, 1947), 24, quoted in Capshew, "Engineering," 144.

22. Nils Wiklander, "From Hamilton College to Walden Two: An Inquiry into B. F. Skinner's Early Social Philosophy," in *B. F. Skinner and Behaviorism in American Culture*, 97.

23. Skinner, *Beyond Freedom and Dignity* (New York: Bantam Books, 1971), 191, 206. See also Mark P. Cosgrove, *B. F. Skinner's Behaviorism: An Analysis* (Grand Rapids: Zondervan, 1982), and the excellent John A. Mills, *Control: A History of Behavioral Psychology* (New York: New York University Press, 1998).

24. Rutherford, *Beyond the Box*, 10, 8.

25. Edward Jones-Imhotep, "Malleability and Machines: Glen Gould and the Technological Self," *Technology and Culture* 57 (2016): 287–321, 289. Jones-Imhotep is particularly interested in the kinds of human that technology produces and in the case of

Gould, how the musician's idealism was not set "against the material world," but rather "in union with it." Jones-Imhotep, "Malleability," 315.

26. Marian Breland, *The Animal Company*, unpublished memoir, c. 1980s, ABE Archives, Box M4307, Folder 8.

27. Andrew C. Godley and Shane Hamilton, "Different Expectations: A Comparative History of Structure, Experience, and Strategic Alliances in the U. S. and U. K. Poultry Sectors, 1920–1990," *Strategic Entrepreneurship Journal* 14 (2019), 1–16, 6. See also Steve Striffler, *Chicken: The Dangerous Transformation of America's Favorite Food* (New Haven: Yale University Press, 2005) and Roger Horowitz, "Making the Chicken of Tomorrow: Reworking Poultry as Commodities and as Creatures, 1945–1990," in *Industrializing Organisms: Introducing Evolutionary History*, ed. Susan R. Schrepfer and Philip Scranton (New York: Routledge, 2004).

28. Animals in Advertising, ABE Archives, Box M4283, Folder 11.

29. Animals in Advertising, ABE.

30. Loose notes (typed) on the history of ABE, n.d., ABE Archives, Box M4308, Folder 5.

31. Loose notes, ABE.

32. Keller Breland, "A Technology of Behavior," China Lake Talk, May 1962, p. 5, ABE Archives, Box M4307, Folder 5.

33. On the creation of hybrid wartime systems that united natural phenomena and machinic behavior, see Jones-Imhotep, *The Unreliable*, 22.

34. Robert E. Bailey, "Training Trainers: From the Days of Yesteryear to the Present," paper delivered at the Association for Behavior Analysis, May 1993, ABE Papers, Box M4312, Folder 2. Keller Breland died in 1965, and Robert Bailey, who had originally worked as a director of training for the Navy marine mammal program, became research director at ABE. He and Marian Breland were married in 1976. See also Robert E. Bailey and J. Arthur Gillaspy Jr., "Operant Psychology Goes to the Fair: Marian and Keller Breland in the Popular Press, 1947–1966," *Behavior Analyst* 28 (2005): 143–59.

35. Breland and Breland, "A Field of Applied," 202.

36. Breland and Breland, "A Field of Applied," 202.

37. Breland and Breland, "A Field of Applied," 203–4.

38. Keller and Marian Breland, "The New Animal Psychology," *National Humane Review* (March 1954): 10–12, 26–27, 12.

39. Breland and Breland, "A New Animal," 10.

40. Breland and Breland, "A New Animal," 10.

41. Keller Breland and Marian Breland, "The Misbehavior of Organisms," *American Psychologist* 16 (1961): 681–84; B. F. Skinner, *The Behavior of Organisms: An Experimental Analysis* (New York: Appleton-Century-Crofts, 1938).

42. Breland and Breland, "The Misbehavior," 681.

43. Breland and Breland, "The Misbehavior," 682.

44. Breland and Breland, "The Misbehavior," 682.

45. Breland and Breland, "The Misbehavior," 683.

46. Breland, "A Technology of Behavior," 17.

47. Breland and Breland, "The Misbehavior," 683.

48. Breland and Breland, "The Misbehavior," 684.

49. "Dynamic Properties of Species Specific Behaviors," ABE, Principal Investigator: Keller Breland. Research proposal, start date: July 1, 1963, p. 24, ABE Papers, Box M4283, Folder 8.

50. B. F. Skinner, "Herrnstein and the Evolution of Behaviorism," *American Psychologist* 32 (1977): 1006–12, 1006–7.

51. B. F. Skinner, "The Phylogeny and Ontogeny of Behavior," *Science* 153 (1966): 1205–13.

52. Skinner, "Herrnstein," 1007.

53. Marian Breland's answers to questionnaire for David Krantz, Lake Forest College, November 1970, Box M4273, Folder 1.

54. Marian Breland Bailey, "Mirror, Mirror on the Wall, Which Is the Smartest Animal of All? Or Intelligence and the Ecological Niche," undated, p. 7, ABE Papers, Box M4307, Folder 12.

55. "A Natural Format for Psychology—The Organization of Organisms," c. 1963, ABE Papers, Box M4309, Folder 14.

56. Gordon M. Burghardt, "Ethology and Operant Psychology," in *The Selection of Behavior: The Operant Behaviorism of B. F. Skinner: Comments and Consequences*, ed. A. Charles Catania and Stevan Harnad (Cambridge: Cambridge University Press, 1988), 414.

57. R. J. Herrnstein, "The Evolution of Behaviorism," *American Psychologist* 32 (1977): 593–603, 598–99.

58. Herrnstein, "The Evolution," 598.

59. Herrnstein, "The Evolution," 602. The work of J. E. R. Staddon, Herrnstein's one-time student, is also important in this regard. Staddon spent a sabbatical working with behavioral ecologists at Oxford University and argued that the superstitious behaviors imputed by Skinner were not idiosyncratic, but food-related instinctual responses. Staddon also identified the importance of the Brelands' work for contradicting Skinner's account of superstitious behavior. See J. E. R. Staddon and Virginia L. Simmelhag, "The 'Superstition' Experiment: A Reexamination of Its Implications for the Principles of Adaptive Behavior," *Psychological Review* 78 (1971): 3–43.

60. W. H. Gantt, "Analysis of the Effect of Person," *Conditional Reflex* 7 (1972): 67–73, 69.

61. W. H. Gantt, letter to Keller Breland, May 2, 1961, W. Horsley Gantt Papers, Alan Mason Chesney Medical Archives of the Johns Hopkins Medical Institutions, Baltimore, Maryland, Box 102.

62. Edmund Ramsden, "A Neurotic Dog's Life: Experimental Psychiatry and the Conditional Reflex Method in the Work of W. Horsley Gantt," *Isis* 109 (2018): 276–301.

63. W. H. Gantt, "Proflex, Schizokinesis, and the Internal Universe," *Pavlovian Journal of Biological Science* 21 (1986): 75–94, 76.

64. A. Rutherford, "B. F. Skinner's Technology of Behavior in American Life: From Consumer Culture to Counterculture," *Journal of the History of the Behavioral Sciences* 39 (2003): 1–23, 17. On the rise of technological pessimism from the 1960s, see Yaron Ezrahi, Everett Mendelsohn, and Howard Segal, eds., *Technology, Pessimism, and Postmodernism* (Dordrecht: Kluwer, 1994).

65. Benjamin M. Braginsky and Dorothea D. Braginsky, *Mainstream Psychology: A Critique* (New York: Holt, Rinehart and Winston, 1974), 70, quoting Breland and Breland, "The Misbehavior," 681.

66. Jones-Imhotep, *The Unreliable*, 15.

67. Skinner, "The Phylogeny."

68. Capshew, "Engineering," 130.

69. Daniel W. Bjork, "B. F. Skinner and the American Tradition: The Scientist as Social Inventor," in *B. F. Skinner and Behaviorism in American Culture*, 42; Jennifer K. Alexan-

der, "The Concept of Efficiency: An Historical Analysis," in *Philosophy of Technology and Engineering Sciences*, vol. 9, ed. Anthonie Meijers (Amsterdam: Elsevier, 2009), 1007.

70. Robert E. Bailey and Marian Breland Bailey, "Uses of Animal Sensory Systems and Response Capabilities in Security Systems," from The Role of Behavioral Science in Physical Security, Proceedings of the Second Annual Symposium, March 23–24, 1977, 50, ABE Papers, Box M4312, Folder 5.

71. ABE, "Dynamic Properties," 14.

72. Breland, "The Philosophy of Organisms."

73. Breland, "The Philosophy of Organisms."

74. Jones-Imhotep, *The Unreliable*.

75. Bailey and Breland Bailey, "Uses of Animal," 60.

76. "A Natural Format for Psychology."

77. "This 'n That at ABE," May 29, 1965, ABE Papers, Box M4291, Folder 6.

78. Breland Bailey, "Mirror, Mirror," 8.

79. "Functions of a Behavioral Museum," n.d., ABE Papers, Box M4309, Folder 4.

80. "Re the article for the Amer. Psych.," n.d., ABE Papers, Box 4307, Folder 2.

Chapter 7

1. Telephone Call from Herb Miller to Kelly McBean—5/20/49, Box 1, Folder 5, Records Group 326, #126477, Program Files Relating to the Development of Enewetak Atoll 1949–1951, National Archives and Records Administration, Riverside, California (Hereafter cited as H&N Program Files).

2. The normative spelling for the atoll is Enewetak. For the sake of fidelity to historical sources, the antiquated American spelling, Eniwetok, appears occasionally.

3. Radiobiological Survey of Bikini, Eniwetok, and Likiep Atolls, July–August, 1949, UWFL-23, 12 July 1950, Box 9, Volume 5, University of Washington, Laboratory of Radiation Biology records, 1944–1970, Special Collections Division, University of Washington Libraries, Seattle, Washington (Hereafter cited as UWLRB).

4. Rough Draft—Notes on Project Manager's Administration at Eniwetok September 1949–August 1950, Box 1, Folder 4, H&N Program Files.

5. Radiobiological Survey of Pacific Lagoons, From: Chief of Naval Operations to: Director, Military Applications Division Atomic Energy Commission, 27 June 1949, Box 6, Folder 5, UWLRB.

6. The disaster has often been told with a focus on the Japanese fishermen harmed by fallout aboard the *Fukuryū Maru*. For this account, see Jacob Darwin Hamblin and Linda Richards, "Beyond the *Lucky Dragon*: Japanese Scientists and Fallout Discourse in the 1950s," *Historia Scientiarum* 25, no. 1 (2015): 36–56.

7. Gabrielle Hecht, *Being Nuclear: Africans and the Global Uranium Trade* (Cambridge, MA: MIT Press, 2012), 6. See also Ruth Oldenziel, "Islands: The United States as a Networked Empire," in *Entangled Geographies: Empire and Technopolitics in the Global Cold War*, ed. Gabrielle Hecht (Cambridge, MA: MIT Press, 2011).

8. Traci Voyles, *Wastelanding: Legacies of Uranium Mining in Navajo Country* (Minneapolis: University of Minnesota Press, 2015).

9. Record of Discussion between Dr. H. L. Bowen of JTF-3 and Mr. D. Lee Narver of Holmes & Narver, Inc. (Job #640), 7 June 1951, Box 1, Folder 4, Record Group 326, #126477, H&N Program Files.

10. Resume of Job 640 History, by D. Lee Narver, 8 June 1951, 1, Box 1, Folder 4, H&N Program Files.

11. Report Atomic Energy Commission Proving Ground Eniwetok, M.I, Vol. 1, II-5,

Box 7, Folder Eniwetok Proving Ground Recon Report Vol. 1 1949, Record Group 326, #7563261, Site Background Files 1943–1971, National Archives and Records Administration, Riverside, California.

12. Ibid., II-13.

13. Qualitative tools like radioautographs did help the biologists understand how radiation moved through the food chain. See Laura Martin, "Proving Grounds: Ecological Fieldwork in the Pacific and the Materialization of Ecosystems," *Environmental History* 23 (2018): 567–92.

14. Radiobiological Survey of Bikini Atoll During the Summer of 1947, UWFL-7, December 1947, 37, Box 9, Volume 2, UWLRB.

15. Radiobiological Survey of Bikini, Eniwetok, and Likiep Atolls, July-August 1949, UWFL-23, 12, Box 9, Volume 5, UWLRB.

16. Eniwetok Radiological Survey 1948, UWFL-19, 8, NV0407849, NNSA/NSO Nuclear Testing Archive, Las Vegas, Nevada (Hereafter cited as NTALV).

17. UWFL-23, 18, Box 9, UWLRB.

18. "Bikini Atoll Food Still Radioactive," *New York Times*, 25 September 1949, ProQuest Historical Newspapers.

19. Task Group 7.5 Historical Installment No. 1 of Operation Castle, 6 November 1953, 9, Box 1, Folder Castle, Operation—TG 7.5 Historical Installment #1 & 2 1953–4, Record Group 326, # 1361627, Project Files of Holmes and Narver, National Archives and Records Administration, Riverside, California (Hereafter cited as H&N Project Files).

20. Task Group 7.5 Historical Installment No. 2 of Operation Castle, 8 February 1954, 4, Box 1, Folder Castle, Operation—TG 7.5 Historical Installment #1 & 2 1953–4, H&N Project Files.

21. Willis Boss to Lauren Donaldson, 19 November 1953, Box 7, Folder 14, UWLRB.

22. Lauren Donaldson to Willis Boss, 5 January 1954, Box 7, Folder 14, UWLRB.

23. Castle Log, 15 March 1954, Box 11, Folder 3, UWLRB.

24. Castle Log, 16 March 1954, UWLRB.

25. Operations Outline for Program 19, Marine Survey Unit, of Operation Castle, UWFL-36, 28–29, 15 February 1954, Box 9, Volume 5, UWLRB.

26. Donaldson Castle Notebook, 21 March 1954, Box 11, Folder 13, Lauren Donaldson Papers, Special Collections Division, University of Washington Libraries (Hereafter cited as MSS Donaldson).

27. Based on their research using X-rays to irradiate salmon during World War II, the Seattle biologists knew exposure to 2500 Roentgens created significant insults to the body and the blood, and induced significant mortality rates. At 2.8 Roentgens per hour, it would take just over five weeks to attain these exposure levels. See: Studies of the Effects of Roentgen Rays on the Growth and Development of the Embryos and Larvae of the Chinook Salmon (*Oncorhynchus tschawytscha*), UWFL-2, October 1945, Box 9, Volume 1, UWLRB.

28. Castle Log, 26 March 1954, UWLRB.

29. Castle Log, 27 March 1954, UWLRB.

30. Donaldson Castle Notebook, 27 March 1954, MSS Donaldson.

31. Castle Log, 13 April 1954, UWLRB.

32. W. H. Adams, "Late Medical Consequences of Exposure to Radioactive Fallout—Rongelap and Utirik 35 Years after 'Bravo,'" *Journal of Radioanalytical and Nuclear Chemistry* 156, no. 2 (1992): 269–90.

33. Holmes & Narver, Report of the Repatriation of the Rongelap People, November 1957, 1–1, Box 1, H&N Project Files.

34. Responsibilities for the Care and Disposition of Native Inhabitants of Rongelap and Utirik Atolls, 6 July 1954, NV0760892, NTALV.

35. Neil Hines, *Proving Ground: An Account of the Radiological Studies in the Pacific 1946–1961* (Seattle: University of Washington Press, 1962), 216.

36. Radiobiological Resurvey of Rongelap and Ailinginae Atolls Marshall Islands October–November 1955, UWFL-43, NV0410696, NTALV.

37. A Radiological Survey of Rongelap Atoll, Marshall Islands, During 1954–1955, UWFL-42, Figure 2, NV0410695, NTALV. This data appears in Appendix One of Hines's *Proving Grounds*, the documentary account of the lab's work in the Pacific.

38. UWFL-42, 23, NTALV.

39. Repatriation of the Rongelap People, November 1957, 1–3, H&N Project Files.

40. Ibid., 1–4.

41. Ibid., 1–32—1–33.

42. Land Crabs and Radioactive Fallout at Eniwetok Atoll, UWFL-50, May 27, 1957, 21–22, NV0407962, NTALV.

43. Note Seattle biologists' at-home demeanor. W-760–12, 7.1, "Scientists in the EMBL Laboratory," 25 July 1957, Box 35, Records Group 326, Jobsite Negatives and Photographs 1953–1959, National Archives and Records Administration, Riverside, California.

44. Repatriation of the Rongelap People, November 1957, 1–30, H&N Project Files.

45. Held to Conard, 16 September 1958, Box 2, Folder 19, UWLRB.

46. Ibid.

47. Potassium and Cesium-137 in *Birgus latro* (Coconut Crab) Muscle Collected at Rongelap Atoll, UWFL-64, 15 January 1960, 7, NV0407867, NTALV.

48. They expanded their work at Fern Lake in Washington State and became involved in Project Chariot in Alaska. See: Matthew Klingle, "Plying Atomic Waters: Lauren Donaldson and the 'Fern Lake Concept' of Fisheries Management," *Journal of the History of Biology* 31 (Spring 1998): 1–32.

49. See: Mary X. Mitchell, "Offshoring American Environmental Law: Land, Culture, and Marshall Islanders' Struggles for Self-Determination during the 1970s," *Environmental History* 22 (2017): 209–34.

Chapter 8

1. Kokubunken dētā setto 国文研データセット, 養蚕秘録 *Yōsan Hiroku* [Sericulture Secrets] Kyōwa 3-nenkan, 1803 http://www2.dhii.jp/nijl_opendata/NIJL0232/049-0075/52.

2. Kamigaki Morikuni 上垣守国, *Yōsan Hiroku* 養蚕秘錄 [Sericulture Secrets] (Tōkyō: Nō-san-gyoson Bunka Kyōkai, 1981), pp. 73–75, 111, 125. Japanese names appear with surname preceding given name, unless the cited literature is published in English.

3. One *chō* corresponds to roughly 99.18 acres. Dainihon Sanshikai 大日本蠶絲会, *Nihon sanshigyōshi dai yon kan*日本蠶絲業史第四巻 [History of Japanese Silkworm Thread Industry Vol. 4] (Tokyo: Dainihon Sanshikai 大日本蠶絲会, 1935).

4. The online silkworm supplier "Everything Silkworms" has also ventured to map the location of mulberry trees in all of Australia. https://everythingsilkworms.com.au/ and https://keyserver.lucidcentral.org/weeds/data/media/Html/morus_alba.htm, accessed 1 Sept. 2019.

5. Robert Reinhold, "Tucson in a Grass-Roots Battle against Allergies," *New York Times*, March 23, 1987, sec. A, 12.

6. Lisa Onaga, "Toyama Kametaro and Vernon Kellogg: Silkworm Inheritance Experiments in Japan, Siam, and the United States, 1900–1912," *Journal of the History of Biology* 43, no. 2 (March 2010): 215–64; Lisa Onaga, "More than Metamorphosis: The

Silkworm Experiments of Toyama Kametarō and His Cultivation of Genetic Thought in Japan's Sericultural Practices, 1894–1918," in *New Perspectives on the History of Life Sciences and Agriculture*, ed. Denise Phillips and Sharon Kingsland, 415–37 (Springer International Publishing, 2015).

7. Investigations into the cultural history of heredity have especially flourished in the early 2000s, as exemplified by works such as Staffan Müller-Wille and Hans-Jörg Rheinberger, eds., *Heredity Produced: At the Crossroads of Biology, Politics, and Culture, 1500–1870* (Cambridge, MA: MIT Press, 2007); and research on the domestication of cattle in works such as Rhoda M. Wilkie, "Domestication to Industry: The Commercialization of Human–Livestock Relations," in *Livestock/Deadstock: Working with Farm Animals from Birth to Slaughter* (Philadelphia: Temple University Press, 2010), 17–42. Indeed, some questions are perennial in the history of biology, such as those that explore the relationship between evolution, artificial breeding, and capital, as discussed of pigeon-fancying in Bert Theunissen, "Darwin and His Pigeons: The Analogy between Artificial and Natural Selection Revisited," *Journal of the History of Biology* 45, no. 2 (2012): 179–21; with regard to animals of labor see Margaret Elsinor Derry, *Horses in Society: A Story of Animal Breeding and Marketing, 1800–1920* (Toronto: University of Toronto Press, 2006); for plant science see Paolo Palladino, *Plants, Patients and the Historian: (Re) Membering in the Age of Genetic Engineering* (New Brunswick, NJ: Rutgers University Press, 2003); and Helen Anne Curry, *Evolution Made to Order: Plant Breeding and Technological Innovation in Twentieth-Century America* (Chicago: University of Chicago Press, 2016).

8. In the example of Nazi-era Germany's desire for self-sufficiency, academic breeders during the 1930s and 1940s developed ration-conscious feeds that would maximize the fat and protein content of pigs. This feed consisted of a balanced mixture of fishmeal, soybean meal, rye, and potatoes. Tiago Saraiva, *Fascist Pigs: Technoscientific Organisms and the History of Fascism* (Cambridge, MA: MIT Press, 2016), 127–35.

9. Lisa Onaga, "Bombyx and Bugs in Meiji Japan: Toward a Multispecies History?" *S&F Online* 11, no. 3 (2013), http://sfonline.barnard.edu/life-un-ltd-feminism-bioscience-race/bombyx-and-bugs-in-meiji-japan-toward-a-multispecies-history/; Lisa Onaga, "The Silkscape of California as a Microcosm of Sericultural Continuities," in *Oxford Handbook of Agriculture*, ed. Jeannie Whayne (Oxford University Press), forthcoming.

10. While processed animal protein was prohibited as a feed source for ruminants according to the Novel Food Regulation in the EU (adopted in 2015, applied in 2018), insect proteins are permitted, suggesting another demarcation within the category of animals from which insects stand apart. Anu Lähteenmäki-Uutela et al., "Insects as Food and Feed: Laws of the European Union, United States, Canada, Mexico, Australia, and China," *European Food and Feed Law Review* 12, no. 1 (2017): 22–36.

11. Terry G. Summons, "Animal Feed Additives, 1940–1966," *Agricultural History* 42, no. 4 (1968): 305–13.

12. Georg Borgstrom, "Food, Feed, and Energy," *Ambio* 2, no. 6 (1973): 214–19.

13. At the turn of the twentieth century, toxins such as sulfuric acid and arsenic generated by copper mining activities killed downwind silkworms in Japan's Shimotsuke Plain and decreased the fertility of longhorn cattle and the viability of their young in Deer Lodge Valley, Montana. Brett Walker, *The Toxic Archipelago: A History of Industrial Disease in Japan* (Seattle: University of Washington Press, 2010), 29–34; Timothy J. Lecain, *The Matter of History: How Things Create the Past* (Cambridge: Cambridge University Press, 2017), 170–81.

14. It is tempting but difficult to equate the first half of the twentieth century in Japan as an explicit rise of fascism like that seen in Western countries. Assumptions about

a lack of ground-up fascistic development are explored critically in Yoshimi Yoshiaki, *Grassroots Fascism: The War Experience of the Japanese People*, trans. Ethan Mark (New York: Columbia University Press, 2015).

15. Teikichi Hotta, *Taxonomical Studies on the Morus Plants and Their Distributions in Japan and Its Vicinities* (Tokyo: Japan Society for the Promotion of Science, 1958); Tian-Tian Zhang and Jian-Guo Jiang, "Active Ingredients of Traditional Chinese Medicine in the Treatment of Diabetes and Diabetic Complications," *Expert Opinion on Investigational Drugs* 21, no. 11 (November 1, 2012): 1625–42.

16. Tadao Yokoyama, "Sericulture," *Annual Review of Entomology* 8, no. 1 (1963): 287–306; Hisao Aruga, *Principles of Sericulture* (Rotterdam: A. A. Balkema, 1994), 296–302; Yataro Tazima, *The Silkworm: An Important Laboratory Tool* (Tokyo: Kodansha, 1978), 85–86.

17. "Dai san shō: Tayō na nōgyō seisan he no tenkan" 第III章 多様な農業生産への転換、桑園の現在の姿 [Chapter III: Conversion to diverse agricultural production, a current view of mulberry plantations], Ministry of Agriculture, Forestry, and Fisheries. http://www.maff.go.jp/kanto/kihon/kikaku/jyousei/28jyousei/attach/pdf/28jyousei-11.pdf.

18. GHQ was convinced that food production would not be affected if food crops were also grown in between rows of mulberry. SCAP, January 14, 1947, "Prices of Cocoons and Its Relation to the Economy of Japan," Military Agency Records RG 331, Records of the General Headquarters Supreme Commander for the Allied Powers (GHQ SCAP), National Archive and Records Administration, Rockville, MD.

19. Nakajima Shige, n.t., *Sansō Yōho* 8 (1939): 155.

20. S. Demianowski, *Arbeiten der Centralen Forschungsstation für Seide u. Seidensapuenzucht* (Moscow) 5, no. 1 (n.d.): 99–142; 143–162; Itō Toshio 智夫伊藤, "Kaiko no jinkōshiryō to daizu" 蚕の人工飼料と大豆 [Artificial Silkworm Feed and Soybeans], *Kagaku to seibutsu* 化学と生物 [Chemistry and Biology] 13, no. 4 (1975): 242–48.

21. Thomas Havens, *Farm and Nation in Modern Japan: Agrarian Nationalism, 1870–1940* (Princeton: Princeton University Press, 1974), 276; Louise Young, *Japan's Total Empire: Manchuria and the Culture of Wartime Imperialism* (Berkeley: University of California Press, 1999).

22. Reformers, bureaucrats, and advocates of farmers together aimed to improve the financial health of extant village communities. Kerry Douglas Smith, *A Time of Crisis: Japan, the Great Depression, and Rural Revitalization* (Cambridge, MA: Harvard University Asia Center, distributed by Harvard University Press, 2001), 11–13.

23. James W. Miller, "Pre-War Nazi Agrarian Policy," *Agricultural History; Chicago* 15, no. 4 (1941): 175–81.

24. Smith, *A Time of Crisis*, 139–49.

25. Smith, *A Time of Crisis*, 260–61; Kerry Smith, "Building the Model Village: Rural Revitalization and the Great Depression," in *Farmers and Village Life in 20th Century Japan*, ed. Ann Waswo and Yoshiaki Nishida (London: Routledge Curzon, 2003), 137–41.

26. Young, *Japan's Total Empire*, 327–33.

27. Young, *Japan's Total Empire*, 175–80, 190; Prasenjit Duara, *Sovereignty and Authenticity: Manchukuo and the East Asian Modern* (Lanham: Rowman & Littlefield, 2004); Mori Takemaro, "Colonies and Countryside in Wartime Japan: Emigration to Manchuria," *Asia-Pacific Journal | Japan Focus* 2, no. 6 (2004): 1–10; Janis Mimura, *Planning for Empire: Reform Bureaucrats and the Japanese Wartime State* (Ithaca, NY: Cornell University Press, 2011), 5–9, 41–106; Aaron S. Moore, *Constructing East Asia: Technology, Ideology, and Empire in Japan's Wartime Era, 1931–1945* (Stanford, CA: Stanford University Press, 2013), 188–224. For additional discussion of agricultural emigration

to Manchuria, see Kevin McDowell, "Japan in Manchuria: Agricultural Emigration in the Japanese Empire, 1932–1945," *Eras Journal* 5 (2003); Gregory Paul Guelcher, "Dreams of Empire: The Japanese Agricultural Colonization of Manchuria (1931–1945) in History and Memory" (PhD diss., University of Illinois at Urbana-Champaign, 1999).

28. S. Hori and U. Bokura, "Soy Bean Cake as a Substitute for Peptone in the Preparation of the Nutrient Media," *Japanese Journal of Phytopathology* 1, no. 1 (1918): 27–31; Okada Azuma 奥田東, *Dojō hiryō sōsetsu* 土壌肥料綜說 [Soil Fertilizer and Its Application] (Tōkyō: Yōkendō, 1951), 151–66; M. Kubo, J. Okajima, and F. Hasumi, "Isolation and Characterization of Soybean Waste-Degrading Microorganisms and Analysis of Fertilizer Effects of the Degraded Products," *Applied and Environmental Microbiology* 60, no. 1 (January 1994): 243–47.

29. G. F. Deasy, "The Soya Bean in Manchuria," *Economic Geography* 15, no. 3 (July 1939): 303.

30. Ichikawa Daisuke, "*Mansan shoten no hiryō seizō to hanro kaitaku*" 萬三商店の肥料製造と販路開拓 [Fertilizer Manufacture and the Expansion of the Mansan Company's Sales Network], *Shakai-Keizai Shigaku* 社会経済史学 [Socio-Economic History] 79, no. 1 (2013): 25–43; David Wolff, "Bean There: Toward a Soy-Based History of Northeast Asia," *South Atlantic Quarterly* 99, no. 1 (Winter 2000): 241; Sakura Christmas, "Japanese Imperialism and Environmental Disease on a Soy Frontier, 1890–1940," *Journal of Asian Studies* 78, no. 4 (2019): 809–36.

31. Kinoshita Asakichi 木下淺吉, "*Shoyu jōzō ni kansuru daizu no kenkyū*" 醤油釀造に關する大豆の研究" [Soybean Studies in Relation to Soysauce Brewing], *Jōzō kyōkai zasshi* 釀造協會雑誌 [Brewing Society's Journal, Japan] 8, no. 8 (1913): 45–50; Sadakichi Sato, "The Proteins of Soja Bean and Their Industrial Applications, Part 11," *Journal of the Society of Chemical Industry, Japan* 23, no. 9 (1920): 905–10; Katō Naosaburō 加藤直三郎, "Untersuchungen Uber Die Wirkung Der Soja-Urease. V.," *Yakugaku Zasshi* 藥學雑誌 [Pharmaceutical Journal] no. 506 (1924): 238–81; Masuno Minoru 增野實, "*Daizu tanpakushitsu no seisei ni kansuru kankyu (dainihō) teikyū arukōrurui ni yoru seisei ni oite suibun no eikyō*" 大豆蛋白質の精製に關する研究(第二報) 低級アルコール類に依る精製に於て水分の影響" [Studies on the Purification of Soybean Protein, Part 2: The Effect of Moisture on the Purification of Soybean Protein by Low-Grade Alcohols], *Tokyo Kagakukaishi* 東京化學會誌 [Journal of the Society of Chemical Industry, Japan] 32, no. 11 (1929): 1042–47.

32. Chikuma Ei. n.t., *Sanshi Kaiho* 蚕糸界報 [Silk Thread World Bulletin] 41, n.9 (1933):75.

33. Kinoshita Jōzōkenkyūsho 木下醸造研究所 [Kinoshita Brewery Research Institute], "*Daizu de yōsan*" 大豆で養蚕 [Sericulture With Soy Beans], *Shōyu to Miso* 醤油と味噌 [Soy Sauce and Miso] 2, no. 6 (June 1933): 40.

34. Kinoshita Jōzōkenkyūsho 木下醸造研究所 [Kinoshita Brewery Research Institute] "*Daizu shiryō no yōsan*" 大豆飼料の養蚕 [The Soy Bean Feed of Sericulture], *Shōyu to Miso* 醤油と味噌 [Soy Sauce and Miso] 2, no. 9 (September 1933): 35.

35. For example, rice in Spain during the Franco regime, and wheat in Mussolini's Italy and Portugal under Salazar. Tiago Saraiva, "Fascist Labscapes: Geneticists, Wheat, and the Landscapes of Fascism in Italy and Portugal," *Historical Studies in the Natural Sciences* 40, no. 4 (2010): 457–98; Lino Camprubi, *Engineers and the Making of the Francoist Regime* (Cambridge, MA: MIT Press, 2014), 77–102. The relationship between self-sufficiency and autonomous rule in relation to agriculture and fascism in Europe is discussed in the volume edited by Saraiva and Wise, 2010. See Tiago Saraiva and M. Norton Wise, "Autarky/Autarchy: Genetics, Food Production, and the Building of Fascism," *Historical Studies in the Natural Sciences* 40, no. 4 (2010): 419–28.

36. Yamamoto Haruhiko 山本晴彦, Manshū no nōgyō shiken kenkyūshi 満洲の農業試験研究史 [Agricultural Experiment Research Station History of Manchuria] (Tōkyō: *nōrin tōkei shuppan* 農林統計出版 [Agriculture and Forestry Statistics Publishing]), p. 93; Itō Toshio 智夫伊藤, "Kaiko no jinkō shiryō to daizu" 蚕の人工飼料と大豆 [Artificial Silkworm Feed and Soybeans], *Kagaku to Seibutsu* 化学と生物 [Chemistry and Biology] 13, no. 4 (1975): 242–48.

37. Nakajima Shige, n.t., *Sansō Yōho* 8 (1939): 155.

38. Patent for 蚕の強化資料 (45.7.20登録、578645号) Itō Toshio; Kameyama Tamiko 亀山 多美子, "*Tokkyo kōkoku kara mita kaiko no jinkō shiryō shiiku ni kansuru gijutsudōkō*" 特許公報からみた蚕の人工飼料飼育に関する技術動向 [Technological Trends in Rearing Silkworms with Artificial Feed as seen from the Viewpoint of a Patent Gazette], *Sanshi kenkyu* 蠶絲研究 [Sericultural Research] 111 (1982): 169–82; Fukuda et al., n.t., *Nihon Sanshigaku Zasshi* 日本蚕糸学雑誌 [Journal of Sericultural Science of Japan] 29, no. 1 (1960), n.p.; and Ito et al., *Nihon Sanshigaku Zassh* 日本蚕糸学雑誌 [Journal of Sericultural Science of Japan] 29, no. 121 (1960), n.p.

39. Hamamura Yasuji 浜村保次, "*Kaiko no Sesshoku kikō to jinkō shiryō*" カイコの摂食機構と人工飼料 [The Feeding Mechanism of Silkworms and Artificial Feed], *Kagaku to Seibutsu* 化学と生物 [Chemistry and Biology] 1, no. 7 (1963): 24–30.

40. Genomic analyses have given additional context for understanding the simple question of why silkworms seem to eat only mulberry. Shimada Toru 嶋田透, "*Kaiko ha naze kuwa wo kūnoka: genomu kara no hinto*. "カイコはなぜクワを食うか：ゲノムからのヒント." *Sanshi konchu baioteku* 蚕糸・昆虫バイオテック = *Sanshi-Konchu Biotec* 78, no. 2 (December 2009): 103–7.

41. Tanaka Yoshimaro 田中義麿, *Idengaku* 遺伝学 [Genetics] (Tōkyō: Shōkabō, 1941); Tazima Yataro 田島弥太郎, "*Shōsai wo kiru to chisha wo kū*" 小顋を切るとチシャを食う [Eats Lettuce When the Small Maxillae Is Cut], *Asahi Shimbun* 朝日新聞 [Asahi newspaper], June 23, 1953, 8.

42. Torii Ichio and Morii Kensuke 鳥居一男・森井謙介, *Kaikoken ihō* 蚕研彙報 [Silkworm Research Bulletin], n.t., 2 no.1 (1948): 3-12 cited in Nīmura Masazumi 正純新村, "*Hatsuiku ni tomonau shikōsei no henka o riyō shita kaiko no jinkō shīkuhō*" 発育に伴う嗜好性の変化を利用した蚕の人工飼育法 [Artificial Breeding of Silkworms Using Changes in Palatability Associated with Development], *Nippon sanshigaku zasshi* 日本蚕糸学雑誌 [Journal of Sericultural Science of Japan] 41, no. 5 (1972): 375–82; Torii Ichio 鳥居一男, "*Sanji wa naze kuwa ha nomi wo shokusuru ka*" 蚕児は何故桑葉のみを食するか [Why Silkworm Larvae Eat Only Mulberry Leaves], *Sanshikaihō*蚕糸界報 [Silk Thread World Bulletin]58, no. 674 (January 1949): 16–20.

43. In 1958, Fraenkel was invited to deliver a paper at the annual meeting of the Entomological Society of Japan, and he and Itō previously published a paper in 1957 on chick embryo development. Ito Toshio, Yasuhiro Horie, and Gottfried Fraenkel, "Feeding on Cabbage and Cherry Leaves by Maxillectomized Silkworm Larvae," *Journal of Sericultural Science of Japan* 28, no. 3 (1959): 107–13; G. Fraenkel, "The Raison d'etre of Secondary Plant Substances," *Science* 129 (1959a): 1466–70; G. Fraenkel, "A Historical and Comparative Survey of the Dietary Requirements of Insects," *Annals of the New York Academy of Sciences* 77, no. 2 (1959b): 267–74.

44. Tazima Yataro, "Alteration of Food Selection Character by Artificial Mutation in the Silkworm," *Bulletin of the International Silk Association* 20 (1954): 27–29.

45. Chromosome 11, position 30.5. Tazima Yataro 田島弥太郎, "*Seibutsu kaizō: watashi no shirukurōdo*" 生物改造: 私のシルクロード [Biological Remodeling: My Silk Road] (Tokyo: Shōkabō, 1991): 79–82; Tazima Yataro, "Alteration of Food Selection Character by

Artificial Mutation in the Silkworm," *Bulletin of the International Silk Association* 20 (1954): 27–29.

46. Yokoyama Tadao 横山忠雄, "*Totsuzenheni kaiko no shokusei ni kansuru kenkyū*" 突然変異蚕の食性に関する研究 [Study of the Eating Habits of Mutant Silkworms], *Sanshi-kagaku kenkyū ihō* 蚕糸科学研究所彙報 [Reports of the Silk Science Research Institute], no. 15 (March 1967): 14–22.

47. Murakoshi, Shigeo. "Effect of Leaf Powder of Various Species of Plant Added in the Diet on Feeding and Growth of the Silkworm, *Bombyx Mori* L.," *Japanese Journal of Applied Entomology and Zoology* 18, no. 1 (1974): 29–31; Shigeo Murakoshi et al., "Effects of the Extracts from Leaves of Various Plants on the Growth of Silkworm Larvae, *Bombyx Mori* L.," *Japanese Journal of Applied Entomology and Zoology* 19, no. 3 (1975): 208–13.

48. Hamamura Yasuji, "Food Selection by Silkworm Larvæ," *Nature* 183, no. 4677 (June 1959): 1746; Hamamura Yasuji 浜村保次, "カイコの摂食機構と人工飼料, *Kaiko no Sesshoku kikō to jinkō shiryō* [The Feeding Mechanism of Silkworms and Artificial Feed] 化学と生物 *Kagaku to Seibutsu* [Chemistry and Biology] 1, no. 7 (December 1963): 24–30.

49. Keimatsu 慶松, "tōmei naru kanten abura no seihō" 透明ナル寒天脂ノ製法 [Manufacturing Method of Transparent Agar Jelly] *Yakugaku Zasshi* 藥學雜誌 [Journal of the Pharmaceutical Society of Japan], no. 269 (July 1904): 587–88; H. G. Schlegel, *Geschichte der Mikrobiologie. Acta Historica Leopoldina 28* (2nd ed.) (Stuttgart: Wissenschaftliche Verlagsgesellschaft, 2004) as cited by Mathias Grote, "Petri Dish versus Winogradsky Column: A Longue Durée Perspective on Purity and Diversity in Microbiology, 1880s–1980s," *History and Philosophy of the Life Sciences* 40, no. 1 (November 29, 2017): 11.

50. *Kanten* (寒天) are cubes of agar jelly prepared in vegetarian dishes or desserts in Japan since the seventeenth century. After agar became an important component of biological research, especially in bacteriology, Japan actively engaged in the export of agar from the late nineteenth century until World War II. J. P. Baumberger and R. W. Glaser, "The Rearing of *Drosophila* Ampelophila Loew on Solid Media," *Science* 45, no. 1149 (1917): 21–22; Taku Komai, "The Culture Medium for Drosophila" *Science* 65, no. 1672 (1927): 42–43; Viesturs Pauls Karnups, "Latvian-Japanese Economic Relations 1918–1940," *Humanities & Social Sciences Latvia* 24, no. 1 (Spring–Summer 2016): 38–50; Howard I. Thaller, "Reclamation of Used Agar," *Science* 96, no. 2479 (1942): 23–24.

51. Grote, "Petri Dish versus Winogradsky Column."

52. G. Fraenkel, "A Historical and Comparative Survey of the Dietary Requirements of Insects," *Annals of the New York Academy of Sciences* 77, no. 2 (1959b): 267–74.

53. Itō Toshio, "A Preliminary Note on the Nutritive Value of Soybean Oil for the Silkworm, *Bombyx mori*," *Proceedings of the Japan Academy* 36, no. 5 (1960): 287–90.

54. Toshifumi Fukuda, Mitsumasa Suto, Tamiko Kameyama, and Shoichi Kawasugi, "Synthetic Diet for Silkworm Raising," *Nippon Nōgeikagaku Kaishi* [Journal of the Agricultural Chemical Society of Japan] 36, no. 10 (1962a): 819–25; T. Fukuda, M. Suto, T. Kameyama, and S. Kawasugi, "Synthetic Diet for Silkworm Raising," *Nature* 196, no. 4849 (October 1962b): 53.

55. Toshio Itō and Narihiko Arai, "An Amino-Acid Diet for the Silkworm, *Bombyx mori* L.," *Nippon Nōgeikagaku Kaishi* [Journal of the Agricultural Chemical Society of Japan] 40, no. 2 (1966): 110–12.

56. Elyssa Faison, *Managing Women: Disciplining Labor in Modern Japan* (Berkeley: University of California Press, 2007), 140, 145–51.

57. Faison, *Managing Women*, 157–60.

58. Itō Toshio calculated that ten kilograms of powdered soybean lees are neces-

sary to generate one kilogram of cocoons. Itō Toshio 智夫伊藤, "Kaiko no jinkōshiryō to daizu" 蚕の人工飼料と大豆 [Artificial Silkworm Feed and Soybeans], *Kagaku to seibutsu* 化学と生物 [Chemistry and Biology] 13, no. 4 (1975): 242–48.

59. Economic Planning Agency, "Income-Doubling Plan," in *Sources of Japanese Tradition*, vol. 2 (2nd ed.), ed. William Theodore De Bary, Carol Gluck, and Arthur Tiedemann (New York: Columbia, 2005), 402–5. The text of this plan was issued by the Japanese government in both English and Japanese. For a discussion of the ideology of productivity in Japan, see William M. Tsutsui, *Manufacturing Ideology: Scientific Management in Twentieth-Century Japan* (Princeton: Princeton University Press, 2001), 147–51, 164.

60. Satoshi Sasaki, "Scientific Management Movements in Pre-war Japan," *Japanese Yearbook on Business History* 4 (1988): 50–76; Tessa Morris-Suzuki, *The Technological Transformation of Japan: From the Seventeenth to the Twenty-First Century* (Cambridge: Cambridge University Press, 1994), 128–36.

61. Aurelia George Mulgan, *The Politics of Agriculture in Japan* (Hoboken: Taylor and Francis, 2013), 73–74.

62. n.a. "*Nippon tenshoku no kumagai kōjō kansei, kaiko no eiyō tenshoku shiryō no seisan kaishi*" 日本添食の熊谷工場完成、蚕の栄養添食飼料の生産開始 [Completion of Kumagaya Factory for Japanese Food Supplements, Production of Nutritional Supplements for Silkworms Begins], *Nihon Keizai Shimbun* 日経産業新聞 [The Nikkei] May 12, 1975, 7; n.a. "*Kaiko no jinkō shiryō jidō kyūjiki Yamanashi ken sangyō shikenjō ga kaihatsu*" 蚕の人工飼料自動給餌機山梨県蚕業試験場が開発 [Automatic Feeding Machine for Artificial Feed of Silkworms Developed by Yamanashi Prefecture's Silkworm Research Institute] *Asahi Shimbun* 朝日新聞 [Asahi Newspaper], April 28, 1976, 21.

63. While this translates to Japan Agricultural Industries, Inc., the corporation name in English is NOSAN Corporation. It was established in 1931 as Nihon Eiyō Shokuryō Kabushikigaisha (Japan Nutrition Food Corporation), until its name changed in 1942. N.a. "*Nōsankō, yōsanmuke jinkō shiryō no kaihatsu ni honkaku sannyū*" 農産工、養蚕向け人工飼料の開発に本格参入 [Full-Scale Entry into the Development of Artificial Feed for Agricultural Industry and Sericulture], *Nihon Keizai Shimbun* 日経産業新聞 [The Nikkei], April 14, 1976, 9.

64. n.a. "*Nippon tenshoku no kumagaya kōjō kansei, kaiko no eiyō tenshoku shiryō no seisan kaishi*" 日本添食の熊谷工場完成、蚕の栄養添食飼料の生産開始 [Completion of Kumagaya Factory for Japanese Food Supplements, Production of Nutritional Supplements for Silkworms Begins], *Nihon Keizai Shimbun* 日経産業新聞 [The Nikkei], May 12, 1975, 7.

65. n.a. "*Kaiko no reitōshoku tema habuke, jinkō shiryō no hanne, ima nōjō de*" 蚕の冷凍食手間はぶけ、人工飼料の半値 いま農場で [Frozen Silkworm Food Saves Time and Effort, Half Price of Artificial Feed, Now at the Farm], *Asahi Shimbun* 朝日新聞 [Asahi Newspaper], July 28, 1976, 15.

66. Especially p. 467 in Hayashiya Keizō 林屋慶三, "kaiko no jinkō shiryō to mukin shīku" カイコの人工飼料と無菌飼育 [Artificial Feed and Aseptic Breeding of Silkworms] *Kagaku to Seibutsu* 化学と生物 [Chemistry and Biology] 7, no. 8 (1969): 466–71.

67. Key fermentation studies on L-glutamate, from which MSG is derived, began publication in 1957 by Shukuo Kinoshita, Shigezo Udaka, and Masakazu Shimono, "Studies on the Amino Acid Fermentation," *Journal of General and Applied Microbiology* 3, no. 3 (1957): 193–205. For more on the history of molecular research about taste reception in humans, see Sarah E. Tracy, "Delicious Molecules: Big Food Science, the Chemosenses, and Umami," *Senses and Society* 13, no. 1 (January 2, 2018): 89–107. Additional discussion of Ajinomoto MSG research and development appear in Sarah E. Tracy, "Tasty

Waste: Industrial Fermentation and the Creative Destruction of MSG," *Food, Culture & Society* 22, no. 5 (July 26, 2019): 548–65.

68. Hayashiya, 1969, p. 467.

69. L. Tong, X. Yu, and H. Liu, "Insect Food for Astronauts: Gas Exchange in Silkworms Fed on Mulberry and Lettuce and the Nutritional Value of These Insects for Human Consumption during Deep Space Flights," *Bulletin of Entomological Research* 101, no. 5 (October 2011): 613–22.

Chapter 9

1. For her astute editorialism and endless generosity, I thank my friend and colleague Beth Semel, who discussed multiple drafts of this chapter with me and commented extensively on the final. I would also like to share my gratitude to the *Nature Remade* editors, particularly Mike Dietrich and Chris Young, for their sharp yet supportive feedback on the first draft. My research would not have been possible without Christine Ortiz and Raúl de Villafranca, who each welcomed me graciously into their design worlds. Thanks also to all the other members of the Ortiz lab and attendees of the Veracruz Biomimicry and Design Expedition, whether here in quotes or spirit. Funding for this trip was provided by the MIT International Science and Technology Initiatives.

2. See Michel Foucault, *The Archaeology of Knowledge* (New York: Vintage Books, [1969] 1982).

3. Manfred E. Clynes and Nathan S. Kline, "Cyborgs and Space," in *The Cyborg Handbook*, ed. Chris Hables Gray (New York: Routledge, 1995), 29–34.

4. I use "life" to refer to the totality of the "organic" subset of nature, its "organisms." The word "biology" may denote either the science of life, or the materiality of life. On classifications in the biological sciences, see Michel Foucault, *The Order of Things* (New York: Vintage Books, [1970] 1994). On organisms versus machines, see Georges Canguilhem, "Machine and Organism," in *Knowledge of Life*, ed. Georges Canguilhem (New York: Fordham University Press, 2008), 75–97; Evelyn Fox Keller, "Organisms, Machines, and Thunderstorms: A History of Self-Organization, Part One," *Historical Studies in the Natural Sciences* 38, no. 1 (2008): 45–75; Evelyn Fox Keller, "Organisms, Machines, and Thunderstorms: A History of Self-Organization, Part Two," *Historical Studies in the Natural Sciences* 39, no. 1 (2009): 1–31.

5. See Donna Haraway, "A Manifesto for Cyborgs: Science, Technology, and Socialist Feminism in the 1980s," *Socialist Review* 80 (1985): 65–108, 72, later republished as "A Cyborg Manifesto" in multiple venues.

6. For a discussion of the dichotomy of nature and artifice, see Bernadette Bensaude-Vincent and William R. Newman, eds., *The Artificial and the Natural: An Evolving Polarity* (Cambridge, MA: MIT Press, 2007).

7. For a fuller explication of this duality, see Ronald Kline, *The Cybernetics Moment: Why We Call Our Age the Information Age* (Baltimore: Johns Hopkins University Press, 2015), ch. 6.

8. Philip Pauly, "Modernist Practice in American Biology," in *Modernist Impulses in the Human Sciences, 1870–1930*, ed. Dorothy Ross (Baltimore: Johns Hopkins University Press, 1994), 272–89, 284.

9. Norbert Wiener, *Cybernetics, or Control and Communication in the Animal and the Machine* (Cambridge, MA: MIT Press, 1948).

10. Geoffrey Bowker, "How to Be Universal: Some Cybernetic Strategies, 1943–1970," *Social Studies of Science* 23, no. 1 (1994): 107–27.

11. See N. Katherine Hayles, *How We Became Posthuman: Virtual Bodies in Cybernetics,*

Literature, and Informatics (Chicago: University of Chicago Press, 1999). See also Kline, *The Cybernetics Moment*, for a detailed historical account of how Shannon's theory of information was selected over Wiener's as the lingua franca of cybernetics.

12. A prerequisite to the analogy of organisms and machines was the construction of the neuronal model of information processing by Warren McCulloch and Walter Pitts. Henceforth, organisms and machines became instantiations of a universal process. See Hayles, *How We Became Posthuman*; Kline, *The Cybernetics Moment.*

For an exceptional critique of the "disembodiment" theory of cybernetics, see Mara Mills, "On Disability and Cybernetics: Helen Keller, Norbert Wiener, and the Hearing Glove," *differences* 22, no. 2–3 (2011): 74–111.

13. Clynes and Kline, "Cyborgs and Space."

14. Donna Haraway, "Cyborgs and Symbionts: Living Together in the New World Order," in *The Cyborg Handbook.*

15. Sherry Turkle, ed., *Evocative Objects: Things We Think With* (Cambridge, MA: MIT Press, [2007] 2011).

16. Martin Caidin, *Cyborg* (New York: Warner Books, 1972), 101, 73.

17. For critical analyses of these and other representations of cyborgs in American popular culture, from Marvel's Wolverine to *The Terminator* series, see Gray, *The Cyborg Handbook*, especially part 4.

18. Kline, *The Cybernetics Moment*, 6; James Baldwin, *No Name in the Street* (1972), rpt. in Baldwin, *Collected Essays* (New York: Library of America, 1998), 349–476, on 384, in Kline, *The Cybernetics Moment*, 7.

19. Hayles, *How We Became Posthuman*; Kline, *The Cybernetics Moment.*

20. Peter Galison, "The Ontology of the Enemy: Norbert Wiener and the Cybernetic Vision," *Critical Inquiry* 21, no. 1 (1994): 228–66. See also Kline, *The Cybernetics Moment.*

21. Klein, *The Cybernetics Moment.* I have retained the gendered pronoun "he" when referring to the subject of cybernetics to maintain the masculinist overtones that numerous authors have detected, and that pervade even Caidin's *Cyborg*. See the prologue of Hayles, *How We Became Posthuman.*

22. Haraway, "A Cyborg Manifesto." See also her discussion of the multiple "chronotopes" of the cyborg in Haraway, "Cyborgs and Symbionts."

23. Galison, "The Ontology of the Enemy."

24. Andrew Pickering, *The Cybernetic Brain: Sketches of Another Future* (Chicago: University of Chicago Press, 2009).

25. Chris Hables Gray, Steven Mentor, and Heidi J. Figueroa-Sarriera, "Cyborgology: Constructing the Knowledge of Cybernetic Organisms," in *The Cyborg Handbook*, 1–14.

26. Timothy Choy, "Articulated Knowledges: Environmental Forms after Universality's Demise," *American Anthropologist* 107, no. 1 (2008): 5–18.

27. Donna Haraway, *Modest_Witness@Second_Millennium.FemaleMan_Meets_OncoMouse: Feminism and Technoscience* (New York: Routledge, 1997), 11.

28. My ethnographic data in this section and the next comes from fieldwork conducted on "bio-inspired design." In the fall of 2014, I spent time with members of a biomimetics laboratory at MIT, which spanned six months intermittently. This consisted of attendance at weekly meetings; recorded and semi-structured interviews with the principal investigator and two graduate students who were then working on the project I studied (one hour each); and a literature review of their papers and dissertations (14 total). One graduate student asked to remain unnamed, whom I name "Anonymous." Subsequently, I conducted four years of intermittent fieldwork with biomimicry designers at various sites across North America. In this paper, I introduce ethnographic data

from a week-long workshop I attended in the summer of 2017. The workshop was organized by the Montana-based Biomimicry Institute and held in Mexico: in the capital and in various towns in the states of Puebla and Veracruz. Participant observation and informal interviewing and conversation constituted the majority of my ethnographic data.

29. Ortiz received tenure in 2010 and held the Morris Cohen professorship in materials science and engineering.

30. *Polypterus senegalus* is alternatively known as the gray bichir, the Senegal bichir, or Cuvier's bichir.

31. Author's interview at MIT with Christine Ortiz, December 8, 2014.

32. Jack E. Steele, "An Interview with Jack E. Steele," 61–69, in *The Cyborg Handbook*, 62. Steele first used the term in 1959, republished as Jack E. Steele, "How Do We Get There?" 55–59, in *The Cyborg Handbook*. For histories of bionics, in particular von Foerster and McCulloch's research on frogs, see Kline, *The Cybernetics Moment*, chapter 6, and Jan Mueggenburg, "Clean by Nature: Lively Surfaces and the Holistic-Systemic Heritage of Contemporary Bionik," *communication +1* 3, no. 9 (2014).

33. Rene Roth, "The Foundation of Bionics," *Perspectives in Biology and Medicine* 26, no. 2 (1983): 229–42, 229. See Kline, *The Cybernetics Moment*, and Mueggenburg, "Clean by Nature," for a discussion of these symposia.

34. Kline, *The Cybernetics Moment*, 167.

35. Jon Harkness, "In Appreciation: A Lifetime of Connections, Otto Herbert Schmitt, 1913–1998," *Physics in Perspective* 4, no. 4 (2002): 456–90, 481. Schmitt first uses the term "biomimetics" in Otto Schmitt, "Some Interesting and Useful Biomimetic Transforms," *Proceedings of the Third International Biophysics Congress* (1969), August 29 to September 3, 297.

36. Mueggenberg, "Clean by Nature." In Germany, *Bionik* continues to describe work that, in the US, would be described as biomimetics.

37. See Johnson, "Reinventing Biological Life, Reinventing the 'Human,'" for a full description of these programs and their relationship to the RMA. For an overview of the Revolution in Military Affairs, see Elinor C. Sloan, *The Revolution in Military Affairs* (Montreal, Canada: McGill-Queen's University Press, 2002).

38. By the time Ortiz arrived, the department already had a distinguished vintage, having received generous funding from the US Department of Defense's Advanced Research Project Agency (ARPA) in June 1961 as part of a nationwide investment in materials science and engineering. See Bernadette Bensaude-Vincent, "The Construction of a Discipline: Materials Science in the USA," *Historical Studies in the Physical and Biological Sciences* 31, no. 2 (2001): 223–48.

39. Research done by the Ortiz laboratory on the material properties of *P. senegalus* has also acquired funding from the MIT Institute for Collaborative Biotechnologies, the US Army Research Office, the US Naval Research Office, and the National Security Science and Engineering Faculty Fellowship Program.

40. John D. Joannopoulos, "Institute for Soldier Nanotechnologies," *MIT Reports to the President* (2011–2012); Edwin L. Thomas, "Institute for Soldier Nanotechnologies," *MIT Reports to the President* (2003–2004).

41. See note 31.

42. Also called the gray bichir for its drab coloring, this seemingly unremarkable species of fish first attracted the attention of biologists in 1829 when Georges Cuvier described the organism. The famed French naturalist named it *Polypterus senegalus*, its Latinate genus name denoting "many fins." The British zoologist John Samuel Budgett later took special interest in the mobility and agility of the heavily scaled *P. senegalus*.

In a diary entry dated 1899, he jotted, "Executing the most lithe and supple movements, turning, twisting, darting and pausing in an extreme graceful manner, they thoroughly justify the native Mandingo name of Sayo or snake-fish." John Samuel Budgett, "Observations on *Polypterus* and *Protopterus*," *Proclamations of the Philosophical Society of Cambridge* 10 (1899): 236–40, 239, quoted in Sven Gemballa and Peter Bartsch, "Architecture of the Integument in Lower Teleostomes: Functional Morphology and Evolutionary Implications," *Journal of Morphology* 253 (2002): 290–309.

43. Bensaude-Vincent, "The Construction of a Discipline." The perspective of the Ortiz group is typical of broader representations of biological materials found in the materials science and engineering literature. For a comprehensive review that exemplifies this approach, see Marc André Meyers, Po-Yu Chen, Albert Yu-Min Lin, and Yasuaki Seiki, "Biological Materials: Structure and Mechanical Properties," *Progress in Materials Science* 53 (2008): 1–206.

44. Bensaude-Vincent, "The Construction of a Discipline."

45. Author's interview at MIT with Anonymous, December 3, 2014.

46. For further treatment of the trope of nature as an engineer, see Bernadette Bensaude-Vincent, "Chemists and the School of Nature," *New Journal of Chemistry* 26 (2002): 1–5; Bernadette Bensaude-Vincent, "The New Identity of Chemistry as Biomimetic and Nano-Science," Paper presented at the 6th Annual Conference on the History of Chemistry (2009); Bernadette Bensaude-Vincent, "A Cultural Perspective on Biomimetics." Bernadette Bensaude-Vincent, "A Cultural Perspective on Biomimetics," in *Advances in Biomimetics*, ed. Anne George (IntechOpen, 2011), https://www.intechopen.com/books/advances-in-biomimetics/a-cultural-perspective-on-biomimetics.

47. Benjamin J. F. Bruet, Juha Song, Mary C. Boyce, and Christine Ortiz, "Materials Design Principles of Ancient Fish Armour," *Nature Materials* 7 (2008): 748–56; Lifeng Wang, Juha Song, Christine Ortiz, and Mary C. Boyce, "Anisotropic Design of a Multilayered Biological Exoskeleton," *Journal of Materials Research* 24, no. 12 (December 2009): 3477–94; Juha Song, Christine Ortiz, and Mary C. Boyce, "Threat-Protection Mechanics of an Armored Fish," *Journal of the Mechanical Behavior of Biomedical Materials* 4 (2011): 699–712; Jorge Duro-Royo, Katia Zolotovsky, Laia Mogas-Soldevila, Swati Varshney, Neri Oxman, Mary C. Boyce, and Christine Ortiz, "MetaMesh: A Hierarchical Computational Model for Design and Fabrication of Biomimetic Armored Surfaces," *Computer-Aided Design* 60 (2015): 14–27, 15.

48. Christine Ortiz, Mary C. Boyce, Juha Song, and Steffen H. Reichert, "United States Patent Application: 13/207681—Articulating Protective System for Resisting Mechanical Loads" (August 11, 2011).

49. Christine Ortiz, "Natural Armor: An Untapped Encyclopedia of Engineering Designs for Protective Defense Applications," Presentation given at the Radcliffe Institute of Advanced Study Smart Clothes Symposium, November 15, 2013. This view is corrobated in other accounts: Stephen Mann, "Crystallochemical Strategies," in *Biomineralization, Chemical and Biological Perspectives*, ed. Stephen Mann, John Webb, and Robert Williams (Weinheim, Germany: VCH, 1989), 35–62, 35.

50. Bruet, Song, Boyce, and Ortiz, "Materials Design Principles of Ancient Fish Armour."

51. Janine Benyus, *Biomimicry: Innovation Inspired by Nature* (New York: Harper Perennial, 1997), 8.

52. Ibid., 2.

53. Ibid., 8.

54. Thomas Gieryn, "Boundary-Work and the Demarcation of Science from Non-

science: Strains and Interests in Professional Ideologies of Scientists," *American Sociological Review* 48, no. 6 (1983): 781–95.

55. I borrow the adjective "designerly" from the British design theorist Nigel Cross, who uses the concept of "designerly ways of knowing" to describe social forms particular to the design disciplines, as opposed to the sciences and humanities. See Nigel Cross, "Designerly Ways of Knowing," *Design Studies* 3, no. 4 (1982): 221–27.

56. "History" here should be read as an instance of what the anthropologist Charles Stewart calls "historicities," or "all the diverse ways in which the past may be construed without importing Western assumptions about history." See Charles Stewart, "Historicity and Anthropology," *Annual Reviews in Anthropology* 45 (2016): 79–94, 83. For me, such histories are read as analytic windows into the skein of cultural meanings wound up in biomimicry.

57. See Benyus, *Biomimicry*, 2.

58. Ibid., 3, 2, 2, 2, 2. Benyus contrasts "learning" with the extractive method of modern science championed by Francis Bacon. For more on this rhetoric, see Evelyn Fox Keller, *Secrets of Life, Secrets of Death: Essays on Language, Gender, and Science* (New York: Routledge, 1992).

59. Benyus, *Biomimicry*, 287.

60. Michel Foucault, "Polemics, Politics, and Problematization: Interview with Paul Rabinow, May 1984," In *Essential Works of Foucault*, vol. 1 (New York: New Press, 1998), 117.

61. Michael Fisch, "The Nature of Biomimicry: Toward a Novel Technological Culture," *Science, Technology, and Human Values* 42, no. 5 (2017): 795–821.

62. Benyus, *Biomimicry*, 3, 291.

63. Benyus, *Biomimicry*, 290, 2.

64. Sympathetic magic is an inference assuming "[one] can produce any effect [one] desires merely by imitating it." James Frazer, *The Golden Bough: A Study in Magic and Religion* (New York: Macmillan, 1922), 11.

65. Benyus, *Biomimicry*, 7.

66. Benyus, *Biomimicry*.

67. Benyus also includes, in a wide-ranging search for material that spans nature and artifice, artificial photosynthesisis, zoopharmacognosy, biological computing, and industrial ecology. Benyus, *Biomimicry*.

68. Felix Paturi, *Nature, Mother of Invention: The Engineering of Plant Life* (London: Thames and Hudson, 1974), 11.

69. World Commission on Environment and Development, *Our Common Future* (Oxford: Oxford University Press, 1987); James Grier Miller, *Living Systems* (New York: McGraw-Hill, 1978); Fritjof Capra, *The Web of Life: A New Scientific Understanding of Living Systems* (New York: Anchor, 1997).

70. Whether they "succeed" at such a repression, a question which is frequently posed to me as an objection, is, of course, a consideration secondary to the ideological work of condemning intentionality as a psychological state.

71. Michel Foucault, *The History of Sexuality*, vol. 3: *The Care of the Self* (New York: Pantheon Books, 1978).

72. As numerous cultural anthropologists have observed, the body is a medium through which diverse traditions aspire toward self-actualization. Saba Mahmood, *The Politics of Piety: The Islamic Revival and the Feminist Subject* (Princeton: Princeton University Press, 2004); Charles Hirschkind, *The Ethical Soundscape: Cassette Sermons and Islamic Counterpublics* (New York: Columbia University Press, 2006); Rachel Prentice,

Bodies in Formation: An Ethnography of Anatomy and Surgery Education (Durham: Duke University Press, 2012).

73. Benyus, *Biomimicry*, 2.

74. A full discussion of what this politics entails is outside the scope of this chapter, but the careful reader might suspect that biomimicry is not without its own set of thorny issues, from the resurgence of the moral authority of nature to the exoticization of wilderness and anti-modern primitivism. For a longer critique of biomimicry, refer to my forthcoming dissertation, "In Life's Likeness: Biomimicry and the Imitation of Nature in the United States."

75. Nevertheless, commentators on biomimetics and biomimicry continue to conflate them. Cf. Elizabeth Johnson, "Reinventing Biological Life, Reinventing the 'Human,'" *Ephemera* 10, no. 2 (2010): 177–93; Roger Stahl, "Life Is War: The Rhetoric of Biomimesis and the Future Military," *Democratic Communiqué* 26, no. 2 (2014): 122–37. On the "cyborg metaphor," see Evelyn Fox Keller, *Making Sense of Life* (Cambridge, MA: Harvard University Press, 2002). For detailed historical and ethnographic studies of the cyborg metaphor in other venues, see Stefan Helmreich, *Silicon Second Nature: Culturing Artificial Life in a Digital World* (Berkeley: University of California Press, 2000); Lily Kay, *Who Wrote the Book of Life? A History of the Genetic Code* (Stanford: Stanford University Press, 2000).

76. Philip Pauly, "Modernist Practice in American Biology," 286. See also Sophia Roosth, *Synthetic: How Life Got Made* (Chicago: University of Chicago Press, 2017); Hannah Landecker, *Culturing Life: How Cells Became Technologies* (Cambridge, MA: Harvard University Press, 2007); Robert Brain, *The Pulse of Modernism: Physiological Aesthetics in Fin-de-Siècle Europe* (Seattle: University of Washington Press, 2015); Marilyn Strathern, *After Nature: English Kinship in the Late Twentieth Century* (Cambridge: Cambridge University Press, 1992); Paul Rabinow, "Artificiality and Enlightenment: From Sociobiology to Biosociality," in *Essays on the Anthropology of Reason*, ed. Paul Rabinow (Princeton: Princeton University Press, 1996); and Sarah Franklin, *Dolly Mixtures: The Remaking of Genealogy* (Durham: Duke University Press, 2007).

77. Michel Foucault, *The Order of Things*.

Chapter 10

1. As Colin Milburn has suggested for nanotechnology, so might we view genetic engineering as "simultaneously a science and a science fiction," where the engineers' "rhetorical strategies intended to distance their science from the negative associations of science fiction . . . end up collapsing the distinction, reinforcing the science fiction aspects . . . at the same time as they rescue its scientific legitimacy." Colin Milburn, "Nanotechnology in the Age of Posthuman Engineering: Science Fiction as Science," *Configurations* 10 (2002): 261–95, on 266–67.

2. As one early twentieth-century writer noted, "The Andromeda nebula [*sic*] is said to be so far away that, though a description of a nearby view of its parts would read like divorce statistics in the United States, no dissolving motion can be seen by observers on the earth." Charles Fort, *The Book of the Damned* (1919), in *The Collected Works of Charles Fort, 1919–1923* (New York: Jeremy P. Tarcher/Penguin, 2008), 814.

3. Fredric Jameson, "Progress versus Utopia; or, Can We Imagine the Future?" *Science-Fiction Studies* 9 (1982): 147–58.

4. In time, its strains were even used to characterize prominent exobiologists themselves. In 1972, Dava Sobel described Sagan himself as a cosmic virus. She had expected "to find Carl Sagan a morose, Earthbound malcontent. How could anyone who appre-

ciates such a cosmic view of the universe, who sees the Earth as a free floating piece of debris in the enormity of space, survive the mundane realities of day-to-day living? That idea was wrong. Sagan's enthusiasm for his work, his excitement over new discoveries, spread like some uncontrollable Andromeda Strain through everything he does. Colleagues call him 'The whirlwind.'" Sobel interview with Sagan, 21 November 1972. For more on Tikhov, see my "Blue Vegetation on the Red Planet: Soviet Astrobotany and Earthly Analogues for Life on Mars," forthcoming.

5. Campos, "Life as It Could Be," in *Social and Conceptual Issues in Astrobiology*, ed. Kelly C. Smith and Carlos Mariscal (New York: Oxford University Press, 2020).

6. Preface to Raymond Z. Gallun, "Seeds of the Dusk," in *Adventures in Time and Space: An Anthology of Science Fiction Stories*, ed. Raymond J. Healy and J. Francis McComas.

7. Paul Davies, *Are We Alone? Philosophical Implications of the Discovery of Extraterrestrial Life*, Lectures given in Milan 8–9 November 1993, 1995, 57.

8. For more on this, see Campos, "Life as It Could Be."

9. National Academies of Science, February 21, 1959. The nature of the potential biological harm was of more direct consequence than any physical harm—as the appendix of a CETEX-1 report published in *Science* in 1958 had noted, "By their very nature, experimental missteps in biology may do irreversible harm; in the physical sciences they may lead at most to exasperation, delay, and waste." Appendix comments on CETEX-1 Report, as published in *Science*, October 17, 1958.

10. Second Meeting of the ad hoc committee on Contamination by Extra-terrestrial Exploration, The Hague, March 9–10, 1959. For more on "contamination" as a metaphor in recombinant DNA contexts, see Wolfe and also Colyvas, cited below.

11. Lederberg, March 10, 1959; see also Wolfe, "Germs in Space: Joshua Lederberg, Exobiology, and the Public Imagination, 1958–1964," *Isis* 93 (2002): 183–205.

12. H. G. Baker and G. Ledyard Stebbins, eds., *The Genetics of Colonizing Species: Proceedings of the First International Union of Biological Sciences Symposia on General Biology* (New York: Academic Press, 1965).

13. Wolfe, "Germs in Space." Wolfe points to the containment of the Communist threat as a key context for planetary protection even on the home front (newly invented devices aimed to "protect Americans from alien contamination in space as well as contain Communism at home"), 195.

14. Harry Schwartz, "Nationalism Is Obsolete in Outer Space," *New York Times*, May 19, 1969.

15. As Steven Dick and James Strick have noted, sterilization, back contamination, "or the inadvertent return of possible cosmic microbes to Earth . . . could perhaps allow *Andromeda Strain* scenarios to develop. Few scientists, surely, have ever seen their objectives, both scientific and policy-oriented ones, converted into reality so completely and so quickly by a government agency as happened with Lederberg and exobiology." *The Living Universe* [NASA Third Semiannual Report to Congress, 1 October 1959–31 March 1960, 90, 158–59; NASA Fifth Semiannual Report to Congress, 1 October 1960–30 June 1961, 133–35.]

16. Michael Crichton, *The Andromeda Strain* (New York: Knopf, 1969); Joanna Radin, "Michael Crichton, Science Studies, and the Technothriller," Histories of the Future: How Do We Make the Future website, available at http://histscifi.com/essays/radin/technothriller. For more on Crichton's later work and its relation to matters of concern in STS, see also Joanna Radin, "Alternative Facts and States of Fear: Reality and STS in an Age of Climate Fictions," *Minerva* 57 (2019): 411–31.

17. http://www.michaelcrichton.com/the-andromeda-strain/.

18. As Joanna Radin has noted, "The question of whether or not Earthly microbes could develop adaptations that would allow them to occupy the upper atmosphere was ultimately to be filed by Simpson 'under the heading of idle speculation.' In the note that amends that statement, Simpson continued: 'or would be a large source of material for science fiction, which has so far been singularly unimaginative and uninstructed in dealing with possible future evolutionary development here on earth,' Simpson, *The Major Features of Evolution* 1953, 212n." Joanna Radin, "The Speculative Present: How Michael Crichton Colonized the Future of Science and Technology," *Osiris* 34 (2019): 297–315, on 305, note 44.

19. As Radin has summarized: "At a time when many were still fixated on the threat of nuclear crises, *The Andromeda Strain* (1969) was a story of biological crisis—the introduction of a mysterious and deadly viral agent upon the return of a space voyage. In the novel, a military satellite returns to earth, landing in northern Arizona. The team sent to recover the satellite breaks radio contact. Because the military suspects an infection, a new team of scientists is dispatched to investigate only to find that everybody in the nearest small town is dead (except for an old man and a screaming infant). The two survivors are taken to an underground laboratory where the culprit, an organism named the 'Andromeda Strain,' is identified; it then nearly kills all of humankind. American citizens never learn of this brush with apocalypse, or that the entire crisis is the product of its government's top-secret germ warfare research." Ibid., 305–6. Elsewhere, Radin has noted that "Crichton soon established his beat as a reporter of the near future, the possible but not yet actual." Radin, "Michael Crichton, Science Studies, and the Technothriller." Calls from a National Academy of Sciences committee in 1974 for a voluntary moratorium on some potentially hazardous recombinant DNA experiments were, in turn, framed by the media in turn as reading "almost like something from a science fiction thriller." "Dangerous Knowledge," *Times* (Trenton), July 25, 1974.

20. Lederberg to William Koshland, Knopf publisher, June 6, 1959.

21. Jay Clayton, "The Ridicule of Time: Science Fiction, Bioethics, and the Posthuman," *American Literary History* 25, no. 2 (Summer 2013): 317–43. Radin has called for not only "biographies of science fiction authors" but "perhaps more urgently, biographies of the products of their creative endeavors." This chapter responds to that call.

22. Michael Meltzer, *When Biospheres Collide: A History of NASA's Planetary Protection Programs*, 74.

23. Zinder to Horowitz, July 10, 1974 Box 183, Norton Zinder Collection, Cold Spring Harbor Laboratory.

24. Statement of Stanley N. Cohen, MD, Assoc Prof of Medicine, Stanford, April 22, 1975. Box 32, Folder 3, p. 6, Maxine Singer Papers, Library of Congress. Joshua Lederberg had raised the same parallel a year earlier, in a letter to the director of NIH: "The hazards of interplanetary quarantine (contamination of the planets and back contamination of the earth). In fact there is a fairly close analogy to the issues raised by the biohazards committee." Lederberg to Robert Stone, August 23, 1974. Folder 608, Oral History Collection on the Recombinant DNA Controversy, MIT Special Collections, MC100.

25. "Shades of Science Fiction," *Salt Lake City Tribune*, July 21, 1974.

26. In the novel, the landing location is in Piedmont, Arizona; in the film, it is in Piedmont, New Mexico. For more on the film's use of the hazmat suit, see Radin, "The Speculative Present," 306.

27. Berg to Stoker, July 21, 1972. Paul Berg Papers, Stanford University, SC0358, Box 3, Folder "1973 S-Z."

28. "Halt to Genetic Manipulation Urged," *C&EN* 52, no. 30 (July 29, 1974): 4.

29. Paul Berg, "Asilomar 1975: DNA Modification Secured," *Nature* 455 (2008): 290–91.

30. "Scientists Worry that New Lab Organisms Could Escape, Do Harm," *Boston Sunday Globe*, November 11, 1973, 66.

31. Gairdner B. Moment, Editorial, "Andromeda Strain?" *BioScience* 24, no. 9 (September 1974): 487.

32. Interestingly, a report from a 1972 symposium on virology reported that the meeting "brought home the fact that a vaccine for human cancer is not imminent—not because it is not technically feasible, but rather because Federal regulations prohibit the culturing of some 'Andromeda Strain.'" Rather than calling for federal regulations to prevent the Andromeda strain, this author felt such safeguards were already in place—and problematic: "As might be expected, cancer virologists are not happy about regulations that stand in the way of making a human cancer virus. Yet more than a few of them are not sure they would change the regulations. Apparently the threat Michael Crichton writes about in his novel, *The Andromeda Strain*—the chance of a virus leaking out of the laboratory and infecting the population—is always a possibility. Some safety mistakes were made in the creation of a polio virus vaccine." "Perspectives on Cancer: Viral Link Elusive," *Science News* 101, no. 8 (February 19, 1972): 118. Two years later, it was reported that "the NAS committee took their case to both the scientific community and the public-at-large, perhaps recognizing that an 'ostrich' attitude could only lead to further mistrust among those who already fear 'Andromeda Strain' effects." Cristine Russell, "Weighing the Hazards of Genetic Research: A Pioneering Case Study," *BioScience* 24, no. 12 (December 1974): 691–94, 744.

33. "Swearing Off Gene Transplantation," *Medical World News*, August 9, 1974. The *New York Times* even noted, "It is not very often that the world gets treated to two events straight out of science fiction in the same week." Tabitha M. Powledge, "Dangerous Research and Public Obligation," *NY Times*, August 24, 1974.

34. "The Andromeda Fear," *Time*, July 29, 1974.

35. Charles Thomas, "Biohazards, Real and Imaginary," February 12, 1976. Folder 599. Oral History Collection on the Recombinant DNA Controversy, MIT Special Collections, MC100. "The disease and the circumstances of its origin are, of course, fictitious. But the circumstances—genetic manipulation of certain bacteria with results reminiscent of 'The Andromeda Strain'—are precisely what molecular biologist had in mind in instituting the unprecedented ban now in effect on such experiments." Michael Woods, "Why Ban Research on Genetic Transfer? Molecular Biologists Fear 'Andromeda Strain' Results," *Toledo Blade*, August 18, 1974. See also Judith Randal, "A New Fear: Building Vicious Germs," *Washington Star News*, July 22, 1974.

36. Michael Rogers' beautifully written article for *Rolling Stone*, "The Pandora's Box Congress," recounts the conference's turn of events more fully and fascinatingly than the many other available secondary sources. See note 48.

37. Robert Cooke, "Scientists to Resume Risky Work on Genes: Danger of 'Andromeda Strain' Posed," *Boston Globe*, February 28, 1975, 1.

38. Stuart Auerbach, "And Man Created Risks," *Outlook*, Sunday, March 9, 1975, B1.

39. "Is Harvard the Place for Frankenstein Tinkering?" *Boston Star*, June 16, 1976.

40. Robert Cooke, *Improving on Nature: The Brave New World of Genetic Engineering* (New York: Quadrangle/the New York Times Book Co., 1977), 186.

41. George Alexander, "Rules Set to Explore Potential, Peril of Genetic Manipulation," *Los Angeles Times*, December 22, 1975.

42. "Asilomar became a scientific version of 'Jaws,' and the public was titillated, but

also frightened." "Recombinant DNA is becoming the Farrah Fawcett-Majors of scientific issues of the day. It is impossible to get through a week's reading of newspapers and magazines (let alone scientific publications) without encountering yet another insider's view of this glamorous subject." Willard Gaylin, "Sounding Board: The Frankenstein Factor," *NEJM* 297, no. 12 (1977): 665. On Frankenstein: "We want to be damned sure the people of Cambridge won't be affected by anything that would crawl out of that laboratory." "Is Harvard the Place for Frankenstein Tinkering?"

43. "The popular term, genetic engineering, might be considered as covering anything having to do with the manipulation of the gametes or the fetus, for whatever purpose, from conception other than by sexual union, of treatment of disease in utero, to the ultimate manufacture of a human being to exact specifications." "Genetic Engineering: Reprise," *JAMA* 220, no. 10 (June 5, 1972): 1356–57.

44. As Charles Thomas explained: "Our journalists (who are now more numerous than ever) prematurely publicize genuine scientific advances in terms that are consistent with those of science fiction. In this process, they often have the full complicity of the scientist himself. Who doesn't like attention? We should not be surprised to see scientists saying irresponsible things in order to enjoy public attention. It is human nature, and scientist are all too human." In "The Fanciful Future of Gene Transfer Experiments," *Brookhaven Symposia in Biology* 29 (April 30 1977): 348–59.

45. Paul Berg interview, May 17, 1975, Box 1, Folder 8, p. 60. Oral History Collection on the Recombinant DNA Controversy, MIT Special Collections, MC100. Mark Ptashne similarly complained that speculation about the creation of Andromeda strains "is not scientific talk—it's science-fiction talk." Arthur Lubow, "Playing God with DNA," *New Times* (January 7, 1977): 48.

46. "Malcolm Martin's personal notes on Enteric Bacteria meeting, August 31, 1976," Maxine Singer Papers, Library of Congress, Box 32, Folder 7, p. 38.

47. Ibid.

48. Michael Rogers, "The Pandora's Box Congress," *Rolling Stone* 189 (1975): 37–42, 74–82. Watson had gathered enemies at home who envisioned an Andromeda-style solution to his antics. Having made few friends in the Harvard department during these years—E. O. Wilson once characterized Watson as the "Caligula" of biology—Watson could scarcely have imagined that one of his "most distinguished colleagues at Harvard privately opined that he would not mind having an Andromeda-strain facility in the biological laboratories—if the nuclear self-destruct device could be planted under Jim Watson's desk." Concerns over engineered epidemics and solutions to problematic colleagues dovetailed in the minds of some. William Bennett and Joel Gurin, "Science that Frightens Scientists: The Great Debate on DNA," *Atlantic Monthly*, February 1977, 43–62.

49. Rifkin, Larry Gordon, and Dan Smith, "DNA," *Mother Jones*, February/March 1977.

50. Watson, quoted in Robert S. Gatsoff, "Watson attacks research critics," folder "Berg et al letters," Paul Berg Papers, Stanford University, SC0358. Zinder, "A Personal View of the Media's Role in the Recombinant DNA War," Box 14, Folder "MS, Media v. & DNA." Norton Zinder Collection, Cold Spring Harbor Laboratory.

51. "Malcolm Martin's personal notes," p. 40. This phrase inspired the title of Susan Wright's history of the early recombinant DNA years, *Molecular Politics: Developing American and British Regulatory Policy for Genetic Engineering, 1972–1982* (Chicago: University of Chicago Press, 1994).

52. Maxine Singer Papers, Library of Congress, Box 33, Folder 6, Page 1, Tape 7a jds, January 21, 1977.

53. Bernard D. Davis, "Epidemiological and Evolutionary Aspects of Research on Recombinant DNA," in National Academy of Sciences, *Research with Recombinant DNA: An Academy Forum, March 7–9, 1977* (Washington, DC: The National Academies Press, 1977), 132, 131.

54. Bernard D. Davis, "The Recombinant DNA Scenarios: Andromeda Strain, Chimera, and Golem: Analyzed in terms of epidemiology and evolution, the risks of working with recombinant DNA in E. coli seem far less than the familiar risks of working with known pathogens," *American Scientist* 65, no. 5 (September–October 1977): 547–55, on 554. [Quoting Handler. 1977. Annual Report of the President, National Academy of Sciences, 26 April 1977.]

55. Bernard D. Davis, "The Recombinant DNA Scenarios."

56. Bernard Davis, "Darwin, Pasteur, and the Andromeda Strain," public lecture at Harvard University January 5, 1977. Reprinted in abridged form in the *Harvard Crimson*, February 2, 1977.

57. "Doomsday: The DNA Furor: Tinkering With Life," *Time*, April 18, 1977.

58. Bernard D. Davis, "Epidemiological and Evolutionary Aspects of Research on recombinant DNA," 131, 124.

59. "Tinkering with Genes: Extreme Hopes, Fears," *National Observer*, March 19, 1977.

60. Folder 260, "Research with Recombinant DNA," Academy Forum, March 7–9, 1977, Introduction, Daniel E. Koshland, Jr. Oral History Collection on the Recombinant DNA Controversy, MIT Special Collections, MC100. See also National Academy of Sciences, *Research With Recombinant DNA*.

61. Folder 540, "NAS: Committee on Recombinant DNA Molecules, Press Conference, July 18, 1974, Wash DC, Transcript of interview with Berg and Baltimore." Oral History Collection on the Recombinant DNA Controversy, MIT Special Collections, MC 100.

62. Charles H. Blake to Sinsheimer, January 7, 1976. Folder 1.2, "Correspondence B-Bl," Robert L. Sinsheimer Papers, CalTech Archives.

63. Folder 224, "Rough draft, re in vivo transmission of plasmids, to NAS Plasmid Subcommittee on Biohazards: Novick, Clowes, Cohen, Curtiss, Data." Oral History Collection on the Recombinant DNA Controversy, MIT Special Collections, MC100.

64. E. S. Anderson to John Tooze, June 7, 1977, re Falmouth, MA, workshop on studies for assessment of potential risks associated with recombinant DNA experimentation (June 20–21, 1977). Folder 266, Oral History Collection on the Recombinant DNA Controversy, MIT Special Collections, MC100.

65. Zinder to Josh Lederberg, Email subject header: "The Truth." Box 178, Norton Zinder Papers, Cold Spring Harbor Laboratory.

66. Carl Sagan was one of many scientists who explicitly linked these considerations, in correspondence with the science fiction author Isaac Asimov: "As with all questions of interplanetary quarantine or recombinant DNA we must ask not only what is the most likely theory but what is the probability that this theory is incorrect" (quoted in David Grinspoon, *Earth in Human Hands* [Ashland, OR: Blackstone, 2016], 376). The anthropologist Ashley Montagu in 1972 even suggested that humans themselves might be seen as unwelcome plague-bearers, and equated the lack of success in locating extraterrestrial intelligence with the idea that such (Andromedan?) superlunaries may have avoided contact "because if they have observed us, they no doubt regard us as we regard rabies or cancer or cholera—in short, as a highly infectious disease that is best quarantined from the rest of the universe." Richard Berendzen, ed., *Life Beyond Earth & The Mind of Man*, Symposium at Boston University, November 20, 1972 (NASA 1973), 21.

67. Maxine Singer, Proposed Guidelines from Potential Biohazards, associated with experiments involving genetically altered microorganism, Asilomar, plasmid group, February 1975. Box 32, Maxine Singer Papers, Library of Congress.

68. Maxine Singer, "Scientists and the Control of Science: Peer Pressure among Scientists Is Sufficient to Ensure Their Public Responsibility," *New Scientist* 74 (June 16, 1977): 631–34. First delivered as a talk at the NAS Forum, "Research with Recombinant DNA (March 7, 1977)," in the session "How Scientists Interact with the Public: Historical Perspective."

69. Joann Rodgers, "Asilomar Revisited,'" *Mosaic* 12, no. 1 (Jan/Feb 1981): 19–26. Boyer experienced similar difficulties with reporters who called seeking confirmation of reports that "a bacterial virus that couldn't be killed—a true Andromeda strain—had been created by accident at the San Francisco laboratory." Boyer grew irritated: "No, that's not true. It didn't happen that way at all. You're not listening. . . . Look, I'm very busy and you're just wasting my time by arguing." John Lear, *Recombinant DNA: The Untold Story* (New York: Crown, 1978), 97.

70. Quoting Toshio Murashige of UC Riverside. Rick Gore, "The Awesome Worlds within a Cell," *National Geographic* 150, no. 3 (September 1976): 386. It's worth noting that Erwin Chargaff, no fan of recombinant DNA research, considered science fiction "one of the most frequent forms of senile delinquency." Chargaff, "On Some of the Biological Consequences of Base-Pairing in the Nucleic Acids," in *Developmental and Metabolic Control Mechanisms and Neoplasia* (Baltimore: Williams and Wilkins, 1965). "Whether Frankenstein's little biological monsters will be grafted on successfully, I cannot say, nor what else may be introduced at the same time unintentionally. Were I not so averse to rancid science fiction, I should say that the spreading of experimental cancer may be confidently expected." Chargaff, "Profitable Wonders: A Few Thoughts on Nucleic Acid Research," *The Sciences* (August/September 1975).

71. Regulation of Recombinant DNA Research: Hearings before the Subcommittee on Science, Technology, and Space of the Committee on Commerce, Science, and Transportation, United States Senate, Ninety-Fifth Congress, First Session . . . November 2, 8, and 10, 1977, pp. 246, 284.

72. Recombinant DNA Regulation Act, 1977: Hearing before the Subcommittee on Health and Scientific Research of the Committee on Human Resources, United States Senate, Ninety-Fifth Congress, First Session, on S. 1217 . . . April 6, 1977, 136.

73. Statement by Philip Handler, President NAS, before Subcommittee on Science, Technology, and Space of Committee on Commerce, Science, and Transportation, US Senate, November 2, 1977; and Recombinant DNA Act: Hearing before the Subcommittee on Science, Research, and Technology of the Committee on Science and Technology, US House of Representatives, Ninety-Fifth Congress, Second Session, April 11, 1978, 7. See also Folder 112, Oral History Collection on the Recombinant DNA Controversy, MIT Special Collections, MC100.

74. Clayton, "The Ridicule of Time."

75. Recombinant DNA Regulation Act, 1977: Hearing before the Subcommittee on Health and Scientific Research of the Committee on Human Resources, United States Senate, Ninety-Fifth Congress, First Session, on S. 1217 . . . April 6, 1977, 90.

76. Carl Cohen, "On the Dangers of Inquiry and the Burden of Proof," in *The Recombinant DNA Debate*, ed. David Jackson and Stephen Stich (Prentice-Hall, 1979), 303–35, on 316.

77. Julie Ann Miller, "Lessons from Asilomar," *Science News* 1278 (February 23, 1985): 122–23, 126.

78. Design of Biomedical Research Facilities: Proceedings of a Cancer Research Safety Symposium conducted at Frederick Cancer Research Center, Frederick, Maryland, 21701, October 18–19, 1979, 76.

79. SUOTL S74–035, Memo to File, 13 July 1976, cited in Jeannette A. Colyvas, "Factory, Hazard, and Contamination: The Use of Metaphor in the Commercialization of Recombinant DNA," *Minerva* 45, no. 2 (June 2007): 143–59.

80. John Jutila, VP for Research at Montana State University and a microbiologist, "Federal Oversight of Biotechnology," hearing before the Subcommittee on Hazardous Wastes and Toxic Substances of the Committee on Environment and Public Works, United States Senate, One Hundredth Congress, First Session, November 5, 1987, 125.

81. Hiromitsu Yokoo and Tairo Oshima, "Is Bacteriophage ΦX174 DNA a Message from an Extraterrestrial Intelligence?" *Icarus* 38 (1979): 148–53.

82. "In other ways, this idea of viruses from space would emerge yet again in 2018 under the more formalized auspices of "astrovirology." See A. Berliner, T. Mochizuki and K. Stedman, "Astrovirology: Viruses at Large in the Universe," *Astrobiology* 18, no. 2 (2018): 207–22.

83. As Colin Milburn has noted, with particular reference to tropes of infection, germs, and alien probes, these many variable uses of science fiction in scientific discourse and debate suggest that "science fiction functions within scientific space, as a repository of modifiable futures." Milburn, "Modifiable Futures: Science Fiction at the Bench," *Isis* 101 (2010): 560–69.

84. "No Sci-Fi Nightmare, After All," *New York Times*, July 24, 1977.

85. Piers Bizony, *The Search for Aliens: A Rough Guide to Life on Other Worlds* (London: Rough Guides, 2012), 21. As others noted, "The decreased level of attention is *not* due to the attainment of a better understanding, but rather to a lack of crises of 'Andromeda strain' events. Any such event, a near-accident, a real accident, an investment fraud scheme, could easily revive public interest, only on a large scale commensurate with the expanded biotechnology operations." Earl D. Hanson, ed., *Recombinant DNA Research and the Human Prospect*, 110. And biotechnology expanded incredibly rapidly in this period, as Susan Wright has noted: "The dawn of synthetic biology coincided with the emergence of a new ethos, one dominated by concerns external to science . . . the value and number of the bioengineering firms rose almost exponentially until 1982. In 1979 alone, the paper value of the four leading firms almost doubled. The number of small firms specializing in the field rose from about twelve in 1978 to about 52 in 1982. The cumulative equity investment in all types of biotechnology companies rose from $50m to over $800m between 1978 and 1981." Susan Wright, "Recombinant DNA Technology and Its Social Transformation, 1972–1982," *Osiris* 2 (1986): 303–60, on 337, 347.

86. "Goodbye to Guidelines," *Nature* 29 (February 4, 1982).

87. "'And look at this nonsense!' he said, almost throwing the papers at her, 'viruses from outer space! What they think they can get people to believe!'" (132). "Their own newspapers and video channels had carried the news but for the time being put the blame onto an entirely new virus, perhaps even an escape from one of the space probes" (140).

88. Kenneth Coleman, "Biotech and Conservation," *Bio/technology* 6 (February 1, 1988): 217.

89. David Baltimore, "Limiting Science: A Biologist's Perspective," *Daedalus* 117, no. 3 (Summer 1988): 333–44.

90. Bernard Dixon, "Don't Believe (All) the Hype," *Bio/Technology* 12 (June 1, 1994):

555, quoting David Pramer and Janet Shoemaker of the American Society for Microbiology.

91. They also repeated the canard that "Public fear was fanned by the popularity of *The Andromeda Strain*, and the myriad 'what ifs' floated by both serious and demagogic commentators." "The Recombinant DNA Controversy," *PNAS* 92 (September 26, 1995): 9011.

92. "If the critics are correct and the Andromeda scenario has even the merest possibility of occurring, or a variation of it, we will have to assume it will occur, based on our experience." Anthony Mazzocchi, "Health Hazards to Labor," in *Research with Recombinant DNA: An Academy Forum, March 7–9, 1977* (Washington, DC: The National Academies Press, 1977), 143. And: "An Andromeda strain (from the 1965 [*sic*] novel 'The Andromeda Strain' by Michael Crichton) is a hypothetical new type of microbe that causes massive destruction of human life. Its impact would depend on its transmissibility, and transmissibility is a neglected subject in microbiology." C. A. Mims, "Special Article: Virology Research and Virulent Human Pandemics," *Epidemiology and Infection* 115, no. 3 (December 1995): 377–86.

93. Joan Slonczewski, "Tuberculosis Bacteria Join UN: WHO Proposes to Include Disinfectant under the Geneva Convention," *Nature* 405 (June 29, 2000): 1001.

94. As Dixon has noted, "Time and again gloomy pessimism about opposition to biotechnology, often linked with anticipated public reactions to books and movies, has proven to be unwarranted. Take only the most recent example. A year or so ago, much of the U.S. biotechnology establishment was in a state of acute alarm over the harm likely to be done by Jurassic Park. In any event, Spielberg's epic did not trigger mass hysteria. The executives panicked. But everyone else remained calm, and the industry was unaffected." Bernard Dixon, "Don't Believe (All) the Hype," *Bio/Technology* 12 (June 1, 1994): 555.

95. Marcia Barinaga, "Asilomar Revisited: Lessons for Today?" *Science* 287 (March 3, 2000): 1584–85.

96. Arthur Kornberg, oral history by Sally Hughes, Stanford University, SC359, ACCN 2002–186, Box 2.

97. "Futures of Artificial Life," *Nature* 431 (October 6, 2004): 613.

98. "There was rampant speculation about laboratory creations run amok, and talk of Frankenstein's monster and *Andromeda Strain* scenarios enlivened the news (*Jurassic Park* had not yet been written)." Henry I. Miller and Gregory Conko, "Should We Make a Fuss?" *Nature Biotechnology* 24 (August 1, 2006): 899–900. See Ginsberg, "Resurrecting the Sublime," this volume.

99. Luis Campos, "Jurassic Ark," forthcoming.

100. William D. Cohan, "Coronavirus Gives Investors an Excuse to Cut and Run," *New York Times*, March 2, 2020. See also my "Tinctures of Time and Schrödinger's Virus," *Science* 368, no. 6490 (May 1, 2020): 478–79, from which this paragraph is adapted.

101. Lane continues: "And 'The Andromeda Strain' is right to prophesy that, should life from elsewhere fall to Earth, it will, as likely as not, comprise a small patch of what appears to be blue-green mold with limited social skills. My main concern, frankly, is not that it could mow us down in droves but that, owing to an unfortunate housekeeping glitch, it might get squirted with bleach and removed with a lemon-scented wipe. Thus would end our only contact with another life-form, although President Trump, of course, would insist on seeing the bright side. 'Me and the mold got on great,' he'd say. 'It had a terrific time. I also think, and I'm not just saying this, that I would make a tre-

mendous mold.'" Anthony Lane, "Our Fever for Plague Movies," *New Yorker*, May 15, 2020.

Chapter 11

1. Jacques Loeb, *The Mechanistic Conception of Life: Biological Essays* (Chicago: University of Chicago Press, 1912), 3–4.

2. Lewellys F. Barker, "Foreword," in *Eugenics: Twelve University Lectures* (New York: Dodd, Mead, 1914), ix–xiii, ix.

3. C. Forbes, "Scientists as God: Our Greatest Challenge," *Australian*, November 28, 1996, 3, cited in A. Petersen, "Biofantasies: Genetics and Medicine in the Print News Media," *Social Science & Medicine* 52, no. 8 (April 2001): 1255–68, https://doi.org/10.1016/s0277-9536(00)00229-x.

4. J. Craig Venter, *Life at the Speed of Light: From the Double Helix to the Dawn of Digital Life* (New York: Viking, 2013), 85, 103, 109, 130, 131.

5. L. M. Adleman, "Molecular Computation of Solutions to Combinatorial Problems," *Science* 266, no. 5187 (November 11, 1994): 1021–24; J. C. Genereux and J. K. Barton, "Molecular Electronics: DNA Charges Ahead," *Nature Chemistry* 1, no. 2 (May 2009): 106–7.

6. See for example, Matt Ridley, *Nature via Nurture: Genes, Experience, and What Makes Us Human* (London: Harper Collins, 2003).

7. E. F. Keller, *The Mirage of a Space between Nature and Nurture* (Durham: Duke University Press, 2010); Brad Weslake, "Explanatory Depth," *Philosophy of Science* 77, no. 2 (2010): 273–94.

8. Amy Harmon, "My Genome, Myself: Seeking Clues in DNA," *New York Times*, November 17, 2007, sec. US; Joel Achenbach, "Pondering 'What It Means to Be Human' on the Frontier of Gene Editing," *Washington Post*, May 3, 2016, sec. Health & Science.

9. I am extending here an argument made in my *Science of Human Perfection* (New Haven: Yale, 2012) and drawing upon a large literature, including Thierry Hoquet, "Biologization of Race and Racialization of the Human: Bernier, Buffon, Linnaeus," in *The Invention of Race*, ed. Nicolas Bancel, Thomas David, and Dominic Thomas (London: Routledge, 2014), 17–33; Charles E. Rosenberg, "The Bitter Fruit: Heredity, Disease, and Social Thought in 19th-Century America," *Perspectives in American History* 8 (1974): 189–235; Peter Conrad, *The Medicalization of Society: On the Transformation of Human Conditions into Treatable Disorders* (Johns Hopkins University Press, 2007); Adam M. Hedgecoe, "Geneticization: Debates and Controversies," in *ELS* (John Wiley & Sons, Dec. 2009) www.els.net [doi: 10.1002/9780470015902.a0005849]; A. Lippmann, "The Geneticization of Health and Illness: Implications for Social Practice," *Endocrinologie* 29, no. 1–2 (1990): 85–90.

10. See J. Reardon, *The Postgenomic Condition: Ethics, Justice, and Knowledge after the Genome.* (Chicago: University of Chicago Press, 2017); Sarah S. Richardson and Hallam Stevens, eds., *Postgenomics: Perspectives on Biology after the Genome* (Durham: Duke University Press, 2015).

11. My influences here include Thomas C. Leonard, "Retrospectives: Eugenics and Economics in the Progressive Era," *Journal of Economic Perspectives* 19, no. 4 (2005): 207–24; Daniel T. Rodgers, "In Search of Progressivism," *Reviews in American History* 10, no. 4 (1982): 113–32; Robert H. Wiebe, *The Search for Order: 1877–1920* (New York:

Hill and Wang, 1967); Richard Hofstadter, *The Age of Reform: From Bryan to F. D. R.* (New York: Knopf, 1955).

12. Statement of Aims, Mont Pelerin Society, accessed Nov. 9, 2020, montpelerin .org/statement-of-aims.

13. Wendy Brown, *In the Ruins of Neoliberalism: The Rise of Antidemocratic Politics in the West* (New York: Columbia University Press, 2019). See also Melinda Cooper, *Family Values: Between Neoliberalism and the New Social Conservatism* (New York: Zone Books, 2017); Pierre Dardot and Christian Laval, *Never Ending Nightmare: How Neoliberalism Dismantles Democracy* (Verso Books, 2019); David Harvey, *A Brief History of Neoliberalism* (Oxford: Oxford University Press, 2007). On anti-democratic neoliberalism as antagonistic to Hayek's thought, see Brown, *Ruins*, 9.

14. Elizabeth Pennisi, "A Low Number Wins the GeneSweep Pool," *Science* 300, no. 5625 (2003): 1484.

15. H. J. Rheinberger, S. Müller-Wille, and A. Bostanci, *The Gene: From Genetics to Postgenomics* (Chicago: University of Chicago Press, 2018), chap. 10; Stuart A. Newman, "The Demise of the Gene," *Capitalism Nature Socialism* 24, no. 1 (2013): 62–72.

16. For first drafts of this history, see Jennifer A. Doudna and Samuel H. Sternberg, *A Crack in Creation: Gene Editing and the Unthinkable Power to Control Evolution* (Boston: Houghton Mifflin Harcourt, 2017); Eric S. Lander, "The Heroes of CRISPR," *Cell* 164, no. 1–2 (January 14, 2016): 18–28.

17. See for example John R. Hibbing, Kevin B. Smith, and John R. Alford, *Predisposed: Liberals, Conservatives, and the Biology of Political Differences*, 1st ed. (New York: Routledge, 2013); Charles Kenny, "Dumb and Dumber," *Foreign Policy*, April 30, 2012; Daniel J. Benjamin, David Cesarini, Vilmundur Guðnason, et al., "The Promises and Pitfalls of Genoeconomics," *Annual Review of Economics* 4, no. 1 (September 2012): 627–62; Daniel J. Benjamin, James J. Choi, and A. Joshua Strickland, "Social Identity and Preferences," 2007, 51, http://www.econ.yale.edu/~shiller/behmacro/2007-11/benjamin .pdf; Arcadi Navarro, "Genoeconomics: Promises and Caveats for a New Field," *Annals of the New York Academy of Sciences* 1167, no. 1 (June 2009): 57–65.

18. Catherine Bliss, *Social by Nature: The Promise and Peril of Sociogenomics* (Stanford: Stanford University Press, 2018); Aaron Panofsky, "From Behavior Genetics to Postgenomics," in *Postgenomics: Perspectives on Biology after the Genome*, ed. Sarah S. Richardson and Hallam Stevens (Durham: Duke University Press, 2015), 150–73.

19. See also Nicholas G. Evans and Jonathan D. Moreno, "Children of Capital: Eugenics in the World of Private Biotechnology," *Ethics in Biology, Engineering and Medicine: An International Journal* 6, no. 3–4 (2015), doi: 10.1615/EthicsBiologyEngMed.2016016594; Ted Peters, "Are We Closer to Free Market Eugenics? The CRISPR Controversy," *Zygon* 54, no. 1 (2019): 7–13; Nikolas Rose, *The Politics of Life Itself: Biomedicine, Power, and Subjectivity in the Twenty-First Century* (Princeton: Princeton University Press, 2006).

20. Johann Caspar Lavater, *Physiognomy, or the Analogy of the Conformation of the Features and the Ruling Passions of the Mind*, translated from the Original by J. C. Lavater, by Samuel Shaw (London: H. D. Symonds, No. 20, Paternoster Row, 1800), 1–4.

21. John Graham, "Lavater's 'Physiognomy': A Checklist," *Papers of the Bibliographical Society of America* 55, no. 4 (1961): 297–308, 298. For more recent scholarship on Lavater, see R. T. Gray, *About Face: German Physiognomic Thought from Lavater to Auschwitz* (Detroit: Wayne State University Press, 2004); John B. Lyon, "'The Science of Sciences': Replication and Reproduction in Lavater's Physiognomics," *Eighteenth-Century Studies* 40

(2007): 257–77; Melissa Percival and Graeme Tytler, *Physiognomy in Profile: Lavater's Impact on European Culture* (Newark: University of Delaware Press, 2005).

22. Lavater, *Physiognomy*, 24.

23. B. A. Morel, *Traité Des Dégénérescences Physiques Intellectuelles et Morales de l'espèce Humaine* (Baillière, 1857), http://books.google.com/books?id=IWaiBXuwT8EC. For discussion, see Daniel Pick, *Faces of Degeneration: A European Disorder, C.1848–1918* (Cambridge: Cambridge University Press, 1993).

24. M. Gibson, N. H. Rafter, and C. Lombroso, *Criminal Man* (Durham: Duke University Press, 2006); Rosenberg, "Bitter Fruit," 216–19; Nicole Rafter, *Creating Born Criminals* (Urbana: University of Illinois Press, 1998); Paul A. Lombardo, "Return of the Jukes: Eugenic Mythologies and Internet Evangelism," *Journal of Legal Medicine* 33, no. 2 (2012): 207–33, 213.

25. Edmund F. DuCane, discussion following Francis Galton, "Composite Portraits, Made by Combining Those of Many Different Persons Into a Single Resultant Figure," *Journal of the Anthropological Institute of Great Britain and Ireland* 8 (1879): 132–44, 142–143. On DuCane and prisons, see Peter Tibber, "Edmund DuCane and the Prison Act 1877," *Howard Journal of Criminal Justice* 19, no. 1–3 (1980): 9–16.

26. The best all-around biography of Galton in my view is Nicholas W. Gillham, *A Life of Sir Francis Galton: From African Exploration to the Birth of Eugenics* (New York: Oxford University Press, 2001).

27. On pseudoscience, see Maarten Boudry and Massimo Pigliucci, eds., *Science Unlimited? The Challenges of Scientism* (Chicago: University of Chicago Press, 2018); Pigliucci and Boudry, eds., *Philosophy of Pseudoscience: Reconsidering the Demarcation Problem* (Chicago: University of Chicago Press, 2013); Michael D. Gordin, *The Pseudoscience Wars: Immanuel Velikovsky and the Birth of the Modern Fringe* (Chicago: University of Chicago Press, 2013).

28. Lavater, *Physiognomy*, 11; "Richard Louis Dugdale [obituary]," *New York Times*, June 24, 1884, 8.

29. Francis Galton, "Hereditary Improvement," *Fraser's Magazine* 7 (January 1873): 116–30, 116; August Weismann, *Das Keimplasma: Eine Theorie Der Vererbung, Trans. as The Germ-Plasm: A Theory of Heredity*, trans. W. Newton Parke and Harriet Rönnfeldt (New York: Scribner, 1893). On Weismann's barrier and complex behavior, see Timothy D. Johnston, "The Influence of Weismann's Germ-Plasm Theory on the Distinction between Learned and Innate Behavior," *Journal of the History of the Behavioral Sciences* 31, no. 2 (1995): 115–28.

30. Lyndsay A. Farrall, "Controversy and Conflict in Science: A Case Study—the English Biometric School and Mendel's Laws," *Social Studies of Science* 5 (1975): 269–301; David MacKenzie and Barry Barnes, "Scientific Judgment: The Biometry-Mendelian Controversy," in *Natural Order* (Beverly Hills: Sage, 1979), 191–210; R. Olby, "The Dimensions of Scientific Controversy: The Biometric-Mendelian Debate," *British Journal of the History of Science* 22, no. 74, pt. 3 (September 1989): 299–320.

31. Lindley Darden, "William Bateson and the Promise of Mendelism," *Journal of the History of Biology* 10, no. 1 (March 1, 1977): 87–106; Garland Allen, "Naturalists and Experimentalists: The Genotype and the Phenotype," in *Studies in History of Biology*, vol. 3 (Baltimore: Johns Hopkins University Press, 1979), 179–209; Robert Kohler, *Lords of the Fly: Drosophila Genetics and the Experimental Life* (Chicago: University of Chicago Press, 1994), 58.

32. For an excellent global survey of eugenics, see Alison Bashford and Philippa

Levine, *The Oxford Handbook of the History of Eugenics* (Oxford: Oxford University Press, 2010). On genetic determinism, see Michel Morange, *The Misunderstood Gene* (Cambridge, MA: Harvard University Press, 2001); John Cleese, *The Scientists*, vol. 32, John Cleese Podcast, 2008, https://www.youtube.com/watch?v=-M-vnmejwXo.

33. Arthur H. Estabrook, *The Jukes in 1915* (Carnegie Institution of Washington, 1916). On these pathological families, see John D. Smith, *Minds Made Feeble: The Myth and Legacy of the Kallikaks* (Rockville, MD: Aspen Systems Corp., 1985).

34. Daniel J. Kevles, *In the Name of Eugenics: Genetics and the Uses of Human Heredity*, 2nd ed. (Cambridge, MA: Harvard University Press, [1985] 1995); Paul A. Lombardo, *Three Generations, No Imbeciles: Eugenics, the Supreme Court, and Buck v. Bell* (Baltimore: Johns Hopkins University Press, 2008); Garland E. Allen, "The Eugenics Record Office at Cold Spring Harbor, 1910–1940," *Osiris* 2 (1986): 225–64; Jason Lantzer, "The Indiana Way of Eugenics," in *A Century of Eugenics in America: From the Indiana Experiment to the Human Genome Era*, ed. Paul A. Lombardo (Bloomington: Indiana University Press, 2011), 28–41.

35. See Nuriddin, "Engineering Uplift: Black Eugenics as Black Liberation," in this volume, for a novel take on this argument.

36. Nathaniel Comfort, *The Science of Human Perfection: How Genes Became the Heart of American Medicine* (New Haven: Yale University Press, 2012), chap. 2–4; Philip Reilly, *The Surgical Solution: A History of Involuntary Sterilization in the United States* (Baltimore: Johns Hopkins University Press, 1991).

37. G. S. Stent, *The Coming of the Golden Age: A View of the End of Progress* (New York: Natural History Press, 1969), ix–xii.

38. See Luis Campos, "Strains of Andromeda," this volume. For an earlier history of genetic engineering dating back to the first half of the twentieth century, see Campos, *Radium and the Secret of Life* (Chicago: University of Chicago Press, 2015) and Helen Anne Curry, *Evolution Made to Order: Plant Breeding and Technological Innovation in Twentieth-Century America* (Chicago: University of Chicago Press, 2016).

39. Sally Smith Hughes, *Genentech: The Beginnings of Biotech* (Chicago: University of Chicago Press, 2011), chap. 1; Daniel J. Kevles, "Ananda Chakrabarty Wins a Patent: Biotechnology, Law, and Society, 1972–1980," *Historical Studies in the Physical and Biological Sciences* 25, pt. 1 (1994): 111–35; Kevles, "From Eugenics to Patents: Genetics, Law, and Human Rights," *Annals of Human Genetics* 75, no. 3 (May 2011): 326–33; Nicolas Rasmussen, *Gene Jockeys: Life Science and the Rise of Biotech Enterprise* (Baltimore: Johns Hopkins University Press, 2014); Ashley J. Stevens, "The Enactment of Bayh–Dole," *Journal of Technology Transfer* 29, no. 1 (January 2004): 93–99.

40. Shelley Page, "The Gene Gun: Advances in Genetics May Mean Simple Injections Can Cure a Host of Ills, such as Cancer," *Ottawa Citizen*, June 28, 1992, E1.

41. Unfortunately, the best history of gene therapy predates the field's crash and partial recovery: Jeff Lyon and Peter Gorner, *Altered Fates: Gene Therapy and the Retooling of Human Life* (New York: W. W. Norton, 1995).

42. Garland E. Allen, "Modern Biological Determinism: The Violence Initiative, the Human Genome Project, and the New Eugenics," in *The Practice of Human Genetics*, ed. Mike Fortun (Dordrecht: Kluwer, 1999), 1–23.

43. See Reardon, *Postgenomic Condition*, 27–28.

44. E. Zerhouni, "Medicine: The NIH Roadmap," *Science* 302 (2003): 63–72, 63.

45. Katie Benner, "Mark Zuckerberg and Priscilla Chan Pledge $3 Billion to Fighting Disease," *New York Times*, September 21, 2016, sec. Technology; "UN Panel Warns

against 'Designer Babies' and Eugenics in 'Editing' of Human DNA," *UN News*, October 5, 2015; "Will CRISPR Gene-Editing Technology Lead to Designer Babies?" *New Scientist* 228, no. 3050 (December 5, 2015): 33–35.

46. David Baltimore, Paul Berg, Jennifer A. Doudna, et al., "A Prudent Path Forward for Genomic Engineering and Germline Gene Modification," *Science* 348, no. 6230 (2015): 36–38; Puping Liang, Yanwen Xu, Xiya Zhang, et al., "CRISPR/Cas9-Mediated Gene Editing in Human Tripronuclear Zygotes," *Protein & Cell* 6, no. 5 (2015): 363–72; Antonio Regalado, "EXCLUSIVE: Chinese Scientists Are Creating CRISPR Babies," *MIT Technology Review*, November 25, 2018. Zhang was censured, fined, and sentenced to jail time: Carolyn Y. Johnson, "Chinese Scientist Who Claimed to Create Gene-Edited Babies Sentenced to 3 Years in Prison," *Washington Post*, December 30, 2019.

47. J. Benjamin Hurlbut, "Limits of Responsibility: Genome Editing, Asilomar, and the Politics of Deliberation," *Hastings Center Report* 45, no. 5 (2015): 11–14.

48. See Campos, "Strains of Andromeda," this volume.

49. See Ross, "Knowing and Controlling: Engineering Ideals and Gene Drive for Invasive Species Control in Aotearoa New Zealand," this volume.

50. Sarah Zhang, "A Biohacker Regrets Publicly Injecting Himself with CRISPR," *Atlantic*, February 20, 2018; Josiah Zayner, "CRISPR Babies Scientist He Jiankui Should Not Be Villainized," *STAT* News, January 2, 2020.

51. E.g., Christopher Gyngell and Julian Savulescu, "The Simple Case for Germline Gene Editing," *Genes for Life: The Impact of the Genetic Revolution* (Albert Park, Victoria: Future Leaders, 2018), 28–45; Matt Ridley, "Davenport's Dream," in *Davenport's Dream: 21st Century Reflections on Heredity and Eugenics* (Cold Spring Harbor, NY: Cold Spring Harbor Laboratory Press, 2008), ix–xi; Julian Savulescu, "Procreative Beneficence: Why We Should Select the Best Children," *Bioethics* 15, no. 5–6 (2001): 413–26.

52. Melissa Healy, "Adam Lanza: Will Genetics Reveal What Sleuthing Cannot?" *Los Angeles Times*, December 20, 2012, https://www.latimes.com/health/la-xpm-2012-dec-20-la-heb-adam-lanza-genetics-20121219-story.html.

53. Just a few examples among many: Amy Chung, "Being a Jerk Could Be in Your Genes," *National Post (Canada)*, November 15, 2011; S. E. Cupp, "Don't Blame Me, My 'Slut Gene' Made Me Do It," Fox News, December 10, 2010, https://www.foxnews.com/opinion/dont-blame-me-my-slut-gene-made-me-do-it; "Lazy? Your Genes May Be to Blame," LiveScience staff, Yahoo News, April 9, 2013, accessed November 25, 2019, https://news.yahoo.com/lazy-genes-may-blame-122638541.html; Scott Johnson, "The New Theory That Could Explain Crime and Violence in America," Medium, March 11, 2015, https://medium.com/matter/the-new-theory-that-could-explain-crime-and-violence-in-america-945462826399; Chris Matyszczyk, "Six IQ Points Smarter? There Might Be a Gene for That," CNET, May 10, 2014, https://www.cnet.com/news/six-iq-points-smarter-there-might-be-a-gene-for-that/.

54. Peter Dockrill, "Scientists Just Found Almost 1,000 New Genes Associated with Intelligence," ScienceAlert, accessed May 22, 2020, https://www.sciencealert.com/scientists-have-discovered-almost-1-000-new-genes-associated-with-intelligence.

55. "Genomic Prediction," accessed May 21, 2020, https://genomicprediction.com/; Antonio Regalado, "Forecasts of Genetic Fate Just Got a Lot More Accurate," *MIT Technology Review*, accessed September 13, 2018, https://www.technologyreview.com/s/610251/forecasts-of-genetic-fate-just-got-a-lot-more-accurate/. See Robert Plomin, *Blueprint: How DNA Makes Us Who We Are* (Cambridge, MA: MIT Press, 2019), for a breathless account of the coming benefits of polygenic scores, but also see Eric Turk-

heimer, "Genetic Prediction," *Hastings Center Report* 45, no. 5 suppl (October 2015): S32–38, https://doi.org/10.1002/hast.496, for a more skeptical view.

56. See, e.g., Kathryn Paige Harden and Philipp D. Koellinger, "Using Genetics for Social Science," *Nature Human Behaviour*, May 11, 2020, 1–10, https://doi.org/10.1038/s41562-020-0862-5; Harden, "Why Progressives Should Embrace the Genetics of Education," *New York Times*, July 28, 2018, sec. Opinion; Plomin, *Blueprint*, 151.

Chapter 12

1. Ronald Kline, "Construing 'Technology' as 'Applied Science': Public Rhetoric of Scientists and Engineers in the United States, 1880–1945," *Isis* 86, no. 2 (1995): 197–99; Thomas C. Leonard, *Illiberal Reformers: Race, Eugenics, and American Economics in the Progressive Era* (Princeton: Princeton University Press, 2016), 34–36. For specifics about euthenics as engineering, see Sarah Stage and Virginia Bramble Vincenti, *Rethinking Home Economics: Women and the History of a Profession* (Ithaca: Cornell University Press, 1997), 56.

2. See Meghan Crnic, "Better Babies: Social Engineering for 'a Better Nation, a Better World," *Endeavour* 33, no. 1 (2008); Maarten Derksen, *Histories of Human Engineering: Tact and Technology* (Cambridge: Cambridge University Press, 2017).

3. Charles B. Davenport, *Heredity in Relation to Eugenics* (New York: Henry Holt, 1911), 1; International Congress of Eugenics, *Scientific Papers of the Second International Congress of Eugenics Held at American Museum of Natural History, New York, September 22–28, 1921*, vol. 1 (Baltimore: Williams & Wilkins, 1923).

4. Daniel Kevles, *In the Name of Eugenics: Genetics and the Uses of Human Heredity* (New York: Alfred A. Knopf, 1985); Diane B. Paul, *Controlling Human Heredity: 1865 to the Present* (New York: Humanity Books, 1995); Diane B. Paul, *The Politics of Heredity: Essays on Eugenics, Biomedicine, and the Nature-Nurture Debate* (Albany: SUNY Press, 1998); Alexandra Stern, *Eugenic Nation: Faults and Frontiers of Better Breeding in Modern America* (Berkeley: University of California Press, 2005).

5. Stern, *Eugenic Nation*, 14–15.

6. Madison Grant, *The Passing of the Great Race* (New York: Scribner and Sons, 1918), 77.

7. Harriet Washington, *Medical Apartheid: The Dark History of Medical Experimentation on Black Americans from Colonial Times to the Present* (New York: Doubleday, 2006); Jonathan Metzl, *The Protest Psychosis: How Schizophrenia Became a Black Disease* (Boston: Beacon Press, 2014); Clarence Lusane, *Hitler's Black Victims: The Historical Experience of Afro-Germans, European Blacks, Africans, and African Americans in the Nazi Era* (New York: Routledge, 2000); Nicole Hahn Rafter, *White Trash: The Eugenic Family Studies, 1877–1919* (Boston: Northeastern University Press, 1988).

8. Gregory Michael Dorr, *Segregation's Science: Eugenics and Society in Virginia* (Charlottesville: University of Virginia Press, 2008); Gregory Michael Dorr and Angela Logan, "'Quality, Not Mere Quantity, Counts': Black Eugenics and the NAACP Baby Contests," in *A Century of Eugenics in America: From the Indiana Experiment to the Human Genome Era*, ed. Paul Lombardo (Bloomington: Indiana University Press, 2011); Daylanne K. English, *Unnatural Selections: Eugenics in American Modernism and the Harlem Renaissance* (Chapel Hill: University of North Carolina Press, 2004); Shantella Y. Sherman, *In Search of Purity: Popular Eugenics and Racial Uplift among New Negroes* (Xlibris Corp, 2016); Michell Chresfield, "To Improve the Race: Eugenics as a Strategy for Racial Uplift, 1900–1940" (master's thesis, Vanderbilt University, 2013).

9. Kevin Gaines, *Uplifting the Race: Black Leadership, Politics, and Culture in the Twentieth Century* (Chapel Hill: University of North Carolina Press, 1996); Victoria Wolcott, *Remaking Respectability: African-American Women in Interwar Detroit* (Chapel Hill: University of North Carolina Press, 2001); Evelyn Brooks Higginbotham, *Righteous Discontent: The Women's Movement in the Black Baptist Church, 1880–1920* (Cambridge, MA: Harvard University Press, 1993).

10. John Harvey Kellogg, "Eugenics and Euthenics," *Good Health Magazine*, August 1921. For more on Kellogg, see Stern, *Eugenic Nation.*

11. Hannah Landecker's concept of plasticity provides a useful way to understand how African Americans make arguments about the heritability and plasticity of racial characteristics. For more, see Hannah Landecker, *Culturing Life: How Cells Became Technologies* (Cambridge, MA: Harvard University, 2007), 8–10.

12. Terence Keel, *Divine Variations: How Christian Thought Became Racial Science* (Stanford: Stanford University Press, 2018), 56–58; Christopher Willoughby, "'His Native, Hot Country': Racial Science and Environment in Antebellum American Medical Thought," *Journal of the History of Medicine and Allied Sciences* 72, no. 3 (July 2017): 328–51.

13. Mia Bay, *The White Image in the Black Mind: African-American Ideas about White People, 1830–1925* (New York: Oxford University Press, 2000), 7–8.

14. Frederick Douglass, "Claims of the Negro Ethnologically Considered," in *The Life and Writings of Frederick Douglass: Pre-Civil War Decade 1850–1860*, vol. 2, ed. Philip S. Foner (New York: International Publishers, 1950), 289–91.

15. Douglass, "Claims of the Negro," 293.

16. H. T. Kealing, "The Characteristics of the Negro People," in *The Negro Problem: A Series of Articles by Representative American Negroes of To-Day* (New York: James Potts, 1903), 163.

17. For more on the history of slave breeding, see David Brion Davis, "Slavery, Sex, and Dehumanization," in *Sex, Power, Slavery*, ed. Gwyn Campbell and Elizabeth Elbourne (Athens: Ohio University Press, 2014), and Gregory D. Smithers, *Slave Breeding: Sex, Violence, and Memory in African American History* (Gainesville: University of Florida Press, 2012).

18. Kealing, "The Characteristics of the Negro People," 173.

19. Kealing, "The Characteristics of the Negro People," 172.

20. Lesley M. Rankin-Hill and Michael L. Blakey, "W. Montague Cobb: Physical Anthropologist, Anatomist, and Activist," in *African American Pioneers in Anthropology*, ed. Ira E. Harrison and Faye V. Harrison (Urbana: University of Illinois Press, 1999), 110–12.

21. William Montague Cobb, "Physical Constitution of the American Negro," *Journal of Negro Education*, July 1934, Cobb papers, Moorland-Spingarn Research Center (MSRC).

22. Cobb, "Physical Constitution," 341, 387.

23. William Montague Cobb, "The Negro as a Biological Element in the American Population," *Journal of Negro Education*, July 1939, 342, Cobb papers, MSRC.

24. Samuel Roberts, *Infectious Fear: Politics, Disease, and the Health Effects of Segregation* (Chapel Hill: University of North Carolina Press, 2009); Lundy Braun, *Breathing Race into the Machine: The Surprising Career of the Spirometer from Plantation to Genetics* (Minneapolis: University of Minnesota Press, 2014).

25. Cobb, "Negro as a Biological Element," 336–37.

26. Ray Spangenburg and Diane Moser, *African Americans in Science, Math, and*

Invention (New York: Facts on File, 2003), 162; see also Christopher Crenner, "Race and Laboratory Norms: The Critical Insights of Julian Herman Lewis (1891–1989)," *Isis* 105, no. 3 (2014): 477–507, doi:10.1086/678168.

27. Lewis originally wanted the book title to be *The Anthropathology of the Negro*. Julian Herman Lewis, *The Biology of the Negro* (Chicago: University of Chicago Press, 1942), x; W. Montague Cobb, "Review: The Biology of the Negro," *Crisis*, December 1942, Cobb papers, MSRC.

28. Lewis, *Biology of the Negro*, x.

29. Lewis, *Biology of the Negro*, xiii.

30. Gregory Dorr, *Segregation's Science: Eugenics and Society in Virginia* (Charlottesville: University of Virginia Press, 2008), 99–100; Gregory Dorr and Logan, "'Quality, Not Mere Quantity, Counts': Black Eugenics and the NAACP Baby Contests," in *A Century of Eugenics in America: From the Indiana Experiment to the Human Genome Era*, ed. Paul A. Lombardo (Bloomington: Indiana University Press, 2011), 78.

31. Thomas Wyatt Turner, "Biological Laboratory and Human Welfare," Thomas W. Turner Papers Box 153–6 Folder 18, Manuscript Division, MSRC.

32. Kealing, "The Characteristics of the Negro People," 183.

33. Martin S. Pernick, "Eugenics and Public Health in American History," *American Journal of Public Health* 87, no. 11 (November 1997): 1767–72; Nathaniel Comfort, *The Science of Human Perfection: How Genes Became the Heart of American Medicine* (New Haven: Yale University Press, 2012), 31–32, 88–90.

34. Vanessa Northington Gamble, *Making a Place for Ourselves: The Black Hospital Movement, 1920–1945* (New York: Oxford University Press, 1995), 37.

35. John A. Kenney, *The Negro in Medicine* (Tuskegee: Tuskegee Institute Press, 1912), 50.

36. Wendy Kline, *Building a Better Race: Gender, Sexuality, and Eugenics from the Turn of the Century to the Baby Boom* (Berkeley: University of California Press, 2001), 11–15.

37. James H. Jones, *Bad Blood: The Tuskegee Syphilis Experiment* (New York: The Free Press, 1981), 16–22; Susan Reverby, *Examining Tuskegee: The Infamous Syphilis Study and Its Legacy* (Chapel Hill: University of North Carolina Press, 2009).

38. John A. Kenney, "Eugenics and the Schoolteacher," *Journal of the National Medical Association* 7, no. 4 (October–December 1915): 258.

39. John A. Kenney, "Syphilis and the American Negro—A Medico-sociologic Study," *Journal of the National Medical Association* 2, no. 2 (April–June 1910): 115–17.

40. Katherine Ott, *Fevered Lives: Tuberculosis in American Culture since 1870* (Cambridge, MA: Harvard University Press, 1996), 101–3; Helen Bynum, *Spitting Blood: The History of Tuberculosis* (Oxford: Oxford University Press, 2012), 181–82.

41. "Mortality from Tuberculosis 1921," *Journal of the National Medical Association* 15, no. 1 (January–March 1923): 47.

42. Susan L. Smith, *Sick and Tired of Being Sick and Tired: Black Women's Health Activism in America, 1890–1950* (Philadelphia: University of Pennsylvania Press, 1995), 36–38.

43. International Congress of Eugenics, *Scientific Papers of the Second International Congress of Eugenics*, 59–60; Charles Davenport to Monroe Work, October 1921 and November 1928, Davenport papers, American Philosophical Society.

44. Smith, *Sick and Tired of Being Sick and Tired*, 45; National Negro Health News 1915–1950.

45. Michele Mitchell, *Righteous Propagation: African Americans and the Politics of Racial Destiny after Reconstruction* (Chapel Hill: University of North Carolina Press, 2004).

46. Daylanne K. English, *Unnatural Selections: Eugenics in American Modernism and the Harlem Renaissance* (Chapel Hill: University of North Carolina Press, 2004).

47. English, *Unnatural Selections*, 41.

48. English, *Unnatural Selections*, 167.

49. George S. Schuyler, "Quantity or Quality," *Birth Control Review* 16, no. 6 (June 1932): 166.

50. Alfred M. Gordon, "Mental Abnormalities and the Problem of Eugenics," *Journal of the National Medical Association* 15, no. 1 (January–March 1923): 7, 9.

51. T.E.B., "Birth Control Gains Sanction," *New York Amsterdam News*, January 24, 1934, 5; Jessie Carney Smith, *Notable Black Women: Book II* (Detroit: Gale Research, 1996), 38–39.

52. Rebecca Stiles Taylor, "As a Woman Thinks," *Chicago Defender*, July 17, 1937, 16.

53. Taylor, "As a Woman Thinks," 16.

54. Mitchell, *Righteous Propagation*; English, *Unnatural Selections*.

55. Brittney C. Cooper, *Beyond Respectability: The Intellectual Thought of Race Women* (Urbana: University of Illinois Press, 2017), 3–7.

Chapter 13

1. J. R. Fleming, *Fixing the Sky: The Checkered History of Weather and Climate Control* (New York: Columbia University Press, 2010).

2. R. T. Pierrehumbert, "Climate Hacking Is Barking Mad," *Slate* (February 10, 2015), http://www.slate.com/articles/health_and_science/science/2015/02/nrc_geo engineering_report_climate_hacking_is_dangerous_and_barking_mad.single.html.

3. J. Henrich, S. J. Heine, and A. Norenzayan, "The Weirdest People in the World?" *Behavioral and Brain Sciences* 33, nos. 2–3 (2010): 61–83; J. R. Fleming, "Excuse Us, While We Fix the Sky: WEIRD Supermen and Climate Engineering," in *Men and Nature: Hegemonic Masculinities and Environmental Change*, ed. S. MacGregor and N. Seymour, *Rachel Carson Center Perspectives* 4 (2017): 23–28.

4. National Academies of Sciences, Engineering, and Medicine, *Climate Intervention: Reflecting Sunlight to Cool Earth* and *Climate Intervention: Carbon Dioxide Removal and Reliable Sequestration.* (Washington, DC: The National Academies Press, 2015).

5. S. Arrhenius, "On the Influence of Carbonic Acid in the Air upon the Temperature of the Ground," *Philosophical Magazine* series 5, 41 (1896): 237–76.

6. N. Ekholm, "On the Variations of the Climate of the Geological and Historical Past and Their Causes," *Quarterly Journal of the Royal Meteorological Society* 27 (1901): 1–61.

7. H. Brown, *The Challenge of Man's Future* (New York: Viking, 1954).

8. W. S. Broecker and Robert Kunzig, *Fixing Climate: What Past Climate Changes Reveal about the Current Threat—and How to Counter It* (New York: Hill and Wang, 2008).

9. J. R. Fleming, "Iowa Enters the Space Age: James Van Allen, Earth's Radiation Belts, and Experiments to Disrupt Them," *Annals of Iowa* 70 (2011): 301–24.

10. For more on the intersections of the Andromeda Strain with the history of biological engineering, see Luis Campos, "Strains of Andromeda," in this volume.

11. W. Ley, *Engineers' Dreams* (New York: Viking, 1954); M. J. Fogg, *Terraforming: Engineering Planetary Environments* (Warrendale, PA: Society of Automotive Engineers, 1995).

12. Fogg, *Terraforming.*

13. J. D. Bernal, *The World, the Flesh & the Devil: An Enquiry into the Future of the Three*

Enemies of the Rational Soul (1929), http://www.marxists.org/archive/bernal/works/1920s/soul/index.htm.

14. Ibid.

15. K. S. Robinson, *Icehenge* (London: Voyager, [1984] 1997).

16. J. Hickman, "Space Colonization in Three Histories of the Future," *Space Review* (Nov. 29, 2010), http://www.thespacereview.com/article/1732/1.

17. Rebecca Reider, *Dreaming the Biosphere: The Theater of All Possibilities* (Albuquerque: University of New Mexico Press, 2010).

18. J. P. Severinghaus, W. S. Broecker, W. F. Dempster, T. MacCallum, and M. Wahlen, "Oxygen Loss in Biosphere 2," *EOS: Transactions American Geophysical Union* 75 (1994): 33, 35–37.

19. S. Höhler, "The Environment as a Life Support System: The Case of Biosphere 2," *History and Technology* 26 (2010): 39–58.

20. AMS Policy Statement on Geoengineering the Climate System: A Policy Statement of the American Meteorological Society (Adopted by the AMS Council on July 20, 2009).

21. National Academies of Sciences, Engineering, and Medicine, *Negative Emissions Technologies and Reliable Sequestration: A Research Agenda* (Washington, DC: National Academies Press, 2019).

22. J. Wilcox, *Carbon Capture* (New York: Springer-Verlag, 2012); F. L. Toth, ed., "Geological Disposal of Carbon Dioxide and Radioactive Waste: A Comparative Assessment," *Advances in Global Change Research* 44 (2011).

23. M. Boettcher and S. Schäfer, "Reflecting upon 10 Years of Geoengineering Research: Introduction to the Crutzen + 10 Special Issue," *Earth's Future* 5, no. 3 (2017): 266–77.

24. P. J. Crutzen, "Geology of Mankind," *Nature* 415 (January 2002): 23.

25. H. J. Schellnhuber, "'Earth System' Analysis and the Second Copernican Revolution," *Nature* 82 (1999): C28.

26. B. Latour, "Anthropocene Lecture," https://voicerepublic.com/talks/anthropocene-lecture-bruno-latour.

27. J. Lovelock, "A Geophysiologist's Thoughts on Geoengineering," *Philosophical Transactions of the Royal Society* A 366 (2008): 3883–90; J. R. Fleming, "Climate Physicians and Surgeons," *Environmental History* 19 (2014): 338–45.

28. Quoted in Fogg, *Terraforming*.

Chapter 14

1. IUCN 2019, *The IUCN Red List of Threatened Species*, version 2019–2, accessed August 30, 2019, http://www.iucnredlist.org.

2. *Resurrecting the Sublime* (2019) is a collaborative artwork by Dr. Christina Agapakis, Dr. Alexandra Ginsberg, and Sissel Tolaas, made with the support of Ginkgo Bioworks Inc. and IFF Inc.

3. Ludwig Radlkoffer and Joseph F. Rock, *New and Noteworthy Hawaiian Plants* (Honolulu: Hawaiian Gazette, 1911), 12–14; Vaughan MacCaughey, "A Survey of the Hawaiian Land Flora," *Botanical Gazette* 64, no. 2 (August 1917): 89–114, 94–95; Joseph F. Rock, *The Indigenous Trees of the Hawaiian Islands* (Honolulu: published under patronage, 1913), 299–301.

4. Carol Ann McCormick, "The Heartbreak of Psoralea," *North Carolina Botanical Garden Newsletter* (September–October 2007): 8; Billie L. Turner, "Revision of the Genus *Orbexilum* (Fabaceae: Psoraleeae)," *Lundellia* 11, no. 1 (2008): 1–7.

5. Robert A. Salisbury, *The Paradisus Londinensis: Or Coloured Figures of Plants Cul-*

tivated in the Vicinity of the Metropolis (London: William Hooker, 1805–1807), 104–5, accessed September 7, 2019, https://biodiversitylibrary.org/page/36898390.

6. Anthony Rebelo, "Wynberg Conebush—Extinct for 200 Years," iSpot, accessed August 10, 2018, https://www.ispotnature.org/communities/southern-africa/view/observation/529785/wynberg-conebush-extinct-for-200-years.

7. "Ginkgo Bioworks: The Organism Company," Ginkgo Bioworks, Inc., accessed September 15, 2019, https://www.ginkgobioworks.com/about/.

8. The origins of the project and Ginkgo's process were explored by Rowan Jacobsen in a feature for *Scientific American* in 2019. Rowan Jacobsen, "Ghost Flowers," *Scientific American* (February 2019): 30–39, accessed September 1, 2019, https://www.scientificamerican.com/article/fragrant-genes-of-extinct-flowers-have-been-brought-back-to-life/.

9. This part of the project is supported by International Flavors and Fragrances Inc. (IFF), a company that supplies flavor and fragrance molecules to industry.

10. Michael Elowitz and Wendell A. Lim, "Build Life to Understand it," *Nature* 468, no. 7326 (2010): 889–90.

11. Christine Riding and Nigel Llewellyn, "British Art and the Sublime," in *The Art of the Sublime*, ed. Nigel Llewellyn and Christine Riding, Tate Research Publication, January 2013, accessed September 15, 2019, https://www.tate.org.uk/art/research-publications/the-sublime/christine-riding-and-nigel-llewellyn-british-art-and-the-sublime-r1109418. For a post-colonial view on environmentalism and the sublime, see Aaron Sachs, "The Ultimate 'Other': Post-Colonialism and Alexander von Humboldt's Ecological Relationship with Nature," *History and Theory* 42, no. 4 (2003): 111–35.

12. Ian McCalman, "The Virtual Infernal: Philippe de Loutherbourg, William Beckford and the Spectacle of the Sublime," *Romantic Spectacle* 46 (May 2007), accessed September 14, 2019, www.erudit.org/fr/revues/ron/2007-n46-ron1782/016129ar/.

13. Andrew Wilton and Tim Barringer, *American Sublime: Landscape Painting in the United States 1820–1880* (Princeton: Princeton University Press, 2002).

14. David E. Nye, *American Technological Sublime* (Cambridge, MA: MIT Press, 1994).

15. Philip Shaw, *The Sublime* (London: Routledge, 2006): 3, 7.

16. Beth Shapiro, *How to Clone a Mammoth: The Science of De-Extinction* (Princeton: Princeton University Press, 2015).

CONTRIBUTORS

DOMINIC J. BERRY / *The London School of Economics and Political Science* / Berry (PhD, University of Leeds) is Research Fellow on the European Research Council–funded Narrative Science Project www.narrative-science.org. He has published on the history of science and agriculture, biology and time, synthetic biology, biology and technology, the philosophy of science in practice, and the history of molecular biology; in 2019 he cofounded the Biological Engineering Collaboratory, an international and interdisciplinary network for study of biological engineering (www.bioengcoll.org).

LUIS A. CAMPOS / *University of New Mexico* / Campos (PhD, Harvard University) is Regents' Lecturer and Associate Professor of the History of Science at the University of New Mexico. His research brings together archival discoveries with contemporary fieldwork at the intersection of biology and society. He has written widely on the history of genetics and synthetic biology and is the author of *Radium and the Secret of Life* (University of Chicago Press, 2015) and coeditor of *Making Mutations: Objects, Practices, Contexts* (MPIWG, 2010). Campos recently held the Baruch S. Blumberg / NASA Chair of Astrobiology at the Library of Congress, and currently serves as Secretary of the History of Science Society.

NATHANIEL COMFORT / *Johns Hopkins University* / Comfort (PhD, SUNY-Stony Brook) is a professor of the history of medicine who writes on genetics, eugenics, and genomics. He's previously served on the editorial board of the *History and Philosophy of Life Sciences* and *Isis*. His books include *The Science of Human Perfection* (Yale, 2012) and *The Tangled Field: Barbara McClintock's Search for the Patterns of Genetic Control* (Harvard, 2001). In addition to scholarly articles, he writes for *Nature*, the *Atlantic*, the *Nation*, and other publications.

MICHAEL R. DIETRICH / *University of Pittsburgh* / Dietrich (PhD, University of California, San Diego) is a professor of history and philosophy of science whose research focuses on controversies in twentieth-century biology. He has previously served as Editor-in-Chief of the *Journal of the History of Biology*. He has coedited four books, including most recently *Dreamers, Visionaries, and Revolutionaries in the Life Sciences* (University of Chicago Press, 2018).

RICHARD FADOK / *Massachusetts Institute of Technology* / Fadok is currently a doctoral candidate in the interdisciplinary History, Anthropology, and Science, Technology, and Society program. His scholarship on biology and design spans the anthropology of science and technology, design studies, and the environmental humanities. His dissertation, *In Life's Likeness: Biomimicry and the Imitation of Nature in the United States*, is a historical and ethnographic study of the moral phenomenology of biomimicry.

JAMES RODGER FLEMING / *Colby College* / Fleming (PhD, Princeton University) is the Charles A. Dana Professor of Science, Technology, and Society at Colby College. He is the author of *Meteorology in America* (Johns Hopkins, 1990), *Historical Perspectives on Climate Change* (Oxford, 1998), *The Callendar Effect* (American Meteorological Society, 2007), *Fixing the Sky* (Columbia, 2010), *Inventing Atmospheric Science* (MIT, 2016), and *FIRST WOMAN: Joanne Simpson and the Tropical Atmosphere* (Oxford, 2020). He served on two National Academy of Sciences study panels, was a contributing author for the Intergovernmental Panel on Climate Change, and is series editor of Palgrave Studies in the History of Science and Technology.

ABRAHAM GIBSON / *Arizona State University* / Gibson (PhD, Florida State University) is a Lecturer in the Department of History at Arizona State University. He has authored more than twenty publications, including articles in *Isis*, *Environmental History*, and *Journal of the History of Biology*. His first book, *Feral Animals in the American South: An Evolutionary History*, was published by Cambridge University Press in 2016.

ALEXANDRA DAISY GINSBERG / *Somerset House Studios, London* / Ginsberg (PhD, Royal College of Art, London) is an artist exploring our fraught relationships with nature and technology, and the human impulse to "better" the world. She was lead author of *Synthetic Aesthetics: Investigating Synthetic Biology's Designs on Nature* (MIT, 2014) and received the World Technology Award for design in 2011. Daisy exhibits internationally, including at MoMA New York, the Museum of Contemporary Art, Tokyo, the Centre Pompidou, and the Royal Academy, and her work is in museum and private collections.

ANITA GUERRINI / *Oregon State University* / Guerrini (PhD, Indiana University) is Horning Professor emerita of history at Oregon State University and Research Professor of History at the University of California, Santa Barbara. She has authored or edited six books, including *The Courtiers' Anatomists: Animals and Humans in Louis XIV's Paris* (University of Chicago Press, 2015) and *Experimenting with Humans and Animals: From Galen to Animal Rights* (Johns Hopkins, 2003, 2nd rev. ed. 2022), and over seventy publications spanning articles, book chapters, and essay reviews. Her research interests encompass the histories of animals, food, the human body, and the environment from 1500 to the present.

JOSHUA MCGUFFIE / *University of California, Los Angeles* / McGuffie is a doctoral candidate in the history department. His research tracks the development of medicine and biology within the Manhattan Project and how those traditions matured under the US Atomic Energy Commission during the Cold War. He is interested in the intersections between the history of biology and environmental history, especially at sites where researchers transitioned from work on the scale of the lab to work on the scale of populations and landscapes. In 2019, he was a Mellon / Excellence in Pedagogy and Innovative Classrooms Fellow in Environmental Humanities.

AYAH NURIDDIN / *Princeton University* / Nuriddin (PhD, Johns Hopkins University) is a Cotsen Postdoctoral Fellow in the Society of Fellows. She received her BA in international studies and history from American University in 2009. She received a dual master's in history and library science from the University of Maryland, College Park, in 2014. Her research interests include eugenics, birth control, disability, scientific racism, and public health.

LISA ONAGA / *Max Planck Institute for the History of Science* / Onaga (PhD, Cornell University) is a Senior Research Scholar in Department III (Artifacts, Action, Knowledge) at the Max Planck Institute for the History of Science, investigating the production of knowledge at the interface of animal and human life in agricultural, laboratory, health, design, and industrial settings. She is completing a book, *Cocoon Cultures: The Entanglement of Biology and Silk in Japan*, which is under contract with Duke University Press. Her research also appears in publications, including *Technology & Culture*, *positions: asia critique*, *Journal of the History of Biology*, and *Studies in History and Philosophy of Science*, and in *New Perspectives on the History of Life Sciences and Agriculture* (edited by Kingsland and Phillips, 2015).

EDMUND RAMSDEN / *Queen Mary University of London* / Ramsden (PhD, European University Institute) is a Wellcome Trust University Award Lecturer in the history of science and medicine in the School of History at Queen Mary. His current research is focused on the history of experimental animals in psychology and psychiatry and on the influence of the social, biological, and behavioral sciences on urban planning, architecture, and design in the United States with a particular interest in public housing.

CHRISTIAN H. ROSS / *Arizona State University* / Ross is a doctoral candidate in the Center for Biology and Society. A National Science Foundation Graduate Research Fellow and 2019–2020 Harvard Science, Technology & Society Fellow, he studies how scientific responsibility and expertise around emerging genome-editing technologies are configured in tandem with visions of public engagement, democracy, and the right relations between science and society.

TIAGO SARAIVA / *Drexel University* / Saraiva (PhD, Universidad Autónoma de Madrid) is Associate Professor of History, coeditor with Amy Slaton of the journal *History and Technology*, and author of *Fascist Pigs: Technoscientific Organisms and the History of Fascism* (2016), winner of the 2017 Pfizer Prize awarded by the History of Science Society. He is interested in the connections between life sciences, crops, and politics at the global scale.

CHRISTIAN C. YOUNG / *Alverno College and the Urban Ecology Center in Milwaukee, Wisconsin* / Young (PhD, University of Minnesota) teaches courses on natural history, climate change, evolution, and science education as a professor of biology. In addition to two books on the history of wildlife biology and ecology, he has coedited a volume on evolution and creationism. His current writing explores the history of the naturalist tradition as a means of connecting urban communities to the natural world.

INDEX

Page numbers in *italics* indicate figures.